EVACUATION

EVACUATION

THE POLITICS AND AESTHETICS OF MOVEMENT IN EMERGENCY

PETER ADEY

DUKE UNIVERSITY PRESS
Durham and London
2024

© 2024 DUKE UNIVERSITY PRESS
All rights reserved
Project Editor: Liz Smith
Designed by A. Mattson Gallagher
Typeset in Garamond Premier Pro and Futura Std
by Westchester Publishing Services

Library of Congress Cataloging-in-Publication Data Names:
Adey, Peter, author.
Title: Evacuation : the politics and aesthetics of movement in
emergency / Peter Adey.
Description: Durham : Duke University Press, 2024. |
Includes bibliographical references and index.
Identifiers: LCCN 2023053862 (print)
LCCN 2023053863 (ebook)
ISBN 9781478030584 (paperback)
ISBN 9781478026396 (hardcover)
ISBN 9781478059578 (ebook)
Subjects: LCSH: Emergency management—Political aspects. |
Emergency management—Social aspects. | Disasters—
Political aspects. | Crises—Political aspects. | Internally
displaced persons. | Forced migration. | BISAC: SOCIAL
SCIENCE / Sociology / Social Theory | SOCIAL SCIENCE /
Sociology / Urban
Classification: LCC HV551.2 .A326 2024 (print) | LCC HV551.2
(ebook) | DDC 363.34/81—dc23/eng/20240416
LC record available at https://lccn.loc.gov/2023053862
LC ebook record available at https://lccn.loc.gov/2023053863

CONTENTS

ACKNOWLEDGMENTS

This book has been a very long time in the making. I first began working on evacuation during a master's thesis in 2002, when I was encouraged by Tim Cresswell and Deborah Dixon (my supervisors) to try to carry out a strange project on airport passenger simulations. This helped me begin to work through some of the creative moves necessary to pursue how evacuation is diagrammed in multiple recursions and articulations. This work owes so much to their initial guidance.

Another turning point in the book happened when Mimi Sheller read through an early proposal and was incredibly encouraging and productive in her feedback. Brad Garrett, Steve Graham, and Angharad Closs-Stephens also helped at proposal stage and inspired me in different ways. Weiqiang Lin, Rachael Squire, and Sasha Engelmann generously read through earlier chapter drafts. And Mor Shilon and Pnina Shukrun-Nagar provided guidance on the disengagement chapter; any errors of fact or interpretation are my own. Moreover, given that the final version of the book went into production in late 2022–early 2023, it is grossly inadequate in its anticipation of or application to the terrible and asymmetric violence of the Hamas attacks of October 7, 2023, and Israel's bombing and invasion of the Gaza Strip. Great caution and care should be taken in trying to apply this book's discussion to those events.

As the final parts of the book came together, I spent several periods in Australia (2017–18, 2018–19, and 2022–23), where I was lucky enough to work in Melbourne University's geography department courtesy of a fellowship and visiting positions organized by the brilliant David Bissell, along with a short visit at the Emerging Technologies Lab at Monash. As I detail in the book, I lived with my family in Kinglake, which had been devastated by bushfires in 2009; thank you to B and Finn Starbright for sharing their house with us, and the community in Kinglake. I also learned so much through the encouragement of colleagues and students in Melbourne's wider community—some of whom were passing through—this included of course

David Bissell but also Kaya Barry, Thomas Birtchnell, Maria Borovnik (who kindly supported a visit to Massey University), Candice Boyd, Elisabetta Crovara (thanks also for the guitar loan), the author Sophie Cunningham, Jane Dyson, Tim Edensor, fellow visitor and office mate Olga Hannonen, Rachel Hughes, Michele Lobo, Adam Moore, Tim Neale, Catherine Phillips, Sarah Pink, Rob Raven, Lauren Rickards, Libby Straughan, Shanti Sumartojo, Ilan Wiesel, and especially so many people who gave me kind advice and support after my keynote talk at the Australian Mobilities Research Symposium (AusMob) in 2017—it was an amazingly formative and affirmative experience.

The research has involved the support of numerous amazing archivists and staff at different locations and institutions, including particularly the National Academy of Sciences in Washington, DC; the Georgia Archives in Morrow, Georgia (including their support in reproducing several figures); Christchurch City Archives and the Akaroa Museum in New Zealand; the National Archives, Kew, and the National Railway Museum, York, in the United Kingdom; and the National Library of Australia, the Public Records Office Victoria, and the State Library Victoria in Australia.

Some of the early and underlying thinking about emergency within this book is indebted to a wonderful collaboration with Ben Anderson and Steve Graham in our Staging Emergencies project (2008–10) and other work. My links with an emergency planning unit in Staffordshire proved incredibly useful, where Andy Marshall and Steve Hill were extremely generous in supporting our work with their time and access and candor about their practice. Conversations with emergency planners have been exceptionally interesting, and I'm grateful for their generosity, especially to Dan Neely, who gave me a driving tour of the Blue Lines in the Wellington area in 2019.

A Philip Leverhulme Prize got me going in 2011, just as I arrived at Royal Holloway University of London, to find a way to organize this material and plot out big chunks of research. My work also overlapped with a "Light" fellowship at Durham University's Institute of Advanced Studies (where Tim Edensor and I were both fellows, and I benefited from many conversations with Tim). I'm especially grateful to successive heads of department David Gilbert, Katie Willis, Phil Crang, and Danielle Schreve for supporting my work on this and periods of travel and research leave. Colleagues at Royal Holloway have proven critical friends along the way, especially Katherine Brickell, Simon Cook, Klaus Dodds, Sasha Engelmann, Harriet Hawkins, Michael Holden, Anna Jackman, Rikke Jensen, Oli Mould, and Alasdair Pinkerton. I've had some excellent research assistance help too, especially from working with and discussing the project with Rachael Squire and Pip Thornton, Phil Kirby, and two undergraduate assistants, Oliva Longley and Louise Isaac, who were incredibly professional and enthusiastic in transcribing, annotating, and organizing some of the Australian archival work in 2018, courtesy of Department of Geography internship positions.

The book has benefited from at least forty talks I have given at different conferences, workshops, and seminar series about evacuation, where organizers, audiences, and friends have given me so much help and advice. I want to particularly thank Caren Kaplan for her friendly and critical comments at the American Association of Geographers Annual Conference in 2019, and especially for challenging me to think harder about the imagery that I used to support my work on the Triangle Shirtwaist fire. Chris Philo also caused me to take the abject, "clonic," or expulsive forms of evacuation much more seriously. Support from the Academy of Mobility Humanities at Konkuk University (Ministry of Education of the Republic of Korea and the National Research Foundation of Korea; NRF-2018S1A6A3A03043497) has also led me to share my research with colleagues and audiences in Seoul and extend some of my fieldwork and thinking; thank you to Jinhyoung Lee and Inseop Shin for their support.

At Duke the patience, enthusiasm, and crucial guidance of Courtney Berger cannot be exceeded. Thanks also to Laura Jaramillo for helping me through the final stages of completion.

Last, thank you to my beautiful family, as always. I love you. To Hayley, who has proven unevacuable from this project. To Victor, who does a reasonable impersonation of a lyrebird's mimicry of "Evacuate now." And to my eldest son, Arthur, for his wonder, for teaching me the etymology of the word *panic*, and for giving me his own take on evacuation and sea creatures.

INTRODUCTION

evacuate, v.t. (-uable) empty
(stomach, etc.); (esp. of troops)
withdraw from (place);
Discharge (excrement, etc.)
Evacuation, n. (vacuum).
It all fits, the withdrawing, the emptiness, the troops, the vacuum—above all
the vacuum. I was evacuated. I won't comment on the excrement.
—JOHN FURSE, IN B. S. JOHNSON, *THE EVACUEES*

SCENE 1: JULIA GILLARD'S SHOE. In 2012 Australian prime minister Julia Gillard
was attending an Australia Day remembrance event at the restaurant the
Lobby, located close to the iconic Tent Embassy in Canberra. Ironically, the
event was the inaugural National Emergency Medal to honor responders to
emergencies such as the Black Saturday bushfires of 2009. Protesters, hearing
of opposition leader Tony Abbott's disparaging words earlier that morning,
about trying to "move on" from the grievances brought by Indigenous political
action, surrounded the venue and began chanting and beating on the glass.
Gillard's security and police escort, unnerved by the events, which they described as "deteriorating," decided to evacuate their VIP Gillard to a waiting car.

Gillard's evacuation became a demeaning but ambivalent signal, at
once asserting the sovereignty of executive authority and also, even more,

undermining it, especially the prime minister's. Close protection security parted and pushed protesters back in a displacing and divisive set of movements. An Aboriginal elder lost his balance in the wake of the evacuation and was forced to grip the stair rail as security stormed by. As they exited, Gillard tripped and lost her shoe. Her bodyguards dragged and then carried Gillard's body—for these seconds an apparently empty vessel of political autonomy and authority—her person buried in her security guards' limbs. On video the moment seems over the top. Protesters, press, and passersby seem bemused, the threat, perceived or real, overly inflated. The response offers a form of protection to a political leader that others have no access to.

In the wake of the evacuation, the Tent Embassy and the Aboriginal Parliament found and retained the shoe, suggesting it could be put into economic exchange. The evacuation becomes a rupture with the single shoe flowing between the evacuation and its response, and the singular evacuation an embarrassment of excessive and unequal protective (im)mobilities.

The shoe moved between different activists and figureheads, such as Paul Coe. The late Indigenous activist academic Pat Eatock, who was given the shoe, stated during the events, "She can't have it, this is going on eBay. . . . We are going to see if we can get some money for the (tent) embassy" (quoted in Wright 2012). The shoe is even imagined, by Eatock, as becoming a future museum exhibit: "I see it sitting like Cinderella's shoe in a glass case in a museum 10 years from now as this is part of the history of race relations in Australia," as if a symbol of racial struggle, First Nations rights and land titles, colonial (dis)possession, and enduring marginality. It is also more: the shoe is a vehicle for the coming together of a diverse collective in emergency in a way that continues the rupture of sovereign power performed in the over-the-top evacuation. The protesters hold on to the shoe and offer it back as if were an item of sovereign property, mocking the practices the Australian state has forced on First Nations or Aboriginal and Torres Strait Islander peoples through land claims, blood rights, and more. In play and creativity, the protesters mobilize and are mobilized by the shoe. Some pose with it and pretend to try it on, which shifts some of the anger performed during the protest into something different (figure I.1). It seems more solidaristic and even sympathetic when the protesters return the shoe the next day.

SCENE 2: EXECUTIVE ORDER. A 2016 executive order given by North Dakota state governor Jack Dalrymple requires the "cleanup" and mandatory evacuation of several thousand protesters, most Standing Rock Sioux, who are encamped to resist the Dakota Access Pipeline. The order is followed by a

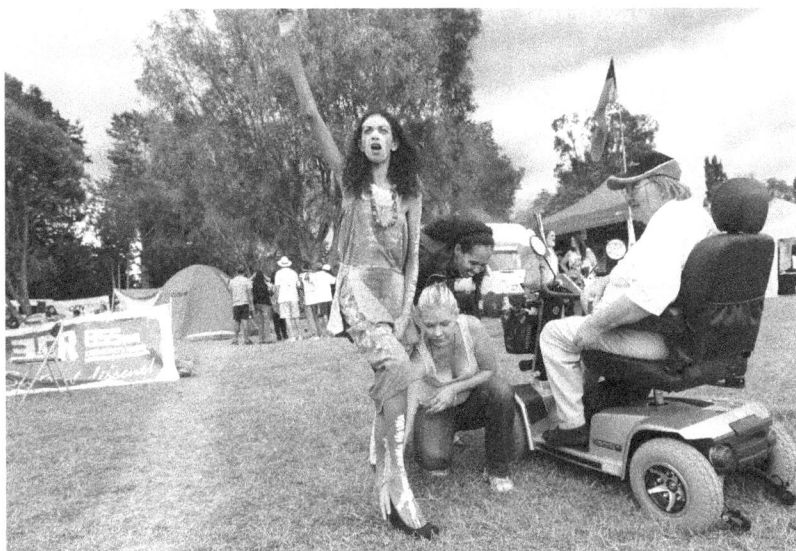

FIGURE I.1. Gwenda Stanley, a Gomeroi woman, poses with Julia Gillard's lost shoe. Newspix/Getty Images, 2012.

further eviction order by the Army Corps of Engineers and a further evacuation order by the new governor in the spring of the next year. One of the many techniques that continued to displace the Standing Rock protesters from their camp, including the dispersal effects of chemical and muscular technologies developed by the police and Israeli military in relation to riots and terrorist and political insurgencies (Simmons 2017), was the use of emergency and evacuation legislation under the 1985 North Dakota Disaster Act. This gave the governor powers to "direct and compel" evacuations of camp activists on the grounds that the "harsh winter conditions" made the encampments unsafe within the evacuation area, which meant the withdrawal of state emergency aid and medical health care (Wong and Levin 2016). Given that fire hoses were turned on the protesters in the cold conditions a week earlier, the evacuation order was a cynical means of removal under the guise of protecting the camp's occupants from the subzero conditions, the potential floodwaters rising in the thaw of winter, and, ironically, the environmental pollution produced by waste from the camp's occupants.

In fact, the coalition of Indigenous American tribes and activists had their own autonomous methods of emergency provision better focused for the camp's care. What's more, the Standing Rock tribal government had a history of declaring their own emergency measures and requesting federal assistance under what has been codified as "tribal emergency preparedness law" (Sunshine and Hoss 2015; see also FEMA 2013; Government Accountability Office 2018).

Between these two scenes, we see a range of fraught relationships within evacuations caught up in contentious histories of colonialism, displacement, occupation, resource extraction, and state and civic power. Standing Rock gives us an insight into the abuses and technicalities of emergency politics, legislation, and evacuation practices and policy that threaten an Indigenous community within an already contracting space of traditional territories under pressure from vast infrastructure expansion with huge environmental implications.

The different evacuations in these scenes show competing emergencies at work. Both sets of protesters interrupt the exertion of state powers in the form of emergency agency in order to lay claim to their own in what Ben Anderson (2017) and Bonnie Honig (2009) might interpret as a civil politics of emergency. A civil politics of emergency criticizes the exceptional and sovereign circling of emergency within executive power, or the bureaucratic and elite spaces of state decision-making, and the more banal violences of emergency practices. And yet both scenes show some promise. We see bodies turning toward and attuning to one another, orienting to one another amid struggle. The protests were collective efforts to stand up to state, logistical, and infrastructural power (Chua et al. 2018) through the temporary bubble-like excursion of civic and Indigenous sovereignties that Kristen Simmons (2017) identifies as the countergeographies of "settler atmospherics." Emergencies can create possibilities of attunement to each other, opportunities to "arc toward one another—becoming-open in an atmosphere of violence" (Simmons 2017).

Such ambivalent senses of evacuation—and what Honig (2009, 2021) conceptualizes as "agonistic" relations in emergency, an approach to politics that values not homogeneity but difference, antagonism, risk, even enmity, and often ambivalence—are central to this book. This follows from Rebecca Solnit's (2010) skepticism toward "catastrophic" renderings of emergency. Her concern, drawn from a patchwork of examples, including New Orleans's experience of Hurricane Katrina in 2005, or that of Californians in the 1906 San Francisco earthquake, is for centralized authorities acting under what she calls a kind of "elite panic," when emergency becomes a way to legitimize the consolidation of power by governments, police, and security agencies especially. Gillard's evacuation and the heavy-handed and disingenuous forms of emergency evacuation at Standing Rock demonstrate the sometimes excessive and pernicious forms of evacuation wielded as state power. But while disaster may lead to the most extreme forms of inhumanity, Solnit also finds in it the possibility of collective and social euphoria and togetherness, generosity, and altruism. Evacuation may show the best and worst of us.

This book seeks to stay with evacuation's agonistic politics and more affirmative versions of emergency as a form of critique and promise. What this book turns on is that when we really and finally underscore what evacuations mean and do, we see how their complexities, ambivalences, and ambiguities are precisely the qualities evacuations are often wielded for: frequently as highly technical achievements, disinterested and yet highly interesting. The many agonisms of emergency drive and constitute so many evacuations. On the one hand, evacuations can perform some of the most unjust, undemocratic, and violent forms of displacement, which build on and exacerbate (neo)liberal individualizing, exclusionary, and normative assumptions. On the other hand, and in spite of these tendencies, they promise human agency, collectivity, compassion, and togetherness.

WHAT THIS BOOK IS ABOUT

This book explores how evacuation has emerged as a concept and a technique of governance to critically explore how and where evacuations are designed and made, and with what effects. Despite the apparent ease with which it is uttered, or called for, or demanded, evacuation is often out of the reach of many. This unevenness of mobility—who can do it and who cannot, and the diverging quality of the experiences of those who are evacuated—is the central concern of this book, as is an attempt to highlight more collective and socially *just* evacuations. How, especially in its governance, might we make evacuation more about *we*? How do we highlight and champion moments, as Solnit (2010) has sought, when evacuations actually collectivize and democratize?

The book doesn't really try to fix evacuation. I don't want to tell planners and consultants and engineers and local authorities how to do their jobs. Rather, it puts evacuation into thought and a wider social and political context of empirical research and conceptual and political discourse, so that it is *not just the job* of planners and consultants to think about it. Part of my claim is that evacuation has tended to be put in the domain of public authorities, legal expertise, technical know-how. Evacuation is frequently designed and planned by people other than those who are made or encouraged to do it. Some communities creatively resist their enrollment within evacuation practices but may demand it in other moments. This is common to emergency governance (Honig 2009). With both proceduralism *and* discretion, emergencies are governed through the elite concentration of knowledge and experience. Honig instead seeks out "new sites of power in emergency settings . . . propulsive generative powers of political action" (10). Of course, ordinary people frequently have to plan for evacuation and help their loved ones prepare for it too. For others this is relatively rare. In some parts of the world, self-evacuations from fire, land denudation, cyclones, and more are much more common, where publics are sometimes involved in drawing up some sort of community-level organization.

It is peculiar that apart from a few isolated places, academia has been relatively blind to evacuation; it is something of an afterthought or a technical process outside of politics. This book seeks to foreground evacuation mobility itself and acknowledge fields like disaster studies and emergency management that explore the benefits and problems of evacuation for policy, planning, and business (Fritz and Williams 1957; Perry 1979; Quarantelli 1980). Several authors, seeking to work across fields such as sociology, human geography (Cutter and Barnes 1982), and disaster management and migration (Zelinsky and Kosinski 1991), have called for the examination of evacuation as an important form of mobility in its own right. It is difficult to claim that this has happened very widely, however, and I worry that the way evacuation surfaces and recedes from view is both a symptom and a condition of its success in the operation of power. In other words: that we don't write and think about it matters and tells us something important about evacuation.

Rosalind Kidd's (1997) *The Way We Civilise* is a classic history of the exploitation and dispossession of Aboriginal populations under the administration of Aboriginal affairs in Queensland. We learn how Australia's North was under threat from Japanese attack during World War II. Some believed the First Nations Aboriginal population to be sympathetic to the Japanese. Thousands were forcibly "evacuated" under the pretense that it was for *their* own protection and the protection of others. The evacuees suffered inadequate provisions and endured extreme poverty, disease, and starvation in Woorabinda, in central Queensland. Kidd shows that the violence and racial injustice performed by the evacuation were part of a much wider bureaucratic and administrative process. And thus the things we call evacuations and the *way* they are conceived, planned, managed, and executed can be drawn into wider contexts, narratives, practices, and experiences of governmental, administrative, and bureaucratic power, in league with colonial and imperial projects. This book attends to the processes, practices, nitty-gritty doings, and performances—the arts, technologies, and experiences of evacuation—as much as it recognizes the strategies, concepts, logics, and rationalities that have driven it.

To take this further, I take inspiration from other genealogies of technology and power. By exploring the emergence of territory as a political concern, Stuart Elden explains that "territory is a word, a concept, and a practice, and the relation between these can only be grasped historically" (2013, 20). Elden bemoans the imprecision of the language of territory, preferring to stick with a more precise historical-geographic understanding. *Evacuation* appears to have been used rather self-evidently, with little sustained examination of its meaning, its historical and geographic character, or the ways in which it has been contested and struggled over. I follow Elden's brilliant genealogy to trace evacuation as a "term," "concept," and "practice"; to investigate the conceptual history of evacuation; and to explore how particular words and their meanings are tied up in evacuation, how specific concepts and practices are named or "designated" as evacuation, and how evacuation is done as a material and embodied practice.

But it is hard to focus on a concept that at its heart is about negation. Evacuation is often about when something or someone is taken away. A denial of the ongoing everyday of life that results from this separation. A severing of relations, temporary and sometimes permanent. Evacuation is often a negative move even if it saves your life. It is emptiness, says the chapter's epigraph, when something is removed from the self. Moreover, evacuation's performance as a "technical" and linguistic object is crucial, to the extent that the word is regularly used and uttered without thought, and its utterance may enclose its properties as much as disclose them, which is part of what this book explores as the aesthetic and affective capacities of evacuation. There has been little attempt to build a theory or conceptual framework with which to make sense of it. This results in an "unproblematic common sense view of evacuation" (Aguirre 1983, 417), which I fear has meant leaving a void in its own name. Evacuation's implicit property is to vacate a position while being literally positioned, poised, in potential. It can name a hopeful practice of escape. Almost entirely different meanings of the term can coexist with one another too, antagonistically or indifferently. The familiar utterance "They have been evacuated" can be ultimately hollow, implying an outside agency pulling one to safety when, instead, a community or group were forced to undertake such movements on their own. Such a treatment is very much in tune with Ann Laura Stoler's (2016, 54) recognition of the "invocations and evacuations" of colonial history, where Stoler evocatively describes the colony as philosophically bypassed, holding little to no weight or conceptual "cachet," but, fittingly, "a political concept in wait, poised to discharge its potentiality," to be called on.

Some peoples have even been moved to death in evacuation's name, given its uneven and unstable relationship to protection. Kidd (1997) recognized the persistent euphemisms of terms such as *dispersal* to describe the displacement and massacre of Indigenous peoples. This book interrogates such practices of naming, pairing those designations and the emergence of evacuation as a concept with "an analysis of practices and the workings of power" (Elden 2013, 8). Precisely as it is taken as self-evident, *evacuation* regularly conjoins with other names, concepts, and practices. Evacuation is promiscuous. This property was realized in one of the only studies dedicated to the idea: Wilbur Zelinsky and Leszek Kosinski's (1991) *The Emergency Evacuation of Cities*, which points out the difficulties in its conceptualization. Evacuation is a "polymorphous" notion that straddles "other modes of movement" (304).

Undoubtedly, it is possible to see evacuation as a technique of protection and a colonial and development technology, sometimes residing in bio- and necropolitical forms of sovereign power. Certainly evacuations arise because of the hazardous nature of contemporary industrial capitalism, extraction, and neocolonialism through chemical spills, nuclear disasters, and building fires; the climatic changes and shocks that go with those processes, such as the increasing regularity of wildfires, cyclones, flood events, landslides, and other extreme weather; and the ways in which our societies are structured and ordered, which places some communities at far greater risk of vulnerability to these events and therefore at far greater risk of being or needing to be evacuated. This is not to assume evacuation universality, however, as an expression of a totalizing apparatus or logic of security, or to grant it simple transferability, as a diagram of power unhooked from one context and transferred into another (Deleuze 2006). It may move through transmissions of emergency practices and safety regulations and standards through global institutions and humanitarian organizations, appearing in comparable conditions and circumstances, but it rarely emerges as quite the same thing.

The focus of this book is to follow the traces, (post)colonial circuits, and heterogeneous conditions of overlapping concerns and sometimes contradictory imperatives and competing technologies and practices (Collier 2009), which somehow hang together through evacuation and what I take to be its "multiple and unstable valences," inspired by Stoler's (2016, 5) recursive approach to the genealogies of the "colony." Stoler's extraordinary approach is to center on how "'knowledge things' are *disassembled*, reassembled, fail or fall apart" (75). I take this to focus on the "tenuous and tenacious qualities," "oscillations in form," and "political rationalities" that might underwrite evacuations, and their deployments in policy and practice, as *evacuation* may oscillate between "charged political concept and innocuous common noun" (76). This is to deploy genealogy as a method, and recursion as a kind of analytics, which means to "'stay on the track of dispersions rather than unities, of the 'dissension of other things,'" urging our attention to "innocuous interstices of words, intimacies, and things" that are repeated in different forms and guises, to the recurrent "visions and practices" of imperial formations and other power structures (24, 6). Evacuation is regularly announced, named, and called, but more regularly, as a "subjacent political concept," it "never stands alone. It is always relational" (117).

Of course, this leaves many gaping holes. Most of the examples and case studies within the book dwell on the Global North, especially, but not exclusively, within the anglophone world, as well as materials available in English for the most part. At the same time, I recognize the interdependency of those worlds with the Global South, with environmental-ecological linkages, transnational movements of people and culture, labor and economic relations and ties, and imperial and neocolonial relations. Indeed, the relics and new formations of colonialism are performed directly through evacuation and emergency planning in liberal humanitarianism, in logistical practices, and even labor industry standards, which are in part transmitted, hybridized, and coconstructed within circuits of influence, aid giving, and sometimes also training and learning.

Evacuation troubles the dichotomies present in the most pressing global and regional concerns and sharp and highly visible displacements, while persisting within the everyday as routine, invisible, and enduring (Cahill and Pain 2019; Christian and Dowler 2019; Nixon 2011). As Yarimar Bonilla (2020) has explored within a context of wider writings on coloniality as itself disastrous, emergencies and their governance cannot be separated from racio-colonial structures and histories (Faria et al. 2021, 89). Such histories may form conditions that set the stage for evacuation and from which it is difficult to imagine or perform evacuations otherwise.

In this highly partial positioning, therefore, the book examines when evacuation becomes another kind of imposition of ideas about how we should deal with emergency, who should be responsible for it, who should be the beneficiaries of such action, and what kinds of lives are valued. Evacuation is entirely uneven, and we must be wary of any attempted universalizations. This book is, in part, a type of critique. *Evacuation* is a simultaneously ethical and political provocation, raising questions about who can evacuate, who wants to evacuate, who can't evacuate, who conducts and plans and orders evacuations, and who else is forced to submit to evacuation.

In parallel with this are questions or rather tensions over what an evacuating subject or evacuee is: whether a body, or an atomistic or fluidic representation of a person; or an assemblage of materials and agencies made corporeally vulnerable through human-animal companionship and human-human solidarities and sympathies. These tensions rub together most around the individual and the collective: the political agency of an individual versus the agencies of a multiple. More often than not, evacuations involve frictions, antagonisms, and agonistic (Honig 2009) relations. This book leans more heavily toward the more collectively embodied forms of evacuation as a way to critique the singular and the exclusive and to foster an ethics of evacuation that, while recognizing evacuation's many ambivalences (and negations), is generous, plural, affirmative, and hopeful.

Some evacuations, and the constellation of actors, technologies, practices, politics, and discourses that shape them, err toward very different forms, experiences, and results from this. Adi Ophir (2007) has identified two different genealogies and versions of modern state sovereignty and their relation to emergencies and disasters, from the "catastrophic" to the "providential." And yet Ophir's typology tends to reduce evacuation to two diametrically opposing types. I am more interested in how these types break down, to move between and complicate these different versions. Ophir's schema could work as a continuum of emergency directions or tendencies—from the caring to the most violent—through which evacuations can be analyzed. Some evacuations, even as they bring populations closer to the care of its processes, registers, and abstractions, may perform various indifferences, abandoning less valued lives, while demonstrating the best of human togetherness, solidarity, and altruism. Evacuation can allow the providential and catastrophic aspects—multiple forms and meanings—of evacuation to overlap, interfere, and coexist. Evacuation might be made of mobility and immobility, care and control, embrace and abandonment, meaning to be simultaneously cared for *and* imperiled (Pallister-Wilkins 2020).

At the heart of the book are important tensions between evacuation and liberal governmentality. Evacuation is perhaps one of the most illiberal things we can do, as we rip ourselves or are ripped from ties to place, to home, to social, familial, and even interspecies networks and relations, destabilizing attachments while new and unpredictable ones may be fostered. Of the millions of people fleeing the increasingly frequent catastrophic wildfires in the western United States and in Canada, in the Amazonian region, in Australia and southern Europe, or the flooding events in Henan province in China and in western Germany, responses have varied wildly: from gratitude when some were evacuated from life-threatening circumstances to disgust at the inadequacies of planning and resources to evacuate those who could not be and were not. There is equal concern over policy responses to the increasing number of these events at the hands of climate change; annoyance at a perceived overreach of governmental and state power to remove people from their homes; and despondency in the habits of those who routinely leave their homes to wait out periods of bushfire threat or high-risk weather and then return to their dwellings. Who could deny that those who were embroiled in the chaos of the Taliban's takeover of Afghanistan in 2021, encamped in or unable to get to the Kabul airport, or who sought evacuation trains in war-torn Ukraine following Russia's invasion and brutal bombardments, have an adequate claim or need to be "evacuated"?

The book's approach to evacuation as an object of governance is to draw on approaches that are sensitive to the study of mobility and emergency and that consider the tensions at play in evacuation, especially around state, organizational, individual, and collective agencies. What kind of subject is imagined or assumed to do evacuation or be evacuated? Liberalism has been characterized by free, autonomous, and individual mobility (contra the collective above), as scholars such as Hagar Kotef (2015) and Tim Cresswell (2006, 166) have argued. The crucial contradiction at the heart of liberal motion is that individuals may move at another's expense, that their freedom of mobility may actually imprison other bodies nearby or far away. In evacuation the (im)mobile body could very easily be simplified within technical procedures, diagrams, and registers to have an assumed capacity for mobility. For example, Zelinsky and Kosinski (1991) and Benigno Aguirre (1983) constructed a kind of taxonomic categorization of evacuation where the evacuee appears as an individual "decision-making unit," acting with their own rationalities and enjoying "some measure of freedom in choosing whether to stay or go" (Zelinsky and Kosinski 1991, 8). This book is cautious of such assumptions, led instead by conceptualizations of the politics and (in)justices of individual and collective forms of (im)mobility (Adey 2016; Cook and Butz 2016; Redfield 2008; Sheller 2013; Sodero 2019).

Evacuation has even been viewed as a type of liberal and economic "adjustment" measure in which communities and societies are meant to respond to an emergency. Gilbert White, who was regarded as the father of floodplain management in the United States (Kates and Burton 2008; Macdonald et al. 2012; Tierney 2014), was highly critical of the interventionist strategies of the Army Corps of Engineers, the martial organization responsible for the building and maintenance of river and lake levees like those that circled New Orleans (Molotch 2014, 157). For White, technocratic infrastructures like this worked to increase risk rather than reduce it. This was the "levee effect" (Collier 2014, 287). A levee might actually encourage development, increasing the risk exposure of people to floodwaters. White expressed a form of liberal, technocratic, and abstract thinking (Kates 1971, 448) to the extent that evacuation and emergency planning could be a basic adjustment people could make—to simply move out of the way. Evacuation is anything but that simple, however, as if an individual can simply act with knowledge and make rational choices. Kenneth Hewitt captures the tonal shortcomings of White's approach: "This cool, reasonable view is, however, not only asking a lot of someone facing 'a roaring typhoon': it is a far cry from the world most of us live in ordinarily. . . . Man may appear in the long run to be a 'manager' selecting certain uses for the 'neutral stuff' of nature. Few men have that opportunity. . . . Does all this mean 'choice' is nonexistent? No . . . insofar as action is concerned, choice is largely regulated by the distribution of power in society" (1980, 310).

This book is highly critical of the persistence of this thinking and seeks to demand more of evacuation concepts and practices.

RECURSIONS AND REPRODUCTION

It is fruitful to consider evacuation recursively. Sometimes this means following how evacuation's different forms turn up again and again, to use recursion as an analytics that allows events to be "understandable in their specificity but not reducible to the uniqueness of time and place" (Stoler 2018, 544). Of the most common recursions followed in the book are forms of social reproduction. Feminist perspectives can also be brought to bear critically on this. This book is attentive to a logic that can go with evacuation as a means to put some valued life or something valued through mobility in order to circulate it back into something. It is a protective and reproductive move, to protect reproduction itself and keep what John Preston calls "the socio-temporal web of Capital's value relationships spinning" (2018, 11). Feminist-informed research on "social reproduction" (T. Bhattacharya 2017; Vogel 1983) has sought to understand the ways in which different oppressions, power relations, and the social-spatial organization of daily life are produced and reproduced. This understands that the "activities involved in sustaining and reproducing daily life" (Braedley and Luxton 2015, vii) are not biological or essentialized characteristics of social difference, even if these forms of reproductive life are stubbornly able to persist. While an evacuation might seem quite alien to these labors, I try to keep less with the clear ruptures and descending lines of evacuation than with the "strange continuities" (Stoler 2016, 28) of recuperation and reanimation, as evacuation rips people from everyday experience and brings them back to it, while reproducing nascent structures of power and inequality, lurking in potential in the background objects and infrastructures of our homes, workplaces, and public spaces.

For me, writing as a British academic, it is difficult to escape some national narratives of evacuation that cement these reproductions. A feminist address of reproduction steers us to evacuation's centrality within imaginations of familial protection, endurance, and even state (re)productivity. The "Blitz experience" and the figure of the evacuated child were a core scene in how I was taught history at school (Welshman 1998). It is understood as a crucial moment of social change in Britain, especially as evacuation mobilities revealed important fractures around class prejudice, racial suspicion, and religious sectarianism, and where evacuees were subject to pathologized stereotypes as the carriers of disease, immorality, and incivilities (Welshman 1999). Mothers and the idea of motherhood came under particular scrutiny (Andrews 2019) within different configurations of classed pronatalism and im/morality too as children and some accompanying mothers were evacuated from cities in several stages. Evacuation became a key prompt for postwar state welfarism, but it was primarily an anticipatory measure to protect different kinds of reproductivity and futurity. The principles for this were partly economic, moral, and affective (Overy 2013). Dead children would distract from the war effort, demoralizing a population on whom the government relied for its legitimacy. The evacuation was a kind of withdrawal, a saving, even if some wondered what they were "being saved for" (John Furse, in Johnson 1968).

Given the discussion above on mobile bodies and liberal autonomy, I am interested in the role of evacuation in embodied and biological notions of reproduction too. Evacuation as a protective concept has been surprisingly central to some debates over women's reproductive rights: as a way to protect the lifestyles and life chances of women—as well as their bodies—*from* motherhood. In this form, evacuation offers a critique *and* an affirmative politics of emergency that seeks to reclaim individual and collective socialities. For instance, Jeffner Allen's feminist ethics argued explicitly for a "philosophy of evacuation" that would "get women out of motherhood" (1996, 315). Allen renders motherhood as a different kind of threat: "In evacuation from motherhood, I claim my life, body, world as an end in itself" (316). Evacuation moves women away from motherhood as the end of female purpose, agency, and social status and away from what she characterizes as the masculine "invasions" of power and patriarchy. Allen's philosophy emphasizes a different version of female futures, futures that emphasize "the power of the possible, and sometimes actual, collective actions" (325). Allen's is an extraordinary clarifying mark of what evacuation can be: a protective and positive measure, a moving toward the safety of the possible, open, collective future. Yet Allen's concept is just so absent from so many ways that evacuations are imagined, planned, and done. Evacuative reproductivity is often marked on bodies to individualize, seclude, and exclude, working precisely against the protective, promiscuous, hopeful, and solidaristic concept of evacuation Allen demanded, albeit with a very different kind of evacuation in mind.

In relation to the discussion of liberal individualistic autonomy, evacuation challenges the boundaries around which the individual is imagined and around where or what a life is. We can even locate some of these dimensions within debates over biological reproduction and the ethics of abortion. Abortion contains a range of practices and constraints on mobility (Cordelia Freeman 2020). Some ethical and moral debates over abortion between pro-life and pro-choice positions have hinged around a medical technique long known as *dilation and evacuation* (D&E). This is the means by which a fetus is removed from the womb by techniques that turn an unwanted or unviable fetus (both linguistically and physical-mechanically)—and the mother—into a potentially abject and abstract status. This practice of the evacuation is precisely what will end the unborn baby's life. In this instance, what protection from evacuation *is*, or for whom, is remarkably bifurcated. The fetus's death by evacuation is for the protection of the mother and/or family, even if this is not without severe cost. In such a relation, evacuation is a way to induce death and is a form of extremely conflicted and contingent care. And it is why, from a famous philosophical argument, many activists and philosophers have not seen the right to abort as a right to kill an unborn baby but rather as the choice to determine whether or not a mother has the right to remove, exclude, or "evacuate" a living thing from inside of her, even if that act will directly lead to a fetus's or baby's death. For some, this could even be considered as a "right of evacuation and not a right of termination" (Kaczor 2005, 107). For others, however, abortion should be considered not only as an issue of bodily autonomy but as "a matter of controlling one's reproductive future" given potential mothers "are also acting on their legitimate reproductive right not to become a biological parent" (Overall 2015, 131).

In more familiar uses of the term, evacuation works against female agency and empowerment, and this book is interested in identifying where and why these kinds of inequalities recur. Cyclone Tracy hit the Australian city of Darwin, Northern Territory, on Christmas day in 1974. The city leaders, led by a civilian and military authority—militaries are often charged with responsibility to evacuate others—decided that the city was uninhabitable for women and children. The largest single evacuation in Australia's history sought to reduce the city's population down to 10,500 people from 45,000. People were evacuated by car and airlift to other metropolitan areas.[1] The decision infantilized women (Cunningham 2014), who were excluded from the task of rebuilding the city, while denigrating the reproductive and domestic labor of which women were traditionally the center. The evacuation was a coerced exclusion. The notion of Darwin as a mobile frontier city, fantastic but dangerous, was rehearsed (Hall 1980). As Jon Stratton notes, the rhetoric of "panic and evacuation sits well with the image of a frontier" (1989, 45).

Communications between Darwin and the Red Cross of Australia reveal how women were rendered out of place. In the desolation of the city, with "piles of rubbish (consisting of fallen trees, twisted galvanized iron, wrecked household commodities) in the streets, light poles askew, leafless trees, and no birds"—important aesthetic sensibilities to evacuation's left-behinds— the Red Cross saw Darwin as a "man's town." In their recruitment of a social worker, "consideration" was to be "given to the appointment being a man," perhaps best able to survive a postevacuation context.[2] The already displaced Aboriginal Larrakia community that lived in the vicinity of Darwin found themselves largely forgotten by the efforts of response and recovery, and their social structures were overlooked in the mass evacuation. The disaster was not a triumph, as the local media presented, but a fracturing around racial division, gender inequality, and more communal ways of dealing with property and landownership, leaving the city highly "vulnerable to emergency" (Day 1975).

It is common for Indigenous, racialized, and nonheteronormative social structures, relations, and bodies to challenge the reproductions of evacuation, and this book is interested in identifying the points of rupture when evacuation is forced to confront the limits of its assumptions. As thousands of women were marshaled to holding areas at Darwin's airport, where they were placed on planes with their babies and pets, stories of Greek men dressing up in women's clothes in order to be allowed onto the evacuation flights surfaced.[3] The rumors contributed to the decision to crack down on population movements, requiring IDs and reentry permits. Ethnic slurs surrounded Greek and other immigrant populations, which saw the imaginary cross-dressers as befitting their characterization as "low-life scum," while continuing the homophobia common within the "frontier 'mentality'" of the city (Kerry 2017).

In the wash of these emergencies, multiple reproductions work through evacuation as cherished, protective features and futures return to essentialisms of biology and social difference. A particular idiosyncrasy of Hurricane Katrina was the so-called evacuation babies or Katrina babies and a higher-than-average birth rate, which the media framed as a story of hope within the "rehabilitation" of the city. The celebration of biological reproduction continued a narrative of heterosocial and normative notions of social futurity in which not only queer but Black lives would not fit, despite the "nonlinear, nonbiological modes of reproduction [that] were available to marginalized populations" (Chapman 2017, 83). There is an almost inevitability about some of this in evacuation's recursions. Evacuation continues to exacerbate highly unequal forms of social reproduction and the disinvestment in reproduction itself, as the "social warrant of hostile privatism" was what turned a flood into a disaster (Katz 2008, 18).

The Grenfell Tower fire disaster in the London borough of Kensington in 2017 suggests symptoms of the same problem. Seventy-two residents who could not leave and had even been encouraged to stay inside (to stay put, or not evacuate, somewhat similarly to the *Sewol* ferry disaster in South Korea, in which 250 schoolchildren drowned and which was seen as a huge national failure, particularly over the failure of the crew, coast guard, and other authorities to initiate and direct an evacuation—shifting "responsibility for evacuation from one to another" [Jin and Song 2017, 232]) suffocated or burned to death on the upper floors. Several babies and small children were believed to have been dropped from higher floors to be caught by bystanders. None were. Some witnesses mentioned seeing a mother holding a baby out of the window to help them breathe. A BBC *Newsnight* (Grossman and

Newling 2017) inquiry imputed that bystanders may have supposed the baby was about to be dropped to safety from the flames. The deaths of children, and tales of some spectacular rescues, see an emergency evacuation amplified by the child's symbolization of reproductive futurity.

These examples of "falling bodies, burning towers" demonstrate some of the uneven "heft" Rob Nixon identifies in the "eye-catching" spectacular emergency of "fast violence" (2011, 3), yet they build on racio-colonial and often highly classed and gendered structures of power that have supported much longer-standing inequalities that tend to reproduce individualizing rather than collectivizing tendencies within a social order. This book challenges how these lie in potential within evacuation governance through a recursive analytics, which is especially sensitive to aesthetic orders through which such tendencies continually recur, and where they might be challenged.

AESTHETICS AND COMMON SENSE

This book foregrounds the aesthetics of emergency evacuation (O'Grady 2018). Evacuation is expressed, represented, and experienced via "aesthetic registers" and particular affects. In this section, we open out aesthetics as a powerful way of addressing the visual and other sensory forms, judgments, and affective experiences that evacuation and emergencies shape. For example, while we have been highlighting evacuation's frequently negative and abstracting and almost desensitized feelings, it also brings certain promises. Evacuation promises. It holds a kind of futural affect that narrativizes aesthetic experience (Ngai 2012) as a hopeful, open address that the future will be taken care of because the present will be reproduced. Yet it comes with closing, disclosing, illusionary, and confusing powers that work precisely because those registers are not so readily distinct.

Pause to consider the 1993 installation by Ilya and Emilia Kabakov, titled *Emergency Exit*. They proposed an overwhelming gallery of artwork, installed in a mazelike pattern in a big exhibition space—the original design was for the Halle Tony Palmier in Lyon. The conception was to provide some release, an escape from the "horror and panic," the seizure of "too much art." For them, evacuation could mean protection, "to slip away, get lost in a corner someplace, catch one's breath, if only for a minute . . ." Evacuation combines its multiple connotations with protective escape and the fantasy and possibility of withdrawal, of affective release. The gallery goer, they gesture, might then see "between the walls of the pavilion a dark crevice . . . and disappear from this 'celebration of art' (Each person knows this desire—to escape)."⁴ Their plan is wicked, an illusion carefully designed to resemble an emergency exit corridor and a glazed doorway enticing the potential evacuee with a glimpse of a sunny autumnal landscape behind it. Except the vision of escape is meant to disappear. The gallery goer soon realizes they are looking at a painting of a parkland landscape in a backstage studio space. The evacuation—at least as planned—is intended as a kind of release from some of the feelings commonly associated with emergency and evacuation—like panic—through humor, playfulness, confusion, and dissonance. And, in this instance, it evokes surprise, disappointment, and then, presumably, indifference or glee at being tricked.

The felt expressive and sensory orders of evacuation are crucial to an aesthetic inquiry on the topic. I build on the development of aesthetic thinking via aesthetic categories, judgments, and affects through the work of Sianne Ngai as well as a wider body of researchers beginning to attend to the aesthetics and anaesthetics of mobile life (Barry 2020; Bissell 2022). Similarly, Ghertner, McFann, and Goldstein's volume on security and aesthetics aims to "comprehend the sensory, symbolic, and affective experiences integral to the regulation of bodies and spaces," which constitute different forms of security (2020, 3). They examine the worlds of security between "affect and order, sense and judgment, and inclination and directive" (4), drawing for the most part on Rancierian aesthetic judgments—an understanding that politics is performed through a distribution of the sensible. I turn particularly to Ngai's investigation of aesthetic categories, some of which are minor and persist between more obvious blocks of sensibility, which is to attend to combinations of judgments and affective experiences. These aesthetic categories, which may come all at once, summon particular capacities to affect and be affected, and presuppose our relations with others. This is to consider the different sides of aesthetic categories, a side that is about judgments and the utterances that form them, but also "the form we perceive, a way of seeing." Both, Ngai suggests, are "sutured by affect into a spontaneous experience" (2020, 1). One example Ngai (2012) examines in her wider exploration of the "interesting," "zany," and "cute"—aesthetic products from Western capitalism—is relevant for our study. It comes from a passage from Herbert Marcuse interpreting the RAND Corporation's development of strategic war gaming, where Ngai notices Marcuse's interpretation of RAND's war-gaming products that bring the perception of conflict into a softening, fun, and domesticated world, alongside "the informational, technocractic style of the interesting" (14). While others have noted the anticipatory affective qualities of RAND's styles of managing and governing catastrophic nuclear futures at length (Ghamari-Tabrizi 2009), Ngai's concerns toward aesthetics help us interrogate those practices within historically emergent aesthetic categories, which contain felt experiences, expressions, and judgments.

Relatedly, some studies of infrastructure have become attuned to the political charge of infrastructure as an affective promise, often through the modulation of an "aesthetic address." These are, for Brian Larkin, the "ambient life that infrastructures give rise to—the tactile ways in which we hear, smell, feel as we move through the world" (2018, 177), and equally the way political force and authority is produced and contested. Part of my approach involves advancing what is probably a more familiar aesthetic critique rendered around infrastructure, in what Kregg Hetherington (2019) and others (Star 1999) have characterized as a figure-ground perspective, where critique lies in the uncovering of previously submerged, "sunk" (Graham and Marvin 2002) materialities and logics we name infrastructure. But so do the multiple and contradictory aesthetic interferences that compose evacuation and emergency (Barber 2019), as Ngai's take on RAND's many products and practices of anticipatory governance in nuclear war gaming combined different aesthetic experiences to bring war into a safe space of fun and "absolving cuteness" (2012, 14). Evacuation infrastructures simultaneously reveal, conceal, and aestheticize different logics. The Kabakovs' is ultimately a capricious and playful experiment with the excesses, disorientations, and aesthetic judgments of evacuation.

A different way into this can be found in the "Blue Lines" project developed in Wellington, New Zealand, by the Wellington Region Emergency Management Office (WREMO). In 2019 I visited the lines and spoke to an official, who kindly took me on a driving tour. The scheme draws on the standardized emergency evacuation signage system for tsunami evacuations, which itself drew on international emergency exit and evacuation signage based on a pictogram for a Japanese Sign Design Association competition. The competition followed a devastating fire in the Sennichi department store in Osaka Prefecture in 1973, which raged through a nightclub on the upper floors of the building (Fujitsuka 2001), where people perished stranded on the roof and jumping from windows. Female cabaret dancers were famously pictured clambering down fire towers or being carried by firemen. In what has become a notable contouring of gendered affect common to emergency, the *New York Times* report made use of familiar aesthetic judgments of mobile bodies, describing a "panicky" stampede and "people tumbling down a stairway like an avalanche" (*New York Times* 1973). Yukio Ota's imagery was submitted as an ISO (International Organization for Standardization) standard and adopted in 1987. Emergency evacuation signage can be used to try to quiet and subdue the heightened feelings of disaster, as a way to direct and channel mobility, to avoid confusion and indecision. As we will see later, the cool, chilling, technical affordances of evacuation governance seem pitted against fears of a volatile, hot, individual and collective physicality of panic that might rupture evacuation. The media reporting in Japan made aesthetic judgments. But the "running man" emergency exit logo of Ota's design is also promissory: it invites the hope of putative safety at the open door. *All* we have to do is to step through it.

Signs and public instructions use particular visual and more-than-visual addresses that try to provoke and prompt aesthetic sensibilities and judgments. In Ngai's terms, aesthetics hail or "call forth not only specific subjective capacities for feeling and acting but also specific ways of relating to other subjects and the larger social arrangements these ways of relating presuppose" (2012, 11). The Wellington Blue Lines are painted directly onto the road surface, and I tried describing this to my four-year-old son on a kitchen table using a blue bag and some cups to demonstrate. The cups were houses. My walking fingers were people escaping. I twisted the bag into a kind of line and fumbled an explanation. The line tells you where the big waves go to. If you live on this side and you run across the line, "Ta Daa! You're hopefully safe." This, I explained, is "what is called evacuation." My son then asked a characteristically tangential but imaginative question: "But if the wave brings sea creatures onto the land, won't they be in their way?" With a bit more discussion, it became clear that he actually meant this from the perspective of the marine life, not the humans. For my four-year-old, sea creatures are much more interesting than the people who are "in their way." His question about nonhuman participants in evacuation will become important through the book and raises a point about who evacuation is meant to be for and how that concern is distributed sensibly.

Evacuation's aesthetics are enfolded within an often unspoken valuation of who or what is seen and counted. One's mobility might depend on another's. This is a political and aesthetic judgment, working with an a priori but politically constructed delimitation of what Jacques Rancière brokers into the "visible and invisible, of speech and noise" (2004, 13). Evacuations are relations between human and nonhuman life moving, mingling, and living among turbulent times. My son's question helps me think how evacuations delimit and include certain lives, to the extent that the nonhuman lives my son wondered and worried about are rarely figured in evacuation apparatus. The lines rely on an arrangement of particular cues, signals, perspectives—

arrangements of the sensory in order to mediate and communicate a number of things to different sets of agents. It is a process that, in relation to public danger signs in Bogotá, Austin Zeiderman has labeled the "calibrations" of "sensory perception to the dangers that inhere in an otherwise familiar milieu" (2020, 72). The Blue Lines came from a community consultation project to devise the simplest way of showing where to move to safety in the event of a tsunami on the roads and in the neighborhoods in Wellington region council districts, oriented toward cars moving at some speed in the hilly and wind-blown neighborhoods. They are background furniture just at the threshold of awareness, hopefully surfacing in significance when emergency threatens.

A debate in the community followed the appearance of the lines as an "aesthetically suspicious thing," to use Ngai (2020, 1) once more. Some were worried about their accuracy. Would they devalue house prices? Others were confused. The lines relied on an understanding of one's correct orientation to them. Unclear of which side of the line they should be on, some residents took to social media. In response to the Wellington residents' concerns, WREMO provided an explanatory diagram—an abstract representation of the sensible thresholds and normative judgments coding evacuation planning (figure I.2).

The Blue Lines debate raised to the surface the planning for tsunami, flooding, and inundation in the greater Wellington area and became a way into the scientific reasoning underpinning the plans. The lines surfaced emergency evacuation possibilities within everyday lives and concerns and struggles, even if those possibilities seemed distant and unbelievable. As Nathaniel O'Grady suggests, "Aesthetics renders future emergencies present . . . on affective and sensorial registers" (2018, 85) but without necessarily presupposing their interest or disinterest.

In emergencies, urgency, panic, and calamity are common affects that are often regarded as causing evacuation as well as being the feelings evacuation is meant to guard against. As examples of the transversal and nonlinear characteristics often attributed to affective life (Massumi 2002), they are conditioned by contextual situational unfoldings of futures and pasts that play in the present. Evacuation leaves a paradoxical remainder, both a surplus and excess of the feelings that it seeks to subdue and that may have precipitated it—especially when evacuation is conjured as a manner of flight. While the book traces a kind of evolution of understanding and management of evacuation, and ultimately of the "attachment" (Anderson 2014, 92) of evacuation to certain kinds of affects and atmospheres, the aesthetics of evacuation can

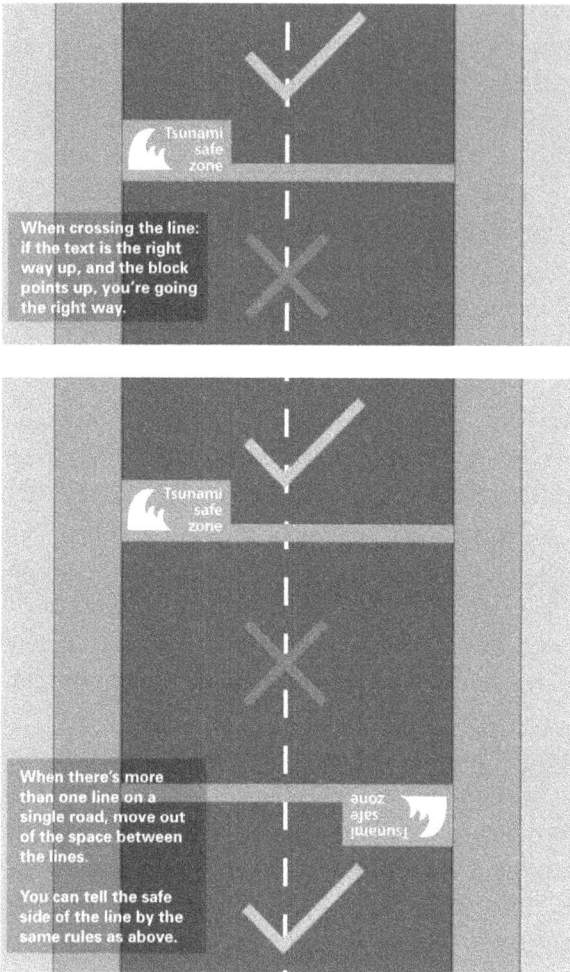

FIGURE I.2. A WREMO graphic introduced to help explain people's perspectival orientation to the lines (Hutt City Council 2017).

mediate and tune in or out of affective life. They shift our capacities to affect and be affected and maybe move and not move.

Today Wellington's Blue Lines project reflects an embodied and mobile orientation in terms of who or what comes to count in evacuation, how evacuation can be sensed and provoked, and how aesthetics might be used "to render articulable and to govern futures whose uncertainty . . . had been considered beyond the realm of the fathomable" (O'Grady 2018, 71). The signs depended on *and* structured a way of seeing and perceiving by the driver or cyclist who is meant to see them via a glancing (Urry 2004), limited, and mobile position, perhaps through glare or rain (figure I.3).

FIGURE I.3. Spotting WREMO's Blue Lines. Photo by the author, 2019.

The Blue Lines are an example of a considered and creative evacuation scheme in which residents were involved in the process of designing the signs, engaging in heated discussions, rejecting them, and coming up with alternatives. The scheme orients people to evacuation, where evacuation doesn't have to mean its implicit or eternal subtraction from the political. The lines mobilized the senses of mobile bodies and subjects who constructively critiqued the tendency to simplify and universalize the bodies expected to do it. They invited public scrutiny of the "commonsense" rationalities that underpinned them, even if they were broadly embedded within a system of private automobility.

One of the most frequently used aesthetic judgments on evacuation is whether it is common sense to evacuate at all. What this book builds on are the ways evacuation seems in itself, and in its relationship to others, to be judged as a negative act—not simply in a sense of the felt or emotional or in the physical moving away from one place to another but as a conditioning property. In such a judgment of what is common sense, and especially in the universalizing maneuvers such a judgment often assumes and performs, something is removed. Perhaps the most famous and explicit of those judgments was given by British Conservative member of Parliament (MP) and former leader of the House of Commons Jacob Rees-Mogg in a radio interview in which he suggested that the victims of the Grenfell Tower disaster didn't use what he called "common sense" to *evacuate* and leave the building. Rees-Mogg was just one of a long line of people to echo the directionality of blame in evacuation and disaster by declaring what is common sense. This is a particular kind of aesthetic closure that renders evacuation as *apolitical* and helpfully forgets that the adding of the dangerous aluminum composite material (ACM) cladding resulted from a process of gentrification designed to improve the aesthetic appearance of the building for wealthy local residents—early planning reports had characterized the tower's appearance as "blight" that decreased land values (Grenfell Action Group 2017). The comments served to render the tower's residents—often multigenerational migrant families— who followed the advice of police, fire officers, and call operators (BBC 2019) as too stupid to know better and ignored the social inequalities that put poorer and Black and ethnic-minority peoples (Hanley 2017; Preston 2018) within the "ordinary verticalities" (Harris 2015; Rosen and Charney 2016) of social housing tower blocks in Britain (Dorling et al. 2007).

Tim Cresswell (2006) has teased out a similar relation in debates during Hurricane Katrina around access to private automobility. Cresswell takes on an article posted in an American magazine by a reporter identifying the lack of automobile ownership as a key barrier for Black Americans to escape the city. Cresswell (2006, 261) quotes from the piece, which claimed at the end of the article that "it was auto ownership, not race, that made the difference between safety and disaster." This is another, similar aesthetic formulation, when mobility is emptied of its social content, and car ownership is divorced from structural inequalities, racisms, and violence (Culver 2018). Evacuation is closed off from its social and economic consequences and the social and economic structures that condition it and, in recurrent moments, absolved from the passions, ethics, and responsibilities that might interfere with its other cool and calculating planning aesthetic. Separating "mobility from race (and class and age, in particular) is simply nonsensical," argues Cresswell (2006, 261), even if it appears common sense to do so.

In other words, evacuation appears at once empty, while also emptying. It appears to lack substance beyond a very technical sense of a process—which of course is not true—while it itself is able to withdraw the relations of other things and events to a technical register as opposed to a social or political one. And yet it is often bound up in very specific feelings-based aesthetic evaluations. These combine Ngai's attention to the kinds of ways aesthetics seems to compel particular codified ways of seeing, talking, uttering, and writing, while aesthetic judgments are bound up in simultaneously normative, discursive-evaluative, and affective moves.

HOW THE BOOK IS ORGANIZED

The book's chapters build and organize different facets of evacuation, establishing connections and lines of association from different contexts, using scenes and juxtapositions to stay with the continuities and discontinuities, dispersions, recursions, and comings-together that Stoler invites within a critical but unconventional and recursive genealogy. The order, though, is broadly chronological, where chapters pick up on evacuation's names, concepts, and practices as they twist and turn, fragment and reaggregate, through space and time.

In chapter 1 we look up. The book examines evacuations as they have developed within high-rise emergencies, especially in North America, where turn-of-the-twentieth-century garment factory fires placed the urban working classes in precarious working and living conditions, especially within high-rise tenement buildings. In the context of women's suffrage and freedom movements, the events coincided with unionized disruption and led to eventual workplace reform. The chapter works backward from the evacuation of the Twin Towers of the World Trade Center by teasing out the development of building and engineering codes and standards. From those standards to the models that have been used to investigate events like the 9/11 evacuations, the chapter examines how certain assumptions and imaginations of the mobile body have been drawn on. These bodies are often rendered in highly normative ways that reproduce raced and gendered assumptions of capacity to move and, indeed, these bodies' culpability for the emergency they are trying to escape. The chapter sets up the particular diagrams and diagrammings that draw evacuation mobility through particular relations and interruptions of power. The chapter alights especially on the ways bodies seem to evacuate otherwise to the diagram, and in a manner that refuses the individualizing assumptions that seek to govern their evacuation and that blame them for moving differently. The intimate and embodied acts of shoe sharing, hand holding, and moving together offer ways in which some have been able to survive and endure emergencies.

Chapter 2 focuses most on vehicles. It continues from the late nineteenth and early twentieth century to examine quite different evacuations during wartime, where evacuation was a way to remove the injured and wounded from battlefields, especially in northern Europe during World War I. This chapter focuses on the "viapolitical" (Walters, Heller, and Pezzani 2021) struggles at stake in the innovation of different evacuation vehicles and technologies that would alter the geo-rhythms of war. The innovations in horse-driven ambulances, motor ambulances, and ambulance trains are explored as battlefields were reconfigured to enable the wounded to be brought back for treatment or put back into martial circulation. At the same time, those very modes of evacuation mobilities were often the symbol of national and highly politicized debates over the state's sense of care and responsibility for the young fighters going to war, where the embodied sensibilities of evacuation mobilities became a proxy for government interests in protecting its fighting forces. The chains and networks of evacuation shuttled the public's concern back and forth to the battlefield, becoming far more intimately involved in the conflict and war's violent consequences for fleshy (non)human bodies and minds. Evacuation involved caring configurations of military and non-military bodies brought closer together, vibrating with affective intensities. Similarly, the vehicles of evacuation are explored as key spaces for the involvement and evolution of humanitarianism, as aid societies and women joined an emergent apparatus for both moving and caring for bodies on the move.

The infrastructures and vehicles harnessed by militaries and medical providers in war are shown to bear out very differently when evacuation is drawn into processes of incarceration, deportation, forced displacement, and even mass murder, and where words become different kinds of vehicles for evacuation politics. In chapter 3, perhaps the most extreme forms of evacuation are explored in the context of World War II and the postwar period of struggle for redress over the lexical meaning and practice of evacuation. In these instances, evacuation emerges as a set of terms—and aesthetic practices—used to disguise the forced mobility of people not away from harm but into it, even in the name of protection and care. Paralleling and juxtaposing the use of *evacuation* to name the systems and practices of expelling Jews from Nazi Germany and occupied Europe, and the incarceration of Japanese Americans in the United States, the chapter wields Stoler's concept of aphasia. Aphasia helps make sense of the disorienting and wicked ways that the words, vehicles and infrastructures, and aesthetic practices of visualizing and enunciating the forms of mobility named *evacuation* disguised them. The chapter explores the layering of meaning as evacuation intersects as an almost palimpsestic device with other practices of forced mobility and displacement. This becomes particularly clear as the sites of Japanese American incarceration are questioned through legal and political redress and the contestation of words such as *evacuation* and *internment* to *American concentration camps*, while the birth and dismantling of some of the camps took place on New Deal territories and reservations designated for Indigenous Native American tribes to resettle.

The urban context of evacuation explored in chapter 1 is returned to in chapter 4 in the situation of the Cold War urban landscape, where evacuation is considered at the scale of the North American city. The chapter examines the development of particular fields of knowledge and expertise around evacuation, tracing the emergence of different ways of thinking about evacuation as an object of concern and, at the same time, evolving organizational forms of government, national civil defense structures and the organization of university departments and research centers around the problem of evacuation in the context of nuclear war. In the middle of other developments in fields such as sociology, urban studies, behavioral psychology, geography, and what became known as disaster studies, complex interdisciplinary engagements evolved. They used fieldwork and theoretical studies on staging various evacuation events and prior knowledge of peacetime evacuations and disasters to anticipate wartime evacuation. The city was the problem within which evacuation was framed and made sense of, and this meant it was bound up in wider sets of antagonistic concerns over race, class, and urban poverty. Evacuation was calibrated under particular fields of affect over which racialized notions of panic were diagnosed.

Chapter 5 follows some of the urban concerns of the previous chapter, rooted in the Cold War, by taking seriously the charges that evacuation is a dehumanizing process. In decentering evacuation's focus on human subjects and extending it to animals—especially when caught up in pernicious and racialized formations of urban security and control—the chapter explores how evacuation has been drawn as a kind of limit. The chapter uses a form of juxtaposition to examine the ways in which animals have been bound up in evacuation practices—and simultaneously tethered to humans. By focusing on the human-animal relations and ethics adopted in animal evacuation schemes in Britain during World War II, and both during and in the aftermath of Hurricane Katrina in the United States, the chapter explicates different bio- and necropolitical rationalities of care and control that evacuation has presumed and performed. By mirroring these different events, the chapter explores the way animals have been brought into the purview of evacuation, its planners, and its plans but often through variously human-centered notions of agency, ownership, and human-societal relations. How animals were treated in an emergency evacuation in World War II exceeds its particularity as evacuation is conceptualized as a kind of functional limit on life. As evacuation has been expanded as a protective mode to apply to animals, the necropolitical choices that are made over animal life have, in some emergency contexts, rebounded. They rebound from nihilistic choices to kill unevacuable animals living unlivable lives once severed from human attachments, to humans themselves, severed from their lives because they are perceived—like their animals—to be unevacuable and, therefore, unable to live.

Chapters 6, 7, and 8 follow closely the post- and neocolonial logics of evacuation. First, in chapter 6, we continue chapter 5's focus to examine a curious and extreme relation between Hurricane Katrina and the highly studied and exposed geographies of the Israeli occupation of Gaza but through the lens of evacuation, its contestation, and a semblance of related terms and meanings that revolve around evacuation's suspension rather than its protection of reproductive futurities. Written prior to the Israel-Hamas war of 2023, the chapter works through Israel's 2005 experience of "disengagement" from settler occupation of Gaza, which involved the highly contested "evacuation" of Israeli settlers from the Gush Katif settlements. Ophir's divine catastrophic conception of emergency is imputed, as the disengagement evacuations are perceived as causative of New Orleans's own evacuations in a form of retributive justice. While demonstrating the workings of colonial structures of occupation and practices of control and displacement, evacuation moves with and against different habits of occupation. The chapter picks up on the recursions of several evacuative tropes that are used to compare and justify Israel's disengagement from Gaza with trauma, present and past, as the Nazi Holocaust and previous recursive moments of Jewish and Palestinian persecution and displacement are remembered and felt. The chapter explores again the fraught politics of evacuation, as evacuation's duplicity surfaces in these moments as a way to displace settlers who had settled territories previously forcibly evacuated of Palestinians.

Chapter 7 engages evacuation's entanglement with liberal humanitarianism and the logistics of evacuation in the viapolitical angle of two disasters: the 2010 Haiti earthquake and the Libyan civil war of 2011. Both events, of course, were preceded and followed by crisscrossing migration flows that paralleled and sometimes came into contact in a confluence of evacuative and displaced mobilities. The chapter builds on chapter 2, on emerging medical-military evacuation and international humanitarian and logistics efforts, and on the legacy of neocolonial policies of trade and global extractive, infrastructural, and logistical pathways that came together to unevenly evacuate citizens. Citizenship is shown to be a key factor shaping powerful and highly uneven propensities-to-be-evacuated by one's own government, taking place via a variety of uneven foreign militaries, diplomatic and medical officers, NGOs (nongovernmental organizations), and humanitarian agencies. The chapter explores the unevenness of these diagrams of mobility viapolitically and at a large scale, enrolling particular *valued* bodies and citizens, some deemed as highly vulnerable—such as adopted children—as well as alternative ways of representing and interpreting evacuation diagrams by publics. The two cases gesture toward the possibility of more civil and less state-directed forms of evacuation.

The relationship between evacuation and (post)coloniality is pushed even further in chapter 8 on bushfires and evacuation in Australia's state of Victoria, where bushfires cannot be separated from the socionatural history of colonialism, Indigenous dispossession, environmental-society relations, and, of course, climate change. Evacuation from fire is part of a longer, interwoven history of colonial practices and logics. Drawing on ethnographic and archival work exploring the Australian Black Saturday bushfire disaster of 2009, the chapter works back through the history of bushfire, colonial settlement, and evacuation policies. In this context, evacuation from bushfires is difficult to separate from the cultural habits and values around masculine notions of defending the home and family from a bushfire instead of escaping it. Conflicting with these values, the experience of bushfires in contemporary Australia coincides with important ways that settler colonial life sought to erode, marginalize, and "preserve" Indigenous life. Recalling the animal evacuations of chapter 5, the chapter concludes with the events at an animal sanctuary, whose animals were evacuated during the 2009 fires. This is set against the background of the sanctuary, whose protected lands were the legacy of an Aboriginal sanctuary brought under eugenicist colonial legislation in the late nineteenth century, which considered First Nations or Aboriginal and Torres Strait Islander communities as less than human and primitive.

Finally, the book's conclusion, "The End," uses perhaps the wildest, most excessive, and most speculative projects for planetary evacuation as a foil with which to reflect on and progress the different themes explored in this book. Expectations of the end of the Earth have given rise to a whole set of planetary evacuation genres, fictions, and plans to evacuate the planet, but they do not move very far away from the problems, pitfalls, and ethical dilemmas explored in *Evacuation*. We use the chapter to return to different aesthetic forms and categories that have recurred within the book—as it were *revacuations*—which cohere around the diagram and diagramming as linking representations and practices of evacuation governance and its resistance; the idea, space, and state of emptiness as animating negativities; and the figure and promise of the future, which seems to reproduce norms of biological and social reproduction. The chapter concludes with a reaffirmation of the agonisms of emergency and evacuation politics as both a mode of critique and a site of possibility.

FOOTSTEPS

DIAGRAMMING HIGH-RISE EVACUATION

It starts with shoes. The footsteps they make; the signature of movement they trace. The space they take up; the bodies that once wore them. It starts with a pair of brown leather oxford shoes once owned by IT recruiter and World Trade Center office worker Fred Segro that were donated to the 9/11 Memorial and Museum in Manhattan and are held in a glass enclosure (figure 1.1). They can be found in one of the see-through cabinets that display the shoes of survivors of the Twin Towers. They allow visitors to the museum to "walk in the same shoes" as the survivors (L. Jackson 2015). The shoes are meant to render an intimate, imaginative, and empathetic embodiment in the underground space that allows the World Trade Center's foundations to be seen. The space is dominated by the slurry wall; steel columns encrusted with rust and graffiti—fossil-like remnants of the building's box columns fixed to the bedrock. Along with the shoes are the Survivors' Stairs, a concrete staircase that had led from the plaza to the street level below it. Visitors to the museum descend in parallel to the eroded staircase.

A guide for parents to assist their children's experience describes the evacuation of the towers as successful. So many people were able to leave because of the "*orderliness* of those evacuating and the courage of the first responders who helped direct their escape" (9/11 Memorial and Museum 2015; my emphasis). The museum's emphasis speaks to problematic assumptions

FIGURE 1.1. Fred Segro's brown oxford shoes, donated to the 9/11 Memorial and Museum (9/11 Memorial and Museum 2010; photo: Jin Lee).

in the development of the World Trade Center buildings and the history of evacuation planning that runs through them. Mobile bodies, with all their proclivities, passions, expressions, and adornments, are often coded with moralistic and normative assumptions (a common concern for mobility; Cresswell 2006) of what *is* efficient, "orderly," and courageous. There are many pairs of shoes in the museum that have been donated by survivors, and several are paired with stories. Footsteps echoing down the stairs of the buildings in smoke and semidarkness; the press of congested bodies. Shoes are often used like this, populating projects of remembrance to signify loss, catastrophic circumstances, and the ordinariness of banal things. They are metonymic and corporeal "contact points" (J. Feldman 2006).

Shoes were shared between those attempting to evacuate the buildings, such as a few office workers tending to a pregnant receptionist, Julie, who was harmed by shattering glass on the seventy-seventh floor of the building. Walking down the stairs, slippery with water, Julie was encouraged to discard her shoes in the stairwell but, reaching the lobby, found glass and sharp debris that promised to lacerate her unprotected feet. Her companion Fred Segro gave her his shoes—brown oxford brogues—before exiting the building. On their walk to a nearby hospital, Julie found a pair of abandoned women's shoes along the way and claimed them for her own, giving Fred his brogues back. The exhibit celebrates this act of generosity.

In this chapter we walk back these moments through a genealogy of high-rise evacuation planning and building design, focusing in particular on the diagrams that imagine and structure and order bodies and materials in anticipation of the event of evacuation. Following the emergence of evacuation plans, architectures, and practices, especially in North America, shows how these diagrams aesthetically distinguish right from wrong evacuations and, by implication, right from wrong subjects and bodies (differently sized, gendered, classed, and raced) that do not fit but wear against evacuation. The chapter explores alternative evacuations that diagram differently, helping us to understand how the evacuees' apparent lack of "fitness," which saw accusations of panic and fault leveled against them, could instead be seen to offer some of the most sympathetic and solidaristic ways of inhabiting emergency.

MODELS

After the fall of the Twin Towers, planners and engineers examined how the fourteen thousand people who were evacuated from the towers did not move as fast as they were expected to. Simulations of different evacuation models used to plan and regulate the emptying of the buildings had them going faster. They also had them going earlier—there was a delay before the building's occupants actually made their way to leave. Billiard ball–like models equated people with atoms. Other models were much more detailed. The evacuees were captured within sophisticated fire-egress simulations that visualize people moving en masse in large buildings. One such system is appropriately called buildingEXODUS. The models were premised on the assumption that the mobile subjects they were simulating looked and essentially appeared to be the same. Same bodies. Same decisions. Same difference. For World Trade Center 1, or the North Tower, it took almost 2.6 times longer for the 7,500 people who did leave the building safely to evacuate than the simulated models predicted. Experts deduce that, had the buildings been full, around twenty-five thousand people at capacity, a lot more people would have died.

The models draw on a long recent history of evacuation science that understands evacuating individuals as "automatons," simulated as if their behavior is an aggregation, an emergent property of simple rules that add up to complex outcomes of mobility (Helbing et al. 2002; Mas et al. 2012). Panic is taken as an emergent but observable property of people as particles, exerting presumed psychological and physical forces (Helbing, Farkas, and Vicsek 2000). Surveys of the World Trade Center evacuation mention the time it took people to shut down computers and perform mundane tasks, such as removing or changing or putting on shoes, before they decided to leave. This was in stark contrast to the one to two minutes many modelers assumed it would take people to decide to leave. The history of evacuation simulation has seen ever-increasing efforts to build more complex variables into the behavior of bodies to more accurately model or simulate the environment they are forced to leave.

According to the Greenwich 9/11 evacuation survey, which used a sample of interviewed evacuees to assess the average body mass index of the population, 69 percent fell into the category of obese or overweight. Some pinned the slow speed of the evacuation on a number of factors but notably on the apparently corpulent bodies (Averill et al. 2005). The slow evacuation was blamed on evacuees who did not match the planning assumptions. They did not fit normative bodily assumptions of size, agility, and speed. They failed as evacuees.

DIAGRAM

Within the media, and the countless reports and investigations that followed, literal diagrams proliferated. They tended to show an awfully simplistic version of the planes hitting the towers; detailed sketches illustrated the weak points of the steel skin that framed the building. Others showed the configuration of the stairways, or the movements of workers and firefighters and police officers moving up and down the building.

The Australian geographer/artist Kaya Barry, in what she calls "transit aesthetics," examines the diagrams designed to shape response to disaster. Common aircraft safety cards are seen as a "fusion of aesthetic, technical, performative and lived experiences" (Barry 2017, 366). We are probably quite used to these representations, often on doors in hotel rooms or in corridors. How they are presented can be highly regulated: only wooden frames and Perspex will do, according to some national standards and guidelines. They are curiously all around us, but we rarely notice they are there. We could think about the evacuation diagram as a "diagram of power" in the way it orders relations between actors and objects in order to marshal movement. The instructions diagram movement and evacuation in particular ways that cast some movements and embodiments as right and others wrong or dysfunctional. This is a helpful way of understanding how evacuations *diagram*.

As an architecture of power, the evacuation arrows and lines gesture toward not only iterative change but reproducibility; they will steer more than one body, and they may be transposed from one context to another, "detached from any specific use" (Michel Foucault, quoted in Deleuze 2006, 30).

As diagrams of power, evacuations—like queues—seem to sequence things into a series or process of movement; as a "strange attractor," they capture "the motion of a multitude and [direct] it into a sequence" (Fuller 2014, 206). They sequence decisions to be made in a particular order; they sequence a preceding series of events leading up to an evacuation across a variety of geographic scales, from the body in a building to millions of people. Gillian Fuller says that queues "packet" and are processes of "tagging and serialising" (2014, 209). This is a common diagrammatic form in an evacuation process. Evacuating subjects and objects are regularly put into the shape of a passenger, a driver, a pedestrian, someone waiting, someone sleeping. They are tagged, perhaps around their neck, on their arm, on their clothes, on their wrist. They might be serialized or put in order. Sometimes this never happens at all.

Barry experiments with art as an "alternative way of knowing" these instructions, to see diagrams as coproductive of sets of relations, giving "rise to an aesthetic resonance" (2017, 371)—the something left over from the diagram, an excess or remainder, something more. Perhaps this is what Brian Massumi means when he calls a diagram a "technique of becoming . . . a technique of bringing to new existence" (2011, 100). In this sense, an evacuation's diagrammatic form holds together quite contradictory things. It is a representation of an ordering of relations to accomplish movement at a particular moment in time. Building regulations in many parts of the world recognize this. Evacuation planners worry about too many changes to evacuation instructions—lest they become confusing. A diagram could be considered as consisting of multiple routes and pathways, perhaps more as a blur. The line may have been reworked just as a river meanders, leaving remnants of its presence. Gazing on an evacuation diagram in one university department, I soon realized with a colleague that the diagram was not within the glass case it was meant to be but had been stuck on top. Peeling it back revealed an evacuation diagram from about ten years ago with an end date as to when it should be reviewed or replaced. Someone could not be bothered to or could not open the regulation glass to replace it.

This moment was not really simply an act, then, of revealing the inner working of an infrastructure but rather one confused with surprise and a kind of disappointment. Our or my aesthetic judgment, resonating for a moment with my colleague, elucidates a kind of ever-present disenchantment with evacuation (Ngai 2020). The rational, efficient, technical performances it promises are suspended—perhaps in cynicism. Like the gimmick, the immediate spontaneous aesthetic experience of an evacuation diagram contains within it the seeds of doubt in ethical, functional, and historical evaluations of its potential, its efficacy, its genuineness. Evacuation might be conceived as a palimpsest, vibrating with expressive frequency and improvisational diversions, still reverberating in the surprise/doubt as we peeled back the paper. In many moments we simply cannot align to evacuation's diagrams; in this moment my colleague and I found ourselves in sympathy through the diagram's revelation.

The method of evacuating the World Trade Center's towers was fraught. Those below where the aircraft struck mostly left the building by the stairs. Those on the floors above this line had no other way of leaving the building, except for one set of stairs, "Stairway A" (Kolker 2011) in the South Tower, which remained intact. Given the experience of the 1993 bombing, when helicopters evacuated some people from the roof, some headed upward only to find the doors locked or malfunctioning. The main stairways throughout the buildings were narrow. They had been constructed according to an evolution of the building codes that governed New York's high-rise construction in the 1960s. The Port Authority architects initially relied on a 1938 edition of the New York City building code, which was eventually surpassed by amendments contained in a new code that was eventually published in 1968. These significantly relaxed the onus on egress or evacuation facilities. They were described by the building's engineers as more "lenient," and by the concerned fire chief John O'Hagan, in 1976, far more gravely. "It is a series of compromises," he explained, "an ideal code would have discouraged construction" (quoted in Goldberger 1976, 22). Compromises gave the building, as the National Institute of Standards and Technology (NIST) report suggests, a "competitive advantage" (Averill et al. 2005), although there is some slippage as to who or what was meant by this within this political economy of evacuation (Lester L. Feld to Robert Linn, memo, January 1987, in Averill et al. 2005, 201). Most of the adjustments made to the 1938 building code were aimed at improving the amount of leasable office space. Evacuation routes and possibilities, including fire retardancy, were offset in favor of the efficiency of the building to raise capital in the form of rent from tenants. Evacuation space was converted into economically productive space in the service of capital. Evacuation lost out to the "moneymaking side of the ledger" (Dwyer and Flynn 2011, 105).

The change in New York's building codes reduced the space for and provision of fire-protected egress routes, corridors, door openings, and stairways. Fire towers, archaic but protected and designated structures within tall buildings required under the 1938 code and designed for quick exit and firefighter entrance into a building, were done away with altogether (Department of Buildings 1968). These were originally known as *Philadelphia fire towers* (Wermiel 2003) as large mill fires such as that at the Randolph Street Mill in 1881 had encouraged Philadelphia's urban leaders to reform building regulations and factory laws. Protected stairways were conceived by the Franklin Institute committee to be far better than the "objectionable" external iron ladders, because of the latter's exposure to the weather and "because women and children can make little use of [them]" (Baird et al. 1881, 410), the committee explained. The Philadelphia fire towers were designed to be smoke-free and accessible from every floor within a building, yet they had long been seen as deducting "from the rentable area where these towers are required" (National Bureau of Standards 1935, 28).

Many inventors had been devising ways to evacuate buildings more effectively; the curiously fixed adornments of fire escapes were initially flexible and mobile. Jennifer Blair's (2008) exploration of early fire-escape designs finds that the word *fire escape* seems to describe an action, as both *fire* and *escape* do too, but when these words are put together, we have something that counters what both mean singularly. The original fire escapes involved moving parts, wires, gears, pulleys. Although these evacuation paraphernalia did not make their way into the World Trade Center's design, something of their aesthetic and affective import did. The diagram of the contraptions concerns the "mutual and indeterminate affectivity of material bodies (human and non-human) as they move and transform in time" (Blair 2008, 55). The main problem of evacuating a building was believed to be the inhabitants themselves. Early designs saw people's "physical and psychological abilities, their sobriety, and their resulting behaviour" as highly contingent variables and "as indeterminate as the movement of the fire itself" (54). To overcome this indeterminacy of involuntary, untrained, or clumsy motions—even terror or panic—the evacuee would become an immobilized unit within the machinery. The so-called horsed escape came into use by the London Metropolitan Fire Brigade in 1897, the very name aestheticizing a complex assemblage of people, horses, and the machinery of steam-powered high ladders. Allowing people to escape burning buildings from high windows in the "twinkle of an eye" (Holmes 1899), it suppressed anxieties on its arrival (*Daily Mail* 1903, 3; *London News* 1898).

Another kind of objectification seems pregnant within these devices whose displays and exhibitions often featured men operating them and women as their objects to rescue. Women were the object of the gaze of technological disciplination and regulation (Crary 2001) and the sexually vulnerable recipients of the heroicized and masculinized firefighters adept at working these technologies (R. Cooper 1995; Maleta 2009). The escape shute was a woven fabric that could be tied to a window and held in tension by people on the ground below, somewhat democratizing the use of the technologies. The escapee could fall within the first part of the shute at a very steep angle, before being permitted to slide more gently to the ground. My own institution's archives at Royal Holloway University of London, which began as a Victorian women's college of higher education, have photos of fabric fire shutes being drilled by the female students from the Founder's Building, once known as the most flammable building in Britain (figure 1.2). While the evacuation of privileged young women may seem incongruous with the many working bodies caught within this discussion, the collusion of both culpability and femininity with vertical high-rise evacuations is once again shown.

The codes shaping the Twin Towers did not require the wrought iron fire escape. The balcony fire escapes that populated tenement buildings offered a different form of escape from the cramped and hot interiors that lacked air and light. Private life spilled out onto these platforms, in children's play, in domestic labor, in access to air, or in respite from the intimacies and even violences of home—some signaled in the exhausted sleeping bodies lying prostrate on the escapes during New York's hot summers. At this scale the materialities of evacuation would affect domestic patterns and routines and the social life of the city—the escapes an indelible urban signature of evacuation's ordinariness. The Tenement House Commission's official study of New York and Brooklyn in 1900 found that almost a quarter of the houses had no balconies at all, and those balconies that did exist were effectively colonized by the building's tenants as living and storage space (figure 1.3). The commission admonished the inhabitants for this dangerous use of an evacuation infrastructure. The fire-escape "encumbrances" were a serious question, yet this challenge undermined the ways of life of the working classes, inhabiting and diagramming the fire escape otherwise (Bonner and Veiller 1900, 18–19). Elizabeth Abel's (2008) study of the segregation practices of the twentieth-century American cinema and theater shows how some fire escapes even became outside staircases for Black Americans to reach the elevated "crow's nest" of segregated seating. Given the prevalence of theater

FIGURE 1.2.
Fire shute drill.
Royal Holloway
University of
London and Bed-
ford New College
Archives, 1928.

fires, which were as frequent and as deadly as high-rise factory fires, the fire escape was inverted as an inadvisable entrance.

Impractical or uneconomic, the Twin Towers egress depended on enclosed stairways centered around the building's core. The number of stairs was reduced from six to three within the new 1968 code and moved from the extremities of the building to its core. They were closer together but also encased in a weak gypsum drywall. People would exit through a lobby rather than directly onto an open street. This was at odds with the earlier codes, which had even worried about the possibility of "panic conditions"

FIGURE 1.3. Fire escape play. Arnold Eagle / New York Tenement Museum, 1935–37.

emerging in lobby spaces—perhaps from social mixing (National Bureau of Standards 1935, 56). The size of the towers' doors was shrunk by eight inches in width. Some of the stairway handrails did not extend to the entirety of the stairs either, and the stairs were not aligned on some floors, meaning that one would have to cross the floor to get to another stairway. A few generous open stairways were built later between floors in a "democratic style" of corporate office at the same time as use of the fire access stairs was discouraged by the automation of elevator operators. This was a "reprogramming" of the high rise and its occupants (Bernard 2014; Trotter 2014), but with consequences for fears of gendered violence to the extent that "walking sometimes seemed safer than riding" (Reiff 1975).

Revisions to the 1968 codes seemed to continue tipping the balance between safety and economics in favor of profit. Even the 1938 code had been designed to free up real estate development and pave the way for slum clearance through relaxing building safety standards, having been drawn up by the Merchants Association of the city. While it reduced regulation for steel-constructed buildings over nine stories, it removed some of the requirements of the perceptibly expensive "fire-proof" buildings too. The reinvigorated "economic life" of the buildings, it was claimed, would far outstrip other building types and facilitate the speedy "cleanup" of slums and tenement housing (*New York Times* 1937).

ALIGNMENT

As Barry asks, "How are we supposed to align our bodies with the implied actions of flowing arrows, hovering above an unfolding situation?" (2017, 372). The workers who evacuated the World Trade Center did not appear to possess the right bodies or were of the wrong sort of fitness. Since the 9/11 NIST report, the width of the stairways—a mere forty-four inches within the 1968 code—has been questioned. Allowing firefighters and emergency personnel to come up the stairs at the same time as people descended limited the channel of bodies to a single file during 9/11. Where did this forty-four-inch standard come from? Sara Wermiel (2003) traces it once more to the history of the fire escape, where we can find forty-four inches recommended by Massachusetts's head of factory inspection, Rufus Wade, then also chief of district police, who in 1880 specified the construction of a fire-escape design that should be at least twenty-two inches wide, with balconies forty-four inches wide. It made its way into the National Bureau of Standards study, which claimed, "The opinion of many who have studied the matter, 22 inches can be taken as the width of a file of people in motion." The minimum was said to come from "experience gained in the Army," so that forty-four inches would permit two files of people "side by side" but only in "occupancies where fire drills are common" (National Bureau of Standards 1935, 59). It is extraordinary that a national committee cannot pinpoint accurately where the standard has come from and that its contingency on the requirement for "drilling" had been somewhat forgotten. Evacuations are a frequent military maneuver, just as moving together in time (McNeill 1997) has been a key measure of the diagram of the disciplination of martial bodies and productive workers under capitalism (Foucault 1977). But why would organized, drilled, and streamlined military bodily movements set the standard against which the eventual occupants of the World Trade Center buildings would be judged?

This is not to suggest that early building codes were completely unsympathetic to different bodies. Some regulations sought to account for bodies with differing capacities. The 1935 national standard weighed up different models of evacuation and egress space within a building, each one checked off with regard to the relative merits of protecting life versus the work that space could do in raising rentable income. Within the so-called capacity model, stairways were additionally a place of potential refuge, calculated to accommodate every person on a floor without movement, a safe channel to wait out an emergency. This relied on dividing the total area of landings and stairways by an "assumed area per person," or by assuming that one person would be standing "every other tread in a 22-inch width" (39). The panel conceded that evacuating from "high buildings is exhausting even to normal persons," while the "knowledge that there is a safe place for everyone is held to minimize the possibility of panic" (39). Having such a space would allow for the "subsequent safe and orderly evacuation of the building without the necessity of forcing aged or infirm people to travel at a rate beyond their physical capacity" (National Bureau of Standards 1935, 39–40). Another way of dealing with this was to alter regulations for the width of stairways according to the building's use. A refuge was deemed unnecessary for buildings that contained work as their populations were considered "necessarily alert and able-bodied" (59). Another method of calculating egress spaces was through a "flow method," but this was seen to be mainly appropriate to theaters, schools, and other low-lying buildings where continuous movement might be possible.

Pedestrian and evacuation scientists have been pushing at these standards for decades. September 11 confronted them visibly and violently. Jake L. Pauls, John J. Fruin, and J. M. Zupan—some of the most cited and influential authors on pedestrian evacuation—have asked similar questions: "To what extent was a minimum width of 1120 mm (44 in.) ever appropriate for coherent crowd flow, for overtaking movement, and for counterflow? What factors were ignored or misunderstood in setting this minimum?" (2007, 57). They also consider the "dubious" standard in the context of "significant changes in people's body size and fitness" (57). These are excellent questions to ask, but the direction in which others have asked them is more dubious precisely because of the questions they ask of the evacuees themselves, imagined as streamlined and asocial kinds of mobile bodies. Geographers Rachel Colls and Bethan Evans make a crucial and critical rereading of the idea of so-called obesogenic environments, so that it might mean those "particular social, cultural, political and economic environments" that can make "living as a fat body problematic" (2014, 735). Lauren Berlant similarly criticizes how political crises are "cast as conditions of specific bodies and their competence at maintaining health or other conditions of social belonging" (2011, 105–6). How do emergency conditions that demand particular forms of mobility make bigger or fat or unfit bodies a problem? Who is able to "judge the problematic body's subjects" (106)? The World Trade Center's towers potentially made big or fat bodies problematic and deadly to themselves. The evacuee is resolved to the categories of both the vulnerable *and* the victim—victim of their own bodily faults, vulnerable to incipient emergency, and somehow culpable for the disaster itself by their own destructive agencies.

And yet the Greenwich study led by the evacuation expert Ed Galea (2012), while looking hard for the correlation of body mass or size with evacuation speed, actually found none. A heavier or unfit person was just as likely to stop as anyone else. One of the major causes of the slowdown was congestion and counterflow, and what others might call the *withness* of mobility, which challenged the atomizing tendencies of the modeling. This is crucial to evacuation and the ways in which its plans, planners, and building engineers have tended to render bodies as individuals. The models see occupants encountering one another as a form of behavioral "conflict resolution" in a limited space. A person or "node" cannot occupy the same space as another. Conflict resolution "time" is added to their total evacuation time according to probabilities assigned within the simulation.

In evacuation and stairway design, the mobile subject is often cast as a singular and territorial animal. John Templer's (1992) monumental history of the staircase sees individuals marked by an ellipse of body-space within which their bodies will rock and sway markedly during descent as they move their body weight between legs and move their arms and other encumbrances. These little atomistic packages of flesh and movement are characterized as in competition with others potentially able to penetrate a wider envelope of interpersonal space—an acceptable buffer zone or bubble, which, "if violated . . . generate[s] defensive patterns of behavior" (69). For Templer, these bubbles resemble organisms vying for adequate space, where the bubbles can expand and contract and come under "tension, harden[ing], regulariz[ing] and sensitiz[ing] the surface so that it becomes a defensive perimeter" (69). That the World Trade Center evacuees collaborated, helping one another, sharing apparel like shoes, shows us something more irrepressible. People were evacuating together, breaking out of the anatomical and atomistic notions of bodies in motion. Some needed help or wanted to evacuate with friends and colleagues, hardly surprising given the events going on around them. What the building codes, plans, and assumptions had not fathomed was that we move and diagram differently and that our diagrams are drawn *with* others.

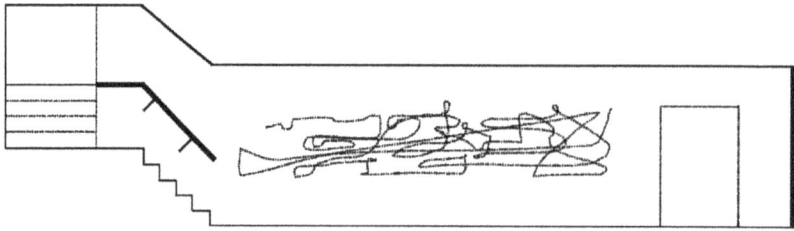

FIGURE 1.4. Diagrams of movement in a white-walled corridor and stairs of Chisendale Dance Space (McCormack 2004, 213).

DIAGRAMMING

The 9/11 World Trade Center evacuation exposes a history of techniques meant to contain certain kinds of bodies from turning awry in alternative vibrations and irritabilities of indeterminate bodies. We might consider that evacuations seek to suppress or quell what David Bissell, Maria Hynes, and Scott Sharpe characterize as a "potentially volatile figure" that becomes conditioned through a "sequence of self-protection rituals demanded by an immanently risky environment" (2012, 695). The diagram works against the volatile body that seems to throw out erratic lines of apparently panicky or maladaptive and even criminalized potentials of movement. But it is equally a line of flight or potential for those who refuse, resist, and generate creative variations. The evacuee may diagram differently. How to consider evacuation diagrammings that are not overdetermined by disciplination or regulation— whose lines work that much more expressively? These are the other lines, such as those that cultural geographer Derek McCormack has traced in another set of stairs (figure 1.4). "Some geometries," he writes, are "partially positive, connective, affective"; the lines he found himself working with were "abstract yet real. Something architectural. Something corporeal. Something incorporeal. Something transversal. Something between" (2004, 212).

Tracing bodies has long been the bread-and-butter work of evacuation pedestrian scientists, following the early technologies and perceptions of time, space, and movement in the first developments in time-lapse photography and industrial time and motion studies that visualized, rationalized, and reconstructed the movement of laboring bodies (see figure 1.5; Cresswell 2006; Solnit 2004; Templer 1974).

Templer showed that stair design had relied on foot and shoe sizes from the 1850s while paralleling nineteenth- and early twentieth-century

FIGURE 1.5. Time-motion studies on staircase movement (Fitch, Templer, and Corcoran 1974, 83).

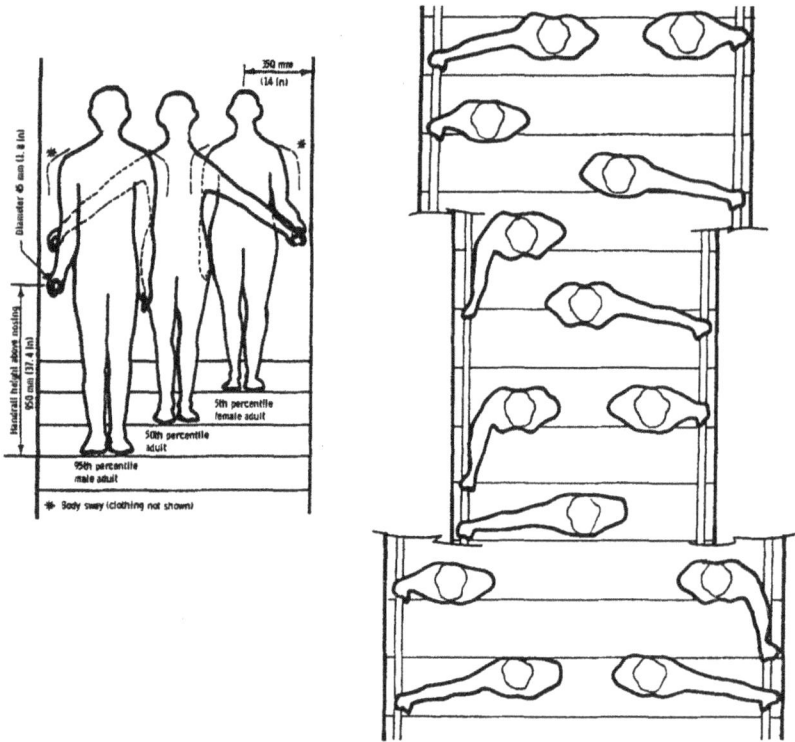

FIGURE 1.6. "Distribution of People on Stairs" (Pauls 1984, 36).

sociological and psychological thought that depicted the crowd as an uncompromising, volatile collective. Various crowd disasters provided the funding for key studies within pedestrian science, like Jake Pauls's *Stair Event* (1979; see also figure 1.6). Theater fires provided some of the earliest disasters to plague urban reformers, whereas sports stadium disasters were as important in the 1970s and 1980s. In response to the Cincinnati rock concert crush of 1979, a small colloquium of experts analyzed the sheer physical pressures involved in the movement of crowds, bending and fracturing steel barriers intended to keep pedestrians in. What they came up with was a highly physicalist version of the crowd, exposing the problems of standards that poorly understood the physical and social forces that were impossible for people moving and dwelling together to resist.

The evacuees of the World Trade Center drew lines to one another through associations, as well as lines to their things and bodily accouterments or prostheses such as shoes. The shoes shared and given by strangers

break out of the impenetrable and defensive social bubbles stairway designers imagined—the "kinespheres" of McCormack (2004) that collaborative
stairway movements ruptured. For evacuation planners, the decision to
change shoes delayed the initial evacuation. "Inappropriate footwear" such
as high heels, slip-ons, and new shoes would slow egress through difficulty
or pain (Fauzi et al. 2014; Gershon et al. 2007). Moreover, several of the
buildings' stairways became congested with discarded shoes, causing people
to trip or fall as they maneuvered around them or awkwardly found their
footing (Corbett 2018). The NIST survey records piles of shoes impeding the
stairways (Averill et al. 2005, 214). The politics, and gender politics, of
the shoe are underappreciated in this context (Margolies 2003) given that
the high heel is an established and tacitly accepted norm of footwear donned
in order to inhabit these settings (Carla Freeman 2000), yet they appear
"inappropriate" for emergency planners indifferent to the gender politics
of footwear in corporate and office life. Women's footwear is often a "poor"
choice, concluded the NIST report. Even if a pair of shoes is recovered as
testament to togetherness by the kindness of strangers giving, swapping, or
lending shoes, evacuation experts advise against this by not accommodating
these alternative diagrammings or the social and economic conditions that
shape them. Indeed, in other settings such as aircraft, it is suggested that high-
heeled shoes be removed prior to evacuating. Evacuation planning appears
to continue to create and then apply fault to the evacuee, while following a
longer pattern of concern especially over dress.

DIAGNOSIS

Jim Dwyer and Kevin Flynn have argued that the fates of the victims of 9/11
were "sealed nearly four decades earlier" (2011, 243) within the building code
changes we have discussed. We know that many decided their only option was
to jump. Evacuation became death. Their bodies even became part of a deadly
rain of metal, concrete, and flesh landing on some below (Mackay 2016).
Within the history of high-rise evacuations, this horror has been more commonplace than you might think. With working-class and immigrant young
women jumping from tall buildings as their only possible mode of escape,
the Triangle Shirtwaist Factory fire in New York City in 1911 became em-
blematic of the poor and gendered working conditions of the working classes
and of just how deadly inadequate fire regulation and emergency protection
measures could be (*Off Our Backs* 1970, 9). The fire, which started on the
eighth floor, saw 149 women killed—most of whom were in their late teens

and early twenties. Under a hundred were killed jumping or falling from the building, and the rest were burned or suffocated within. One fire escape ended in midair on the second floor, which led the women back into the building. Others buckled. The doors to the factory floors had been locked in order to stop the women from taking informal breaks and to prevent their movement between floors. About 150 workers left the tenth floor of the building and escaped onto the roof before crossing to the roof of the building next door. Those on the ninth floor were trapped. By 1911 almost half the factory and garment workers in New York were working above the seventh floor in loft factory spaces (Pence et al. 2003), usually on the upper two to three floors, where light was plentiful, above the tallest ladder of firefighters.

Several kinds of diagrams inhabited this moment of high-rise evacuation, oscillating between the positions mentioned at the start of this chapter and between the aesthetic exposure of fault or cause. One way of representing Shirtwaist was through Brown Brothers' photography, which ranged from imagery of the fire-stricken building and twisted and distorted shapes, including the fire escapes, to photos of the broken bodies of the garment workers lined on the sidewalk (figure 1.7). Passersby go on their way. Others crane their necks upward, presumably looking at the blaze or other workers trying to leave or jump from the building (Mackay 2011).

Ellen Wiley Todd (2005) has argued that different uses of the Brown Brothers photographs can help us understand a gender politics and ethics embedded within the representation of the young women. Different levels of lay and direct authority over the fallen bodies are present. The photos document the processing of bodies on their way to the morgue, a diagrammatic aesthetic to package, to serialize, and to tag, albeit after the event of deadly evacuation. The images are shocking, but I want to explicitly call out the unspeakability of evacuation in events such as this, and the ways in which migrant and working-class women, even in death, were not permitted to adequately voice and contest the structural inequalities that led to the failed evacuation. The tendency within this genre of imagery is perhaps a *forensic* and diagnostic character of evacuation diagrams. One of the most interesting is a pseudo-photograph/montage, which sees a Brown Brothers photo combined with sketches and maps, a synoptic view of the park, and an oblique sketch of Grace Church. The *New York Tribune* (1911) called this a "Diagrammatic Sketch of the Surroundings of Yesterday's Horror," which crawled over the front page of the newspaper (figure 1.8). Other newspapers featured similar imagery with labels, arrows, and numbered legends. Except here the tendencies divide. The forensic diagramming sees culpability within

FIGURE 1.7. Bystanders and police look on from Greene Street watching the
Triangle Shirtwaist factory fire, 1911. Brown Brothers / Kheel Center for
Labor-Management Documentation and Archives, International Ladies
Garment Workers Union Photographs (1885–1985), Cornell University,
Ithaca, NY.

the building, its owners, the regulators; this is at odds with another target
of blame: the evacuee.

The "sketch" was a diagram of power and movement. Subheadings described how several men had used their bodies as a bridge to a neighboring
building on Greene Street, but the weight of the collective bodies was too
much for them to continue. It tries to find causality for the event by using a
mixture of testimony, although very few statements come from the surviving
garment workers themselves but instead from police officers, building inspec-
tors, and other workers from the building. When a garment worker is interviewed, we are unsure as to her reliability as she "swooned" at the "sight of the
bodies hurtling through the air to the passageway below," and on coming to,
"she leaped from one of the windows." An assistant superintendent of buildings suggests that the problem was with the building fire codes themselves.
Fire Chief Edward F. Croker blamed the lack of fire escapes. The Tenement

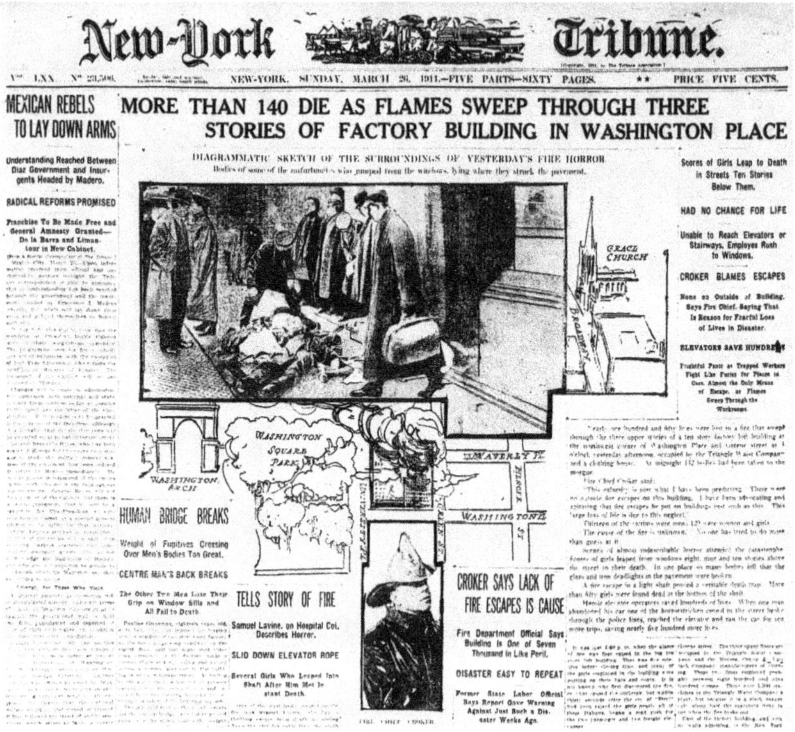

New-York Tribune.

NEW-YORK, SUNDAY, MARCH 26, 1911.—FIVE PARTS—SIXTY PAGES. PRICE FIVE CENTS.

MEXICAN REBELS TO LAY DOWN ARMS

MORE THAN 140 DIE AS FLAMES SWEEP THROUGH THREE STORIES OF FACTORY BUILDING IN WASHINGTON PLACE

Understanding Reached Between Diaz Government and Insurgents Headed by Madero.

RADICAL REFORMS PROMISED

DIAGRAMMATIC SKETCH OF THE SURROUNDINGS OF YESTERDAY'S FIRE HORROR

Scores of Girls Leap to Death in Streets Ten Stories Below Them.

HAD NO CHANCE FOR LIFE

Unable to Reach Elevators or Stairways, Employes Rush to Windows.

CROKER BLAMES ESCAPES

ELEVATORS SAVE HUNDREDS

HUMAN BRIDGE BREAKS

Weight of Fugitives Crossing Over Men's Bodies Too Great.

CENTRE MAN'S BACK BREAKS

TELLS STORY OF FIRE

Samuel Lavine, on Hospital Cot, Describes Horror.

SLID DOWN ELEVATOR ROPE

CROKER SAYS LACK OF FIRE ESCAPES IS CAUSE

Fire Department Official Says Building Is One of Seven Thousand in Like Peril.

DISASTER EASY TO REPEAT

FIGURE 1.8. A "diagrammatic sketch" of the Shirtwaist fire. From the *New York Tribune*, March 26, 1911.

House Commission had only recently identified the Shirtwaist factory as one of seven thousand in the city requiring greater fire escape capacity. Yet the inquest's finding summarily failed to apportion guilt. The *Literary Digest* reported, "The monstrous conclusion of the law is that the slaughter was no one's fault, . . . or, in the fine legal phrase which is big enough to cover a whole multitude of defects of justice, it was 'an act of God'" (1912, 6).

Women's bodies are silenced as failed evacuees, animated instead by bystanders, elevator operators, and other male workers as precipitously unstable and at fault. Gendered and classed, the workers are characterized as "girls nearly all of them Italians," innately passionate and melodramatic according to national stereotypes. They are observed to have broken windows with a "frenzied blow" and to have begun a "mad rush for the two passenger and two freight elevators." The intersection of their heritage and gender signals their "lack of capacity for survival in the industrial order" (McEvoy 1993, 637–38). The elevator operators—cast as heroes by the press—worked as long as the elevators still operated (Bernard 2014; Diffrient 2018) and described a scene of "young Italian girls, their eyes starting from terror," fighting with "insane strength and savagery to gain the elevators." Some "screamed for help" or made "flying leaps" into the elevator cars (*New York Tribune* 1911, 1). The accounts describe a dehumanized or animalistic selfish self, with the girls throwing themselves at and clinging to the elevators' barriers—some with their teeth. Apparently some girls died in this violence of escape: "dead and mutilated bodies" who were "not killed by fire, but torn to pieces, almost, by frenzied human hands. . . . It was a mad fight for life." The newspaper speculates that it was the way in which the girls tried to leave that killed them: "It is certain that many of the unfortunate creatures were killed not by fire, but in the mad trampling of many hundreds of feet" (1). The evacuee becomes a susceptible figure falling into an animalistic-like potential to panic and harm themselves. The crowd below even catches a kind of hysteria.

Applications of panic to evacuation are commonplace. John Protevi suggests that the triggering of rage and panic might even be understood as "an evacuation of the subject as automatic responses take over" (2009, 50). And yet the attribution of panic to emergency evacuations (discussed in chapter 4) is rarely unproblematic. Panic becomes a way of understanding how bodies seem to break out of the atomistic assumptions modelers and building designers have imagined, but this equally subjects the panicking body to classed, gendered, and racial prejudices. Even the jury members to the inquest complained, "I can't see that any one was responsible . . . it must have been an act of God. I think the factory was well managed, and was as good or better than many others. I think that the girls, who undoubtedly have not as much intelligence as others might have in other walks of life, were inclined to fly into a panic" (quoted in McEvoy 1995, 637).

Leon Stein repeats accusations of panic seizing the young women, resulting in a violent selfishness: "Panic stricken girls battled each other on that rickety, terrifying descent" (2011, 57). The girls fly into a panic. Panic pushes them into flight as they are reduced to an animalistic mass. A fire insurer and consultant described "human bundles self-flung to the pavement as a choice to roasting in the flames behind. Bridges and chains of stout limbs and bodies are constructed where apparatus does not avail"; those found trapped were "doomed like trapped beasts in the jungle" (McKeon 1910, 343). The enchained worker bodies seem evidence not of empathy and solidarity or strength but of primal, maladaptive vulnerabilities.

The previous year, the Newark Factory Fire of November 26, 1910, had likewise seen numerous young female factory workers jumping to their deaths

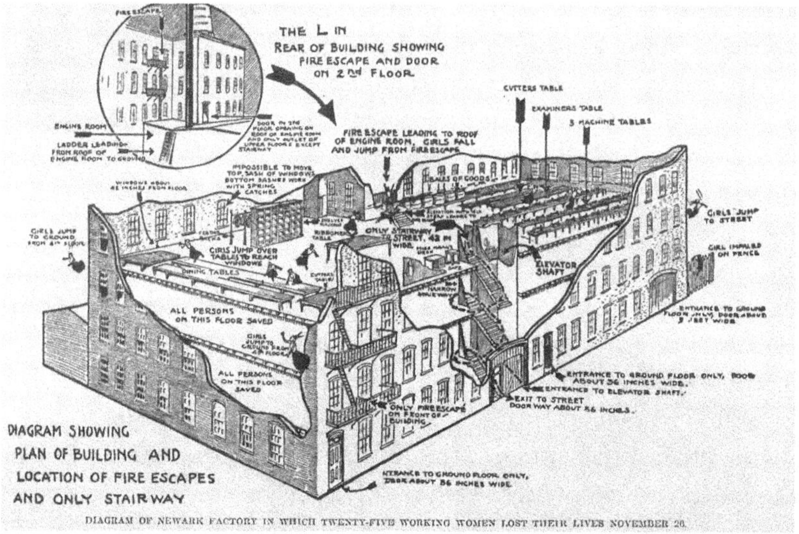

THE L IN
REAR OF BUILDING SHOWING
FIRE ESCAPE AND DOOR
ON 2ND FLOOR

DIAGRAM SHOWING
PLAN OF BUILDING AND
LOCATION OF FIRE ESCAPES
AND ONLY STAIRWAY

DIAGRAM OF NEWARK FACTORY IN WHICH TWENTY-FIVE WORKING WOMEN LOST THEIR LIVES NOVEMBER 26.

FIGURE 1.9. "Girls" leap from the Newark Factory Fire (McKeon 1911, 532).

from the higher floors (*Newark Evening News* 1910). An extraordinary diagram published in the *Survey* (McKeon 1911, 532) shows a detailed cutaway look showing the outside and inside spaces of the factory in one image (figure 1.9). The sketch is annotated with text showing the placement of escapes, windows, interior stairs, elevator shafts, and the machine tables. The annotations are diagnostic and accusatory, identifying window sash locks that made the windows impossible to open. The windows' distance from the floor required the girls to jump onto the tables to reach them. At first glance, it is easy to miss the drawn figures of jumping/falling girls, sketched in different postures and forms of bodily expression of hot desperation, clutching at something, their hair trailing behind. Skirts billow in the rush of air. Some are impaled on spiked railings below that were intended to keep workers in and others out, as was also the case at the Triangle Shirtwaist fire.

Again, the representations of factory workers are resolved as victims of inadequate factory regulation and building codes, as well as culprits harmed by their own inadequacies. The coroner's jury concluded that blame could not be pinned on the factory owners; the workers' deaths were accidental, caused by their maladaptive actions: they jumped. The jury identified an individual worker, Carrie Robrecht, who became the "fall-girl," who "came to her death by misadventure and accident caused by a fall, and not as the result of a criminal act" (*Survey* 1911a, 520). Labeling the cause of death as "misadventure," the findings shifted culpability. Sharing considerable indignation at these findings, the *New York Times* complained, "It will not do to say that the innocent victims lost their self-control and caused their own destruction. Few if any of them jumped from the windows until all hope of rescue was gone and their clothing was in flames. It seems that the accumulation of horrors would have been sufficient if a spike gate had not been left standing open to impale many of them as they fell" (1910, 8).

The Shirtwaist story is bizarrely echoed in a later high-rise fire at the Shirokiya Department Store in Tokyo in 1932. The ladders were once again not high enough to reach the upper floors of the building, but the most notable narrative that emerged from the fire was that some of the victims, which included eight "salesgirls," refused to jump from the building into fire nets or to descend via fire escapes. The traditional kimono was not worn with underwear, and the women's modesty was believed to have prevented them from making their escape. While the event has recurred as an almost urban legend, particularly for encouraging more Western-style dress by Japanese women in high-rise buildings (Suzuki 2023), it discursively constructs the female body as once again made vulnerable to the contemporary urban social order but somehow culpable for the failure of the evacuation practices and technologies that might have saved them.

DRILL

The Triangle Shirtwaist and Newark Factory fires show the evacuees, in refusing the individual limits of body-space inscribed on them, even in the form of an industrial labor force whose movements and practices were already rationalized and made efficient as a material-energetic abstraction of productive processes and capital (Cresswell 2006; Rabinbach 1992), to be found wanting. Their movements were superfluous under conditions of emergency and the demands of evacuation.

The labor reformer, feminist, and suffragist Mary Alden Hopkins found that the workers' disciplination to their tasks may actually have prevented them from leaving. As a worker explained, the attentive economy of the work meant that they were not even aware of the fire until it was too late: "A piece-worker must keep her eyes on her machine if she wants to make out, and I didn't know anything was wrong until I happened to look up and saw all the girls running to one end of the room" (quoted in Hopkins 1911, 665). We might compare this to the last-minute email checking and even monetary trading that delayed many of the workers in the Twin Towers in leaving their desks.

Stein accounts for one woman who "punched" or "clocked out" during the chaos of the Shirtwaist fire. The workers were valued for their productive power as semidisposable, and, as Arthur McEvoy argues, "the Triangle fire wrote out the power of employers to extract wealth from their workers on the bodies of the victims themselves" (1995, 630). The female worker as evacuee is reduced to an "abstraction—a contractor of labor power or an append to the machine—rather than as the complex, material construction of biology and culture that it is" (630–31) and, as Berlant suggests, is blamed as a "failing" and falling "will and body" (2011, 109).

A key response to the Shirtwaist fire was to inscribe those logics back onto the body. Many of us are probably used to seeing our buildings emptied with semiannual regularity in the form of fire drills and exercises, staged or enacted emergencies (Anderson and Adey 2011; Davis 2002). While the World Trade Center's inhabitants were used to performing these kinds of mock evacuations, they did not perform them as fully as you might expect. Local law 5, adopted by the city in 1973, meant several fire drills were conducted each year, but workers were not required to actually enter a stairwell during a drill or pass through an emergency exit doorway—even though that forty-four-inch standard prescribed it. The NIST report wonders whether companies were worried about losing worker productivity. Half of the building's workers are estimated to have never entered the stairs. The 9/11 commission similarly eulogized the lack of evacuation practice, the lack of knowledge about the stairs, and the fact that the doors to the roof were "kept locked," given there was no rooftop evacuation plan (Keane and Hamilton 2004, 281).

A little like the mechanical fire-escape contraptions we considered at the beginning of this chapter, the fire drill enrolled tenement and factory workers into the kinds of army file imagined in the 1935 building standards for forty-four-inch wide stairways. Drills were the common recommendations of many fire experts, consultants, and insurers in the wake of many terrible fire disasters. In Britain one of the first national committees charged with the investigation of fire and fire prevention was the British Fire Prevention Committee, led by architect Edwin O. Sachs. It was formed in 1897, with a testing station in Regent's Park. In 1919 the committee sought to explore the role of fire-prevention activities in an English factory and published its findings in a "red book." Women were the disciplinary objects of this regularizing gaze and practice in a study invited by Kodak Ltd., which had photographic negative and paper production works based in Harrow, Middlesex. Kodak commissioned their fire brigade chief officer, Stanley Thorpe, to focus on both fire prevention and "Self Help."

"Order and cleanliness" were essential to fire prevention in the plant given the flammable litter of rubbish and film on the floors. Windows were built with steel casements. Keys were kept in boxes by the doors. Fire escapes were hinged from the first floor, preventing them from taking up valuable space in normal use. It was deemed "essential that the employees should have a thorough training as to the proper and correct method of leaving the building" (Thorpe 1919, 19). The red book provides an image of two evacuations. One shows the "incorrect method" in the form of "girls"—workers dressed in the familiar white shirt and black skirt, poised on some emergency stairs. "It will be noticed that the skirts of the girls trailing on the steps are apt to be trodden on by the girls following, and therefore are liable to cause one to trip over in the press of others following, and so is likely to cause disaster. This can be easily overcome by means of a little training" (Thorpe 1919, 19).

Women were not always entrusted to perform escape drills. This had given trainee male firefighters the practice and embarrassment of wearing women's clothes to imitate the women they were to "rescue"—to the mirth of firemen appearing in a "long flowing skirt," looking awkward "in their unaccustomed garments" (Holmes 1899). At Kodak, using the "correct method," as shown in the other image (figure 1.10), the "girls" are more disciplined, instructed to march—like the army personnel that set the forty-four-inch standard—"two by two." Interlocking, one was to hold the handrail with "the right hand and her skirt round her with the left, and the girls next should hold the arm of the girl on her right, and her skirt with the left" (Thorpe 1919, 11). The woman who occupied the bottom step before has lost her smile, but "with a little practice it will be found that this can soon be perfected and a large building speedily emptied" (6).

Thorpe's guidance seeks to restrict and discipline evacuation by marshaling dysfunctional skirts, dress, and body movements in many ways at odds with the social conventions and prescribed movements governing female workers in industrialized societies in the early part of the twentieth century. At the same time, the representations of the "correct" manner of evacuating the building show women in a more solidaristic configuration, even if they look less at ease. They have been forced to come together, moving almost as one unit, holding each other's arm, as the one closest to the rail grasps it. They look uncomfortable, conforming to prescribed and more efficient martial forms of a "rank-and-file" mobility, yet they are in a social configuration unbounded by the social norms and conventions of work and evacuation, which have erstwhile imagined them as atomistic units of body-space and labor power. Perhaps there is some promise in this.

FIGURE 1.10. The correct method of evacuating down the fire escape at Kodak (Thorpe 1919, 10).

The Newark diagram tended to show the individual in flight, as opposed to the panicking, frenzied crowd-like pack of young women rendered in the narration and testimony of the events. What was constructed as panic and maladaptation was actually powerful sympathetic gestures, movements drawn by people moving together, sharing their fears (maybe shoes), coaxing, touching, calming, or exciting one another in their efforts to escape. We might say they evacuated otherwise, wearing away at the atomistic and individualistic versions of emergency politics we are more used to considering. McCormack identifies moving bodies forming shapes, seething, expanding, and contracting "as elemental variations excessive of the category of entity" (2018, 29). The Newark and Triangle Shirtwaist disasters saw envelopments of bodies seeking solace and escape from the fire, heat, and smoke around them. They pushed against the atomistic envelopes of bodily space or kinespheres, the restricted factory conditions that were regulated in New York City by the volume of factory space rather than the floor area,

FIGURE 1.11. "The Locked Door!," 1911. Kheel Center for Labor-Management Documentation and Archives, International Ladies Garment Workers Union Photographs (1885–1985), Cornell University, Ithaca, NY.

and the attempts to disperse and isolate the same standing and marching interlocking female bodies—arm in arm like the Kodak employees—that had symbolized the solidarities of the Shirtwaist labor force, which had been on strike only the year before for better working conditions, paid holidays, and fixed working hours (Llewellyn 1987; Mayerson 1910).

Almost everything was working against these more solidaristic configurations, including the evacuation practices, protocols, and even technology and infrastructure—exemplified in the Shirtwaist factory's locked doors and its defunct fire escapes (figure 1.11). Consider the excessive entanglement of bodies and elements with a technology used in both the Newark and Shirtwaist fires: the "life-net." These nets were used by metropolitan fire departments to catch leaping people trying to escape building fires, but they could not cope with the garment workers' embodied and intimate sociality, pressed together by heat and violence, unraveling the structural assumptions of fabric under pressure—bodies that came "down with arms entwined—three or even four together" (Stein 2010, 51). The nets involve material tensions to produce an elastic space-time permitting a soft landing. Fire departments liked demonstrating the nets as muscular and heroic events, popular within dramatic circus and theme park reenactments of rescues from fire—some of the garment workers may have seen these shows at Coney Island (Peiss 1986). Drilling with nets was supposed to improve physical and social coordination, an esprit de corps common to moving in time in militaries (McNeill 1997). Taking five to ten men to hold it, the net decelerates the body. In the coronial inquiry, Chief Croker explained that because the workers jumped—"all went in a pile together" (quoted in Llewellyn 1987, 9)—the nets were just not strong enough. The evacuees diagrammed in social and bodily configurations at odds with the assumptions of the life-net apparatus.

CONCLUSION: THE FABRIC OF EVACUATION

This emergent fabric of evacuation in law, construction materials, and building assumptions cannot match the diagrammings of the bodies and subjects who evacuate in creative, improvisory, and solidaristic ways. Following the Rana Plaza factory collapse in Dhaka in 2013, which killed over 1,300 people, the Shirtwaist fire and its problematic evacuation have been echoed elsewhere as a kind of ghostly afterlife of labor made vulnerable in high-rise disasters. This is probably because of the comparison of a predominantly female workforce seen jumping from the building, the scale of the event, and the poor provision of building regulations and fire safety and evacuation planning. One extraordinary image populated much of the Bangladeshi and Western press, showing female workers sliding down a colorful fabric in a curious echo of the female students practicing and witnessing the evacuation shute drill at Royal Holloway from the 1920s, shown earlier in this chapter. It was an improvisation, a creative variation using the fabric—the cloth from which garment products were made—to spin a new line out of the building in the absence of adequate means to do so in the crumpled building. One outcome of the Rana Plaza disaster has been the rush through global labor alliances involving the United States, the International Labour Organization, and the Bangladeshi government to improve the building safety and working conditions of garment workers, particularly through fire inspections and evacuation drills.

If these evacuation improvisations constitute alternative diagrammings, they signal ways of inhabiting space that see evacuees finding ways to live and endure. Bonnie Honig (2014) has explored different versions or types of emergency politics in what she calls a politics of "promiscuity" in the giving, sharing, and loss of intimate relations. In gay activist and scholar Douglas Crimp (2004), Honig finds a celebration of an embodied politics of promiscuity as a way to counter the emergency legislation, policy, and practices designed to limit AIDS in the 1980s. As opposed to the "sacrifice" of promiscuity within emergency politics, Crimp valued the "excess, sharing, risk, seeking out new relations and realities, creating new forms of life" (Honig 2014, 56)—precisely the encounters and encountering that gay and queer lives were being told to deny.

There is a politics of mobility and evacuation diagramming otherwise that we have seen in this chapter, which could solidify as a kind of promiscuity, constituted by the sororal agencies of smaller-scale, generous, intimate, conspiratorial, and creative acts—such as the wearing and sharing of shoes—that trouble some of the individualizing assumptions that underpin evacuation. Social and embodied promiscuities are precisely one of the things that evacuation plans and guidelines and regulations seem to try to eliminate—finding deviance in the common apprehension of different and alternative evacuations that perforate the bubble-like boundaries and singular lines of evacuation's diagrammatic imaginations—when it might be exactly this kind of relation that allows people to survive these moments. As Alix Chapman (2017) has shown through different forms of Black and queer relations of kinship in surviving, evacuating, and returning to New Orleans after Hurricane Katrina, queer networks of relations were able to form solidarities and pathways of escape and survival just as they were criminalized and excluded from state and law enforcement protections. One of Chapman's participants describes the positive contingencies and solidarities of queer kinship, which meant attempting to evacuate to Baton Rouge by taking an abandoned truck and ferrying more than twenty people with animals and oxygen tanks before they were stopped by police and arrested for looting. Chapman suggests that while "black queer sociality" as a mode of production was rendered "nonproductive" (82), it produced "affinities" or forms of reproduction "determined by corporeal possibilities set within a terrain of scarcity" (81) that helped to adjust to evacuation and the outfall of disaster. As Honig summarizes, promiscuity can be "life supporting" (2014, 55) in its celebration of forms of embrace and excess.

The evacuees of the Triangle Shirtwaist factory and the World Trade Center moved and found little pockets of togetherness and solidarity within a moment of crisis. In their stuttering footsteps and tragic leaps, they trace a longer genealogy of ideas, infrastructures, practices, and regulations. The promise of promiscuity that the evacuations display is another way of reading that seething image of women's bodies in fire and smoke in the Triangle Shirtwaist fire. It is how bodies, in this moment of violence and resistance, can come together.

MOBILE MEDICAL-MILITARY MACHINES

During the medical and military response to Hurricane Katrina, some noted the disturbing use of military helicopter technologies to patrol New Orleans (Graham 2005, 2011). The other but potentially quieter incarnation of aerial helicopter prowess came in a different collision of militarism, not with armed patrols, but with the medical pathways through which people were able to eventually leave the city. Louis Armstrong New Orleans International Airport became a major triage and casualty collection center for over twenty-four thousand aeromedical evacuations. Many of those evacuees were relocated from the city's twenty-three hospitals, most of which lacked power, water, or sewage facilities and were effectively islanded by the floodwaters. As a key infrastructure of mobility (Fuller and Harley 2004), airports seem well disposed to the management and modulation of bodies but not necessarily the susceptible and vulnerable bodies of the injured and acutely ill. The initial 2,000 to 2,500 patients they were told to expect was massively outstripped by the 24,000 that flowed to the airport, while "an evacuation plan" for patients or disaster medical assistance team members was nonexistent (K. Klein and Nagel 2007, 57). The airport's architectures, spaces, and technologies were described as "hostile and austere," (60) oddly in step with the literature on air travel as dehumanizing (Rosler 1988; Schaberg 2015).

The airport became the city's main medical evacuation center for hospitals such as Memorial Medical Center, which is discussed in chapter 5. The less serious cases were helicoptered to lilypad-like staging areas dotted around the city, before they could be moved elsewhere (Beriwal 2006; Lee 2006). At the airport, fears of violence grew, similar to the myths surrounding the Superdome as a hostile and unsanitary environment. Bodies, machines, and militarism were seemingly conflated. A doctor expressed these fears, describing "stabbings, rapes and people on the verge of mobbing" and being most "frightened whenever we entered the sea of displaced humanity that had filled every nook and cranny of the airport" (Vankawala 2005). Both the upper and lower levels of Concourse D, which had suffered the least of the brunt of the storm, became the main area of operations during the evacuation. A bar on Concourse D was converted into a pharmacy. Medical triage tents were set up. A corner of the concourse adjacent to the pharmacy served as the logistics depot for matériel. Medical crews innovated by transporting patients on luggage carts.

Photographs show patients with their legs dangling over baggage carts on the airport apron, while a helicopter hovers, perhaps to land in the background. Baggage claim became a storage/staging area. News reporters comment on the unfamiliarity of the conditions, seeing the indignity of baggage carts being used as makeshift stretchers and hospital beds. A CNN reporter, Soledad O'Brien, narrated people "traveling like baggage, put on one of these big baggage carts. These are the evacuees making their way out of New Orleans. Many things lie ahead for these folks" (CNN 2005). The scene is hard for the reporter to describe. In the aftermath of Katrina, the Federal Aviation Administration noted from its records that at its height 150 helicopters landed and took off per hour (Sanford et al. 2007) from the airport. The feat was widely noted as the biggest helicopter evacuation in history, widely compared to the Vietnam War (K. Klein and Nagel 2007) and the American evacuation from Saigon (Sanders 2005).

This chapter traces a prehistory of military and medical practices, logics, and mobilities colliding in evacuation. Inspired by military historian Mark Harrison's (2010) claim that World War I led a major innovation in evacuation techniques and collusion between military and medical logics and techniques, it combines interest in this period with William Walters's (2015) concept of "viapolitics." Walters has sought anew the politics of mobility and migration through the angle of the vehicles that mediate them. For Walters, the practices and ethics of contemporary migration regimes are not only performed by vehicles but communicated through them via visual economies that bring mobility practices to public attention and potential critique. It is precisely in the configurations of bodies, machines, and infrastructure that debates over mobility are staged, governed, and put into debate at the same time (Walters, Heller, and Pezzani 2021). Taking our starting point from this discussion of the medical evacuations of New Orleans during Katrina, and the strangeness of luggage carts being used as stretchers and hospital beds, raises to the surface particular visual, ethical, and political sensibilities over life, its wounding, and its care. We will examine battlefield evacuation techniques by expeditionary forces (from South Africa to northern Europe) as advances in military and medical aid to triage, treat, and remove the wounded from the battlefield evolved through new mobile machines. With the emphasis on the total war of motorized artillery and its interface with medical aid brought closer to the fighting, evacuation became both machinic (see Gregory 2015a, 2019a) and public. First, the chapter explores the aesthetic logic and concept of battlefield evacuation through the diagram of the "chain." It then, second, explores the tensions between the abstract chain and the realities of inhabiting the cramped and uncomfortable evacuation vehicles and architectures of conflict. Finally, the chapter explores how bodies were brought into closer sympathy with one another even as they were trained to perform almost inhuman and machinic postures and shapes; they showed degrees of tenderness, trust, and solidarity.

THE CHAIN

Between the African Boer Wars and World War I, evacuation became closely entwined with the deployment of modern medicine, what Derek Gregory (2013) has called, following Mark Harrison, a medico-military apparatus of delivering aid on the battlefield and moving the wounded from it. This marked a combination of technologies and practices, a merging or convergence of modern rationalities at work within medicine and military institutions, an advancement of a way of thinking that was "rational, processual, calculative and integrative" even if it "did not simply follow some preordained and disembodied logic of modernization" (Cooter, Harrison, and Sturdy 1998, 15). This meant a burgeoning machinic aesthetic of technologies of modern war and medicine, a reformulation of the conduct and ethics of modern war, a reimagination of the wounded soldier body, a widening field of volunteers and women brought close to or onto the battlefield, and a whole network of infrastructures, vehicles, and personnel.

The French surgeon Dominique Jean Larrey (1766–1842) would bring his surgeon's concern for the vicissitudes of bodily recovery following conflict to the rapid evacuation of bodies from the battlefield. In the spatiotemporal thresholds of war, the battlefield was a space of abandonment from which aid was distanced and in which the wounded would remain until the battle was over. Larrey complained how the "wounded were left on the field, until after the engagement, and were then collected at a convenient spot" (1814, 23). The solider body and the efficiency of the soldering fighting force was at stake. Instead of waiting for a battle to end before the wounded could be attended to on the field, Larrey designed an ambulance that could retrieve bodies from the battlefield and remove them immediately to care. This form of medical mobility marked a key shift in the nature of conflict, accelerating a form of modern humanitarianism seeking to remake those boundaries. The British professor of military surgery Thomas Longmore, who experimented widely with a range of evacuation techniques, was one of the key interlocutors. Longmore is credited as the force behind the dividing up of the battlefield into medical zones that would be protected by international law from opposition forces and violence (Meyer 2019, 35). But it was World War I that marked the wholesale collusion of mechanized warfare with modern military medicine, performed on a massive and machinic scale. As Mark Harrison notes, soldiers were seen as "the human wreckage of routine warfare" (2010, 1); as if damaged car parts, a system was devised to move them back into circulation—a "medical machine" for the reproduction of a fighting force. It was a machine that did not seem to run so smoothly, though. The war was marked by a political and public inquisitiveness and dismay at the slow and tawdry ways in which the wounded were treated and evacuated back home.

The machinery of war treated the wounded evacuees of the British Expeditionary Forces as material for disposal, as excess—swarf to the engines of conflict and remainders to the economics of keeping fighting bodies fit. In the UK parliament, questions were asked of the government about the speed of evacuation and what this really said about the ethical treatment of Britain's troops. In a surely viapolitical stance, Lord Robert Cecil raised the following comment:

> After a man leaves a clearing hospital he is brought down to another hospital, and an immense amount depends upon the rapidity with which that is done. If there is poison in the wound it is working all the time, and it may not have been possible to have it entirely dealt with up to the time of leaving the clearing hospital, and it may be fatal to have delay. It is of vital importance that the wounded man should be treated with the greatest possible speed at a hospital properly so-called

under conditions such as I have tried to describe. I should like to know if the right hon. gentleman is in a position to tell us whether everything is being done—I am sure it is—to secure the greatest possible speed in bringing these men down from the clearing hospitals to the base hospitals. I wish to know whether every use is being made of the motor ambulance, whether, so far as it is in our power to control it, the train service is working properly in those respects, and whether, in fact, there is a reasonably short interval between the infliction of the wound and the treatment of the man in the base hospital?[1]

Medical military evacuation was conceived according to the partition of battlespace, organized into the collecting zone, the evacuation zone, and the distribution zone. The collecting zone marked the area of active combat, where stretcher-bearers would carry the wounded, first to shelter—perhaps a regimental aid post—and eventually to the rear. Forms of movement and transport were required to move casualties through these zones. Bearers would be tasked with taking the wounded from aid posts to ambulance wagons or dressing stations for immediate care. The evacuation zone and distribution zone formed a column of infrastructure of resting stations, casualty clearing stations, and lines of communications. The clearing station appears as the apex of these feeds and redistributions of movement, diverging from the converging collecting zone. This kind of zonation was very much about supporting a system of flows through those demarcations by "freeing up the front of the army from all [the] sick and wounded" for the efficient distribution of military labor (Evatt 1884, 7).

The spatialities of care were complex and in motion in a microgeography of transport vehicles, hospitals, and intimate spaces. Jessica Meyer writes how multiple vehicles, such as hospital barges, stretchers, or trains, were, "in turn, affected by factors such as distance, landscape, and weather conditions, which influenced the particular cultures of caring to be found within them" (2019, 87). A *British Medical Journal* report of 1917 articulated a medical-military sensorium, a more-than-technical, infrastructural space of cars or stretchers, humming with expressions of intense presence and abstraction to the system of medicine and conflict. A clearing station was said to have "an atmosphere all its own—bracing, suggestive, thrilling, yet curiously solvent of illusions and of personal petty ambitions" (217). The spatiality and intensity of evacuation mobility was unfolding within an emergent aesthetic of mechanized efficiency to recirculate the wounded back into military action. Others traveled along an imaginative and literal "road home,"

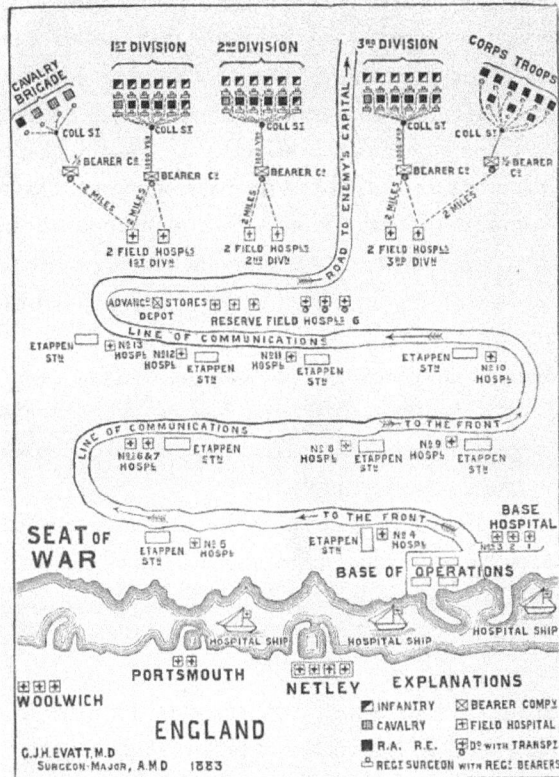

FIGURE 2.1.

G. H. J. Evatt's
illustration of
the chain of
military-medical
evacuation
(Evatt 1884, 2).

PLAN OF THE AMBULANCE ARRANGEMENTS OF AN ENGLISH
ARMY CORPS; STRENGTH, 36,000 men, 12,900 horses, 90 guns, 1153
waggons.

strung along a chain of links between different vehicles and personnel. This
was the so-called chain of evacuation portrayed in numerous sketches and
maps. It gestures toward the rationalism at play in the entwining of modern
medicine with war fighting and in the performative narratives and silences
that overload the diagram's pretentions to remoteness (McCormack 2010).
The diagram seems in tune with other representations of war, from the aerial
photo to even the poem, which seemed to evacuate from them all bodies.
Once again, evacuation is an abstracting aesthetic, removing, partitioning
"flesh and blood," all personhood (Favret 2009; Kaplan 2017).

This endlessly repeated diagrammed chain of evacuation is technical, abstract, *and* moral, branching out like a cladogram of phylogenetic relations,
or a genealogical family tree populated with the inscriptions of functions and
relative locations. The plan of the chain in figure 2.1 is by George Evatt, surgeon major of the Army Medical Department of the Royal Military Academy

at Woolwich, and was presented to the 1884 International Health Exhibition. The view is in part literal, expressing the number of units in a division and brigade along with the spatial relation between functions and field hospitals. Other parts of the diagram show a metaphorical winding road portraying the "line of communications," meandering like a river from the base of operations to the battlefront itself and then across a metaphorical sea to England, to Portsmouth or Netley Hospital. The winding road, Evatt writes in perfect brevity, is "so drawn to save paper" (1884, s1). As Evatt explained to the 1884 exhibition goers, the organization of evacuation arrangements was about purpose, ethics, and empathy. As Evatt claimed, "To-day we are at a threshold of the true method, for we are teaching the people how to be truly humane" (2). Total war dramatically systematized these approaches, made them far more extensive, and related the form and success of war fighting to systems of evacuation and resupply.

The chain was primarily about the redistribution of resources at the battlefield within ever more efficient networks or chains of mobility. Surgery really was preferable so long as the conditions of combat enabled that. Otherwise, operations would happen farther away from the front, and some patients were evacuated back to Britain for recovery. On hearing of early difficulties as the British entrenched on the Aisne in France in 1914, Lord Kitchener was sent a series of private reports from Arthur Lee, who was trusted as Kitchener's personal commissioner. For Lee, the representations of the chain tended to simplify the issues: "In surveying the scene from London, or studying it upon a map, questions of transport present no very serious difficulties. But the actual problems are complicated."[2] It was an understatement to be sure. The abstractions of evacuation did not imply solidity; the chain was considered pliable, warping and wending with the dynamics of combat and the conditions of the day. The keynote of the "successful working of the system for the disposal and treatment of the wounded in war-time," published the *Hospital*, was to "evacuate, evacuate, evacuate" (Officer of the R.A.M.C. 1915b, 307). Evacuation meant the literal mobilities of wounded soldiers within the complexities of the chain; it meant clearance and circulation within a moving network of field ambulances, dressing stations, casualty clearing stations, and transport links. Field ambulances had to be "perpetually 'evacuated' of patients" in order to "conform to the movements of the troops," even if this meant that some distance could then develop between the ambulance and the clearing station. To "avoid the unnecessary evacuation of cases that can be treated at the front" (*British Medical Journal* 1917, 222), Lieutenant Colonel Charles Myers (1873–1946), consulting psychologist to the British Expeditionary Force, sought ways to treat or prevent "so-called evacuation syndromes," which meant losing personnel physically and mentally to the system of evacuation, and preferred treatment close to the trenches in order to preserve "the mental atmosphere of the Front" (Myers, quoted in E. Jones and Wessely 2001, 94).

The system, however, was in almost constant tension. Arthur Sloggett and Sir Almroth Wright disagreed over the emphasis on an aesthetics of speed, so "hypnotised with the plea of 'Military Exigencies,'" Wright observed.[3] At a meeting of the Army Medical Services (AMS) to discuss Wright's claims, the director general was suggestive of the public's admiration of the "celerity with which the wounded are made from France to England, but you and I would very much desire to keep them on the spot and not move them at all."[4] Wright criticized the emphasis on the through-flow of numbers, with the effect of "hustling the wounded from hospital to hospital without regard to the well-being of the wounded."[5] For Sloggett, Wright's concerns were inaccurate but also unrealistic. Military exigencies should take priority: "Evacuation must be carried out if the Lines of Communication and the rear of Armies are not to be clogged to the detriment of military operations."[6] The spat was acrimonious. Sloggett was especially worried about the way Wright might be taking the debate—and evacuation—into a wider public field of (via)politics (Walters 2015), accusing him of an orchestrated media campaign and suggesting he offer his resignation.[7] Wright responded, trying to resolve the situation in private correspondence, "man to man."[8]

The chain's contradictory aesthetics distributed perception of the carnage and chaos of evacuation to disparate points. This meant segmenting the perception of the system into individual parts, not to be bogged down or distracted by the other parts of the chain. It reflected the specialized segments of production within a Taylorist production line, closing off other parts to avoid confusion and maybe shock. A medical officer stationed on a hospital ship, the *St David*, in 1915 paints a kaleidoscopic portrait of the perspectives out from the hospital ship through the evacuation chain. The conflict is seen but only in its "aftermath" by a medical officer who "literally has his finger on the pulse of those whose wounds necessitate a journey home" (Medical Officer 1915, 57) but not metaphorically on the war as a whole. The embarkation officer works through the chain of communication to make arrangements with base hospitals to determine the size and nature of the convoy of motor ambulances to the port, before liaising with the naval officer as to the tide and the schedule of their departure. From the position of the hospital ship, "all that is seen is a long streak of Red Cross motor ambulances threading their way quickly and busily through the streets of the port" (57), the only apprehension of the chain an impressionistic blur. He goes on, "Of the actual worry of transference at the base hospitals and its crowded details we on board know little" (57). Depending on one's position, the evacuation chain was aesthetically blinkered, a limited porthole into the conflict and its wounded bodies.

In diagrams like Evatt's, the color of conflict is mostly drained away within those thick lines and the yellowing and white gaps between them. Film historian Rebecca Harrison (2015) has called this a kind of structure of feeling of "whiteness." The white that stands out "from the brown sepia in every frame" (568) of public information films dominates representations of military medical evacuation. It is to be found especially in bandages, highlighted out of the gray murk of green-painted vehicles and green and khaki uniforms, the shocks of white standing out from the skin of the shirtless wounded. Bare arms are punctured with bullet wounds painted with iodine, and white cigarettes are puffed between white teeth. The ubiquitous enamel paint on the interior surfaces of the ambulance trains shines brilliantly in the train brochures and pamphlets, in contrast to the dark wooden paneling of previous designs. Whiteness denoted cleanliness and technological efficiency and signaled spaces eradicated of germs or dirt. The Quaker's Friends Ambulance Unit mentions the whiteness of surfaces in a poem ironically: "Higgledy, piggledy, my white paint, / White it ought to be but ain't" (*A Train Errant* 1919, 33). Black-and-white newsreels accentuate the contrast. An ambulance train door opens, its white interior paneling clashing with the gray/black of the exterior, to allow stretchers being unloaded from a motor ambulance and several wheeled stretchers to be loaded directly onto the train. For Harrison, "whiteness was a cocoon that neutralised" confusion, dirt, and blood (2015, 568). Denoting a regime of comfort, whiteness glinted out even from the enameled tin cups the Tommies drank from.

Contemplate *Travoys Arriving with Wounded at a Dressing Station at Smol Macedonia, 1916*, painted by Stanley Spencer in 1919 (figure 2.2). Spencer served with the 68th Field Ambulance of the Royal Army Medical Corps (RAMC) in the Salonika Campaign. The perspective is from overhead. The wounded who have arrived at a dressing station on the edge of battle are on trolley stretchers, marshaled by stretcher-bearers grasping their litters with one hand, their other hands stretched back to grasp another. The horses pulling the stretchers are at the front, peering into the Greek Church and the white fabrics lit up by the operating theater within it. The image reflects the kinds of medical-machinic visual economies captured so fully in the evolution of the evacuation chain. It is a line of perspective into the machinic light of war and medicine.

FIGURE 2.2. Stanley Spencer, *Travoys Arriving with Wounded at a Dressing Station at Smol Macedonia, 1916*, 1919. Oil on canvas, 182.8 × 218.4 cm.

Several sets of flows interlock closely, almost like gears. The animals (more on this later) are the closest witnesses. The scene portrays the serious compression of space and time that evacuation logistics would bring, as wounded bodies are brought so close together with infrastructures of care and surgery.

At another end of the chain stood the monumentally grand Netley Royal Victoria Hospital, on Southampton Water. During World War I, evacuation trains tended to embark in Boulogne, before reaching Dover or Southampton, and could end at the military hospital, which Philip Hoare describes as a shimmering "magnificent delusion" of empire, grandeur, and Victorian industrial and architectural skill (2001, 4). Netley was certainly the heftiest structure in the evacuation chain, a vast redbrick structure a quarter of a mile long, made up of enormous internal corridors and a facade looking out over Southampton Water. At first, patients from the hospital boats arriving in Southampton would be entrained to a nearby station and taken by motorcar to Netley. Later, a train station was installed at the hospital, producing a far more efficient interchange for broken bodies on their way to the hospital, as the train would come down the branch line from Southampton, pull "slowly into the hospital platform behind the chapel," and then "disgorge its load into the dark interior" (177). For Hoare, the station and the hospital become almost one, the hospital taking on more of the characteristics of the railway and the logistical chain of movement that provided the means and logic for its existence. The aesthetics of evacuation are pulled through to the hospital, which seemed for some a metaphor for empire: to "look down the ward and we see in imagination the great British Empire stretching out. . . . It is like having a map before us" (135).

Hoare sees the imprint of medical rationalism and military industrial efficiency in the rooms and technologies, in the bodies and their routines, "sleeping, eating and dying to the subconscious rhythm of military drill, compressed and quickened by the urgency of war" (2001, 179). The wards resembled the pace, space, and rhythm of the infrastructure that served them, Hoare concludes, "while limbless soldiers on crutches reached one end of the building and had to turn and face the other, as if they hadn't left the corridors of the trains that brought them here" (181). The wounded themselves arrived "labelled like luggage" (188) with their regiment number, wound, and name. Yet there was more to the logistical rationalism of Netley as one of the final nodes of the evacuation chain. Hoare's evocation of the ambivalence of Netley is helpful, just as Derek Gregory has so vividly shown in a series of remarkable blog posts on these systems of casualty evacuation. Both suggest that the diagrams of the evacuation chain might take us only so far if we think of the "system of evacuation as a linear geometry—an abstract grid of transmission lines," as pictured in a vision of "'the cogs of the evacuating machine,' beautifully oiled and running smoothly" (Gregory 2019b). Gregory points us beyond the "paper landscapes" of the space of evacuation planning, to see not just the geometry of evacuation but a geography, a "bio-physical terrain through which the wounded were moved, and threatened by the savage continuity of military violence." He explains that it is a "corporgraphy" (Gregory 2015a), engaging a "vital reciprocity between those journeys and the bodies that made them" (Gregory 2019b). This placed the vicissitudes of the evacuation aesthetic in particular tension.

VEHICLES OF MODERNITY

The logic of these emerging medical military evacuations was permeated by the rationalism and machinic bureaucracies of militaries, medical services and sciences, and mobile machines. The military's emphasis on the speed and mechanism of evacuation mobility echoed the linguistic momentum of the machines they were increasingly involving. Larrey's (1814) innovations in battlefront evacuation were a plan of "rapid evacuation of wounded soldiers from the battlefield during an engagement" (Skandalakis et al. 2006, 1394), which meant the *ambulances volantes*, or flying ambulances. Ambulance divisions included light vehicles and mobile equipment. They were tested successfully at Metz and came into official use in Italy during the 1796–97 campaign under Napoleon Bonaparte. They were, for Larrey, a design that united "solidity with lightness and elegance" (82). Later, the Paris Exposition of 1878 was a key space in which to debate medical bodies and services within the space of conflict and the ingenuity and technological development of machinery and apparatus intended to whisk the wounded from the battlefield. Dr. Aimé Riant was the organizer of the ambulance display designed for the "rational and rapid transport and hospitalization of the sick and wounded in times of war and epidemic" (quoted in Hutchinson 2018, 165).[9] The Geneva Convention secured the neutrality of Red Cross badged volunteers and personnel and the wounded as well the vehicles and matériel that moved them. To take this matériel away, argued Longmore, would be to reap the consequences of depriving the soldiers of "their first means of safety, and add greatly to their sufferings."[10]

Moving the wounded over vast distances, and at speed, demanded modern techniques of mobility. The relationship with railways began during the British experience in South Africa before the so-called railway war (Colette Hooper 2014) of World War I was structured by the momentum of the railway assemblage and its timetabling (D. Stevenson 1999; A. J. P. Taylor 1969). The British Expeditionary Force in South Africa used the Cape Colony's railway system as a conduit via which to move troops, while Red Cross trains brought the wounded back. Following Netley's gothic, mechanistic vitalism, Rudyard Kipling—who accompanied an ambulance train from Cape Town before publishing his account in letter form in the *Daily Mail*, titled "With Number Three"—renders the train as charismatic, a "little world on wheels," yet a system, rational, disinterested, "her business to get up, load, and get away again" (Kipling 1900). In this narration, the machines of military medical evacuation are given voice; the train is a she: "The rail takes the badly wounded to Cape Town and the sea which leads to"—the behemoth on Southampton Water—"Netley" (1900). The ambulance train framed military mobility logistics and its nonhuman agencies in a much more provisional and corporeally vulnerable way, especially for the bodies they would move and house. "This is the system, said the wagons." Not a smooth or cushioned modernity, it involved shocks and sensory accelerations and rapid decelerations, swaying and noise (Schivelbusch 1986). The wagons were "rained upon, thundered over, and lightened about, jolted and jerked, and jarred" (Kipling 1900). It was "far from positive discomfort," wrote one "South African campaigner" to the *British Medical Journal*. By contrast, traveling by wagon over the roads and tracks of the country entailed such a "suffering" (South African Campaigner 1899, 1485), what Longmore would call the "evils of transportation" in dealing with the wounded.[11] Some of the converted ambulance train coaches used in Britain and later in France would be known as "Netley Coaches."

The RAMC set up base hospitals at rail junctions like Deelfontein, almost thirty miles south of the rail junctions at De Aar, almost five hundred miles from Cape Town. The hospital was just 450 yards from the railway station, before a siding was built right in front of the hospital itself.[12] The field hospitals were intended to be light and mobile too, while stationary hospitals were dotted along the "the line of communication, for its rapid and frequent evacuation" (Royal Commission on South African Hospitals 1901, 5). The British worried about transporting the wounded back to the railhead, which could be many hundreds of miles away (South African Campaigner 1899, 1485). Several royal commissions worried over the vehicles in which the wounded were moved. The British ambulance wagons "jolted and were old fashioned" and "hardly fit to transport the sick," or only suited to "a country with good roads" (*British Medical Journal* 1903, 486–87). They were "very heavy . . . and very jolty and uncomfortable," nor had they been "materially changed or improved upon for many years."[13] The scheme of railway ambulance trains came under significant criticism (Royal Commission on South African Hospitals 1901). Messages did not move freely up and down the line of communications. Ambulance trains would evacuate the wounded far later than expected; the wounded lay by the tracks awaiting the train, while others were forgotten at a station. Compared to the guerrilla war they were fighting, the mechanistic form of evacuation seemed inhumane and inflexible. The Boers, writes Richard Gabriel, "went into battle in pairs, often with brothers or other relatives assigned as buddies," so that "the other was responsible for ensuring that the wounded man was saved from capture and transported on horseback to receive medical attention" (2013, 197).

World War I continued these long and convoluted chains and interchanging modalities of movement. Evacuation was painful and in public became an indictment of an uncaring government, a sign of unnecessary processes of duress, where—suggested Douglas Hall, a Red Cross volunteer in Parliament—

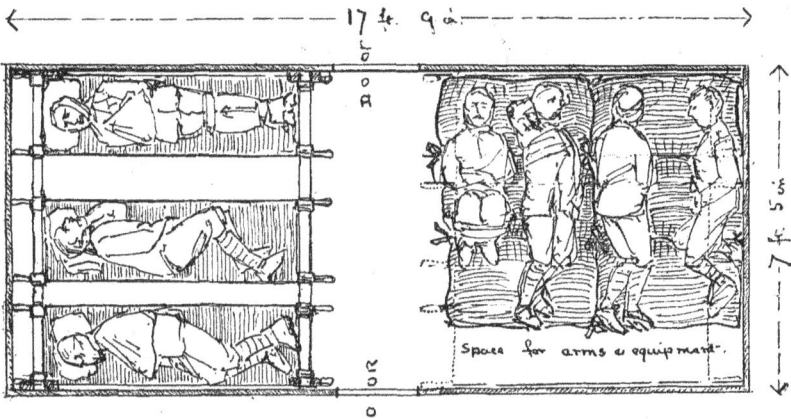

FIGURE 2.3. The wounded on stretchers or six bags of hay on the floor of a wagon (James and Pollock 1910).

"all those movements are a source of great pain to him and should be made as quickly as possible."[14] Hall compared the wounded's treatment to that of animals in the inadequate and improvised hospital trains, having observed "poor men still lying in horse boxes on straw."[15] These were contingencies the RAMC had already foreseen, improvising goods and passenger trains as a necessity; even if they were certainly "not the best type of vehicle," they were realized as "most certainly obtainable" (James and Pollock 1910, 279). Plans were published to determine the most efficient and pragmatic way of stowing the wounded in improvised goods vehicles and to "minimise the shock caused by the jarring of the wagon." The "simplest plan," they concluded, was to "spread straw, hay, or brushwood thickly over the floor" (278). Better still, "if sacks could be found." Two men would require three sacks to make an effective floor mattress (figure 2.3).

The political resonances of discomfort, the sensed and felt characteristics of mobility in cramped and inhospitable spaces, were important in themselves (Walters and Lüthi 2018). Such issues were not unique to the Western Front or the British Expeditionary Force. In the United States, the Civil War had proven a bloody testing ground for evacuation and ambulance operations, or the lack thereof. Pleas for an ambulance system were put forward after reports made much of the personal suffering of the wounded as much as the incompetence and "inhumanity" of the army and civilian drivers who would take the wounded for treatment (Bowditch 1863, 24). For critics, the neglect was equivalent to being "needlessly tortured." Inaction, a bereft father and Harvard professor suggested, was a challenge to the legitimacy of the US government (Bowditch 1863, 15). The Act to Establish a Uniform System of Ambulances in the Armies of the United States put in the most humane terms that a captain of the army would be responsible for "instructing his men in the most easy and expeditious manner of moving the sick and wounded," and in the simplest of terms, "require[d] in all cases that the sick and wounded shall be treated with gentleness and care."[16]

The vehicles of military evacuation that were criticized for being old-fashioned and unsuitable were also valued for their adaptability and efficiencies. Alison Kay, a historian at the National Railway Museum in York, England, suggests that it was possible to unload a train of 123 patients in nineteen minutes (quoted in Spillett 2014). In the leadup to 1914, the British government decided to organize and standardize the hospital train by requesting the main railway companies to design and refit train carriages for the purpose. Medical publications followed closely and fetishized the development of these vehicles as the "best means for speedy evacuation" (*Lancet* 1914, 873). As with other kinds of evacuation seen in this book, evacuations reverse and invert habits, practices, technologies, and infrastructures of mobility that may be about moving us in one way or another, in comfort, at a particular pace, for a particular purpose. Adapted vehicles were where commentators and the public would identify poor provision. The letters that badgered Lord Kitchener referred to the "use of cattle-drawn and horse drawn trucks to transport" the wounded; yet to a "self-consciously modern age such methods seemed primitive" (Carden-Coyne 2014, 24). The adaptation of existing train stock, and even motorcars and vehicles, was situated as a modern and creative response to the needs of war. Criticism that the trains and wagons were not intended for human bodies damaged by war is a reflection of the industrial purposes for which the trains had been designed. A useful paradigm in the confusing confluence of technologies and forces breaking down or collapsing within the ambulance system, modernization looked back at ugly and irrational technologies and practices, it marveled at adaptation, and it positively bristled with excitement at the invention of the new. As Philip Hoare (2001) elicits the rational and the uncanny within Netley, the railway somehow holds together the ambivalent feelings of industrial means, engineering prowess, and something mythical and irrational, a rather "monstrous embodiment of industrial modernity" through its "animalistic noises and jerky accelerations" (S. Bhattacharya 2015, 414).

A "fool's paradise," the same was true for automobility. Comparing a motor lorry ambulance to a "meat lorry," Lieutenant Colonel W. C. Beevor of the RAMC described a lorry "loaded with meat—some frozen, wrapped in salty muslin and sacks (always a foul smelling texture), some fresh and oozing sanious fluid from all sides" (1914, 66–68). If soldiers might be compared to meat in modern warfare by this association, Beevor anticipated the "impossible" problems and trauma of motorcar congestion during evacuation (67). The car's personal autonomy worried observers with its apparent unruliness. As repurposed goods vehicles, the cars were requisitioned from elsewhere. Some of the early motor ambulances were believed to have been repurposed from what H. Massac Buist called the "pleasure car chassis," adapted for the "needs of the field," but where those who "have to travel on stretchers in them are to be pitied" (1914, 544). British roads were compared by some "foreigners" to the "surfaces of the billiard table variety" (544). The Western Front would provide a surface of an entirely kind: "abominable roads," "torrid heat," "heavy rain and snow," not to mention the "barbarous" and dreadful state of French rolling stock (Officer of the R.A.M.C. 1915a, 351). This view would change. Buist's series of reports for the *British Medical Journal* waxed lyrical about the promising changes to the motor ambulance design during the course of the war. Flexibility and adaptability were important. In just thirty seconds, a car could be converted from accommodating lying-down to sitting cases. It had detachable back doors. And "in a race against time," the "motor ambulance" would be "a machine of the compromise sort" (Buist 1916, 18). In the context of overcoming the early evacuation problems of the war in 1915, the motor ambulance moved with "celerity, certainty, and comfort for the wounded results which were quite impossible in the earlier stages of the war" (Officer of the R.A.M.C. 1915b, 308). The motorcar ambulance had "solved the problems of evacuation completely" (308).

Caring for the complex units of energy, wastage, morale, and susceptibility of the body's emotion and senses, the simplest vehicle was the stretcher, an immutable but interchangeable unit around which the modern apparatus of modern military evacuation revolved. Even hospital ships developed lifts designed to load stretchers with their wounded cargo into the bowels of a ship. So as to avoid discomfort, stretcher-bearers walked "silently in rubber-soled boots" and up the "gentle incline of a long, canopied gangway." The stretchers could be attached to a wheeled runner, where "spiral springs between the frame and the wheels break vibration" and relieve pain (*Manchester Guardian* 1916a, 10). The aesthetics of speed and efficiency could go together with comfort and control of the machine and the soldier's own body (10). As Gregory (2019a) has argued, here the machinic is constantly adapted to meet the "stubbornly, viscerally bio-physical," the wounded bodies that "did not present themselves as pristine plates in a medical atlas" or as in the evacuation diagram. The rhetoric that imbues the vehicles of military medical evacuation fulfills a modernistic narrative that sees mobile machinery and organizational form absorb the body's excesses, cries, and wounds.

HOLDING HANDS: BONDS AND CHAINS

The film *The Care of Our Wounded* (Ministry of Information 1918) shows some of the forbidding and unforgiving terrain the wounded had to pass through as regimental and RAMC stretcher-bearers struggled to maintain the balance of a stretcher and its occupant over the rocky and uneven slopes of a recently captured trench on the Western Front. Even from the distance of the camera, their body movements are tentative—treading carefully, inching along. The stretcher seemed to be able to slot relatively cleanly into a range of other machines and infrastructures designed and improvised for war. Abstraction sits alongside the sensuous and visceral nature of war and evacuation. A witness from an ambulance train watches a stretcher-bearer coming from the casualty clearing station down to meet the hospital train, "slithering and stumbling in the watery mire," lit by feeble lamps and a solitary flare (*A Train Errant* 1919, 91). Rhythmic, repetitive motions show the wounded's face, read out the name and number; "another stretcher enters the circle of light; the same words pass, the same motions, and it too moves on blotted out as suddenly as it appears" (91). The mechanistic feeling is one of a motion picture, as the bearers disappear into the dark, unloading their burden onto the train; "time after time, you feel as if a picture was being cast on a screen and flashed off, over and over again; for there is something cruelly mechanical about it all" (91). In the zigzag trenches, things were not so fluid. The stretcher was an immutable and regular object, and especially when carrying a body, bearers were sometimes forced to leave the trenches during the dark (*British Medical Journal* 1917).

As objects and machineries were adapted to fit the conditions, people and bodies were interlocking and interfacing in different ways too. The hospital trains themselves became islands or bubbles, what an observer from the Friends Ambulance Unit train 16 called a "serene and monastic detachment from outside affairs" (*A Train Errant* 1919, 93). The trains, like the Friends unit, devised ways to avoid boredom and strengthen social ties. Ambulance train 39, which served in both France and Italy—and carried twenty-six thousand patients between January and June 1918—had its own concert and variety performance show given by "The Queries"—including humorists, female impersonation, dance, and cockney monologues, performed occasionally on the move.[17] The Friends unit produced its own newsletter designed to amuse, "as a relief for some, after the strain of tending patients and the toil of scrubbing floors, to transport themselves from the world of grim fact to the sphere of ideas, and there to linger awhile with their friends" (J. L. H., "An Introduction to the General Reader," in *A Train Errant* 1919, b). In one passage, a night guard contemplates sleeping in the empty beds of a coach, imagining the "tortured bodies that have tossed and suffered here." The feelings seem potentially (homo)erotic; describing "the groans which the pale lips of this one could not stifle" and signing off on the "very detrimental effect upon the moral instinct," the orderly meditates on the sleeping men on their beds, one "with raven locks wandering in gorgeous disarray about his snowy pillow," and "H——with muscular lips moving perseveringly even in his sleep" (*A Train Errant* 1919, 43).

The collisions of gender, conflict, and medicine surfaced through evacuation and troubled the categories and dichotomies that overdetermined them. Women ambulance drivers were made famous by Munro's Flying Ambulance Corps and the Voluntary Aid Detachments to the British Expeditionary Force. Some were upper-class, landed, and notable writers (Prieto 2015), serving just behind the front lines in Belgium. Ambulance driver Helen Hayes Gleason and her husband, the journalist Arthur Gleason, caricatured female drivers as pioneers at the frontier: "now the modern woman emerges from her protected home, and pushes forward, careless and curious," their potential for automobility set free. They described feats of driving the evacuating wounded through inhospitable terrain when a male soldier would not, undoing stereotyped assumptions of gender, automobility (Gleason and Gleason 1916, 205) and the masculine taming of the passions, the wayward vehicle, and the danger of conflict. It was a challenge of landscape: "the treacherous terrain" compared to foodstuffs, "scummy with mud. It is like butter on bread," and the ambulance skids in the "paste." They observe Elsie Knocker, who would become famous with Mairi Gooden-Chrisholm (already a motorbike rider for the Women's Emergency Corps) within the British and Belgian public as "the Two" heroines of Pervyse (Prieto 2015). Knocker and Gooden-Chrisholm brought immediate comfort to the wounded even before they were evacuated or removed from the battlefield. The Gleasons observed Knocker overcoming this "geography of hell" (Gómez Reus 2012) and her apparent gendered physical limits and social status, while caring for the vulnerable bodies tentatively suspended within the interior geography of the car: "Motion tore at their wounds. Above all, they must not be overturned.

An overturn would kill a man who was seriously wounded. Driving meant drawing all her nervous forces into her directing brain and her two hands. A village on fire at night is an eerie sight. A dark road, pitted with shell holes and slimy with mud, is chancy. The car with its human freight, swaying, bumping, sliding, is heavy on the wrist. . . . Safely home in the convent yard, the journey done, the wounded men lifted into the ward, she broke down" (Gleason and Gleason 1916, 214).

The general experience was more ambivalent. As Laura Doan and others have suggested (Doan 2006; Gómez Reus 2012; Prieto 2015), the female ambulance drivers may have experienced automobile exhilaration, autonomy, and the thrill of danger and speed in driving the ambulances they loved, gaining deep technical knowledge. Yet the experience was physically and emotionally tortuous and constricting because of the muddy geography and masculine hierarchies, which meant some women were never allowed to go out on an ambulance. Doan finds little to suggest that ambulance drivers were emancipated by their practice; perhaps the "drivers were less intoxicated by the experience of speed than fatigued and wracked, physically and emotionally, by the work's pounding, mundane repetitiveness" (2006, 33).

The machinic does not mean that the practices of removing bodies from war zones had to be inhuman in the sense of the distant or abstract, as cold and uncaring as the diagrammatic sketches, but equally monstrous, sympathetic, and susceptible. Helen Hayes Gleason wrote, "Life at the front is not organized like a business office. . . . War is raw and chaotic" (Gleason and Gleason 1916, 192), Gleason rejects the masculinized version of femininity—the ambivalence continued into the diaries of women like Knocker and Gooden-Chrisholm, considered by Teresa Gómez Reus (2012) as "cross-dressed texts," blurring conventions through apparently masculine performances of dress and habit. Men's own versions and narratives of masculinity, brotherhood, sexuality, and femininity were also confusing (Das 2006). Gleason renders vivid a raw-nerve attunement to the war. "War," she wrote, "made itself felt, still more" (Gleason and Gleason 1916, 194). As Ana Carden-Coyne describes so beautifully with regard to Gleason's experiences, "Driver and wounded inhabited a world of pain together and in the closed intimate space of her ambulance the pain of another became her pain too" (2014, 56).

As vehicles of sorts, the bodies that disintegrated, vaporized, and broke down into the mud and sludge of the terrain of war (Gregory 2015b) were nonetheless brought together in evacuation mobility. The chain of evacuation was about "caregiving" that drew other lines of intimacy and sociality, supporting different "narratives of male comradeship and bonding that

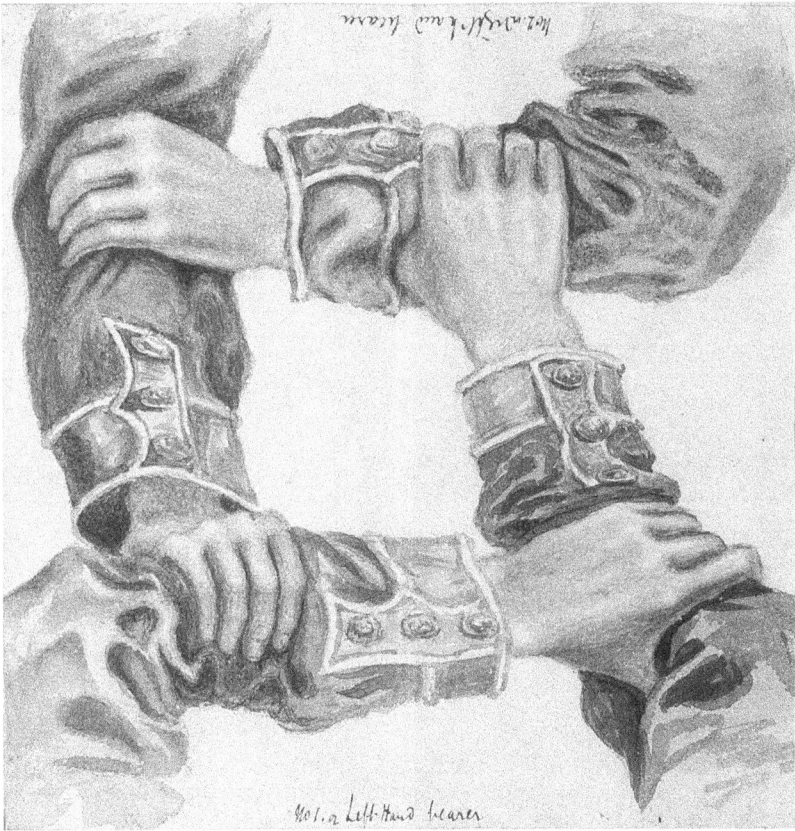

FIGURE 2.4. Left-hand four-handed seat carry for "mutual support." From Thomas Longmore's pamphlet collection (1869b), Wellcome Collection.

developed in the physically and emotionally intimate spaces of the trenches" (Meyer 2019, 88). The "strain" of the burning and aching muscles observed in women drivers gripping the wheel, along with Longmore's experiments at Netley and the battlefield of the stretcher-bearer, constituted other kinds of intimate, affective relations and attachments in motion. The focus on the bodily strains, sprains, and exertions of the bearers show up consistently, as do the hands, "flesh rubbed raw" (Mayhew 2013, 21) by friction, penetrated by splinters and wire. The pain of the bearers constituted forms of heroic masculinity, working "[their] hands off" (Clair and Clair 2004). Yet the touching proximity of male bodies moving other bodies out of danger (see the hand positions required by the "sedan chair" method in figure 2.4) compelled other attunements, as Santanu Das considers the "nakedness of

FIGURES 2.5 AND 2.6. The sedan chair bearer lift. From Thomas Longmore's pamphlet collection (1869a), Wellcome Collection.

the hand" and the "raw" gestures of soldiers' limbs and bodies entwining in a chain stumbling along, gassed and blinded, to a dressing station (2006, 1).

Sometimes it took bodies acting mechanistically to do this. The Netley professor Thomas Longmore's (1869) monumental treatise on stretcher carrying shows something of this as soldiers were taught to comport themselves in odd ways to act like a machine, to become a stretcher, to become an animal bearer in form of the "piggy-back." Arm in arm. Arm over a shoulder, two people moving as one to carry another. Arms forming a seat in the form of the "sedan chair" position (figures 2.5 and 2.6). Children's games and postures, which many of us will remember from when we were younger, became vital to saving life. Bodies literally became another's vehicle by intimately chaining themselves together.

Longmore's concern is for the fatiguing work on the muscles in terms of the forces bearing down on a bearer's forearms, the stress on the fingers, interlocked, noticing only the "bemusement" of the men at adopting certain poses. The bearers became incredibly inured to their proximity to terrible wounds, the pain of those they cared for as well as their own. Longmore's sketches and cuttings show something more tender. To "hold the hot hand"—to take

W. H. Atkins's poem of the "The R.A.M.C." (Meyer 2019, 183)—speaks to the physical intimacies and bonds (many of them homosocial) that became part of evacuation mobilities. The bearer's tactile contact appears as one of many ways that soldiers were pressed together, "but in opposition to and as a triumph over death," as those intimacies helped to move, however slowly, bodies out of the battlefield (Das 2006, 118).

CONCLUSION: BRINGING WAR HOME

In a famous scene in *Gone with the Wind* (Fleming 1939), Scarlett O'Hara searches the train sheds of the Atlanta main city railway station yard. The station was a crucial node within the Civil War for moving supplies and war material within the Confederacy, as well as being a hospital base. O'Hara steps out into the open as stretchers are being offloaded onto horse-drawn ambulances. The wounded, writes author Margaret Mitchell, are "disgorged" (the same term Philip Hoare used) from the trains (1936, 146). The camera begins to move out and up to reveal, everywhere, stretchered wounded men before pulling back from a Confederate flag gently fluttering in the wind. The Civil War could not quite bring war home because it never took it away from communities and towns and cities like Atlanta. Between the Boer War and the Western Front, the medical-military mobilities of evacuation could be a function of what Gregory (2011) has called the "everywhere war." Boundaries between war and peace, battlefront and homeland, were befuddled by evacuation, expanding conflict into a wider ethical, medical, and political space.

Thinking viapolitically through this chapter, we have seen that evacuation mobilities drew war into a public imagination, politicizing the wounded and the fate of the armed forces. Longmore believed that public and political sympathy for the wounded was increasing in an industrializing world "of continued watchfulness," with regard to the army and the "personal concerns, the health, and the welfare of all the individuals composing it."[18] Longmore's basis for the expansion of public scrutiny over the wounded seems chiefly an expression of a global civil society, what he called "the rapid communication of intelligence to great centres almost irrespective of distance, and its immediate circulation through whole communities."[19] "The wounded officers and soldiers will be watched," he claimed, "from the place of fighting to the hospitals, and anxious attention will be given."[20] The vehicles and mechanisms of evacuation, in their tendrils of visibility, were conduits through which war could be witnessed and debated. Medical evacuation meant that the war could be felt viscerally and materially, "where men arrived with

FIGURE 2.7. Boer War soldiers relaxing by Southampton Water with a view across the estuary to Netley Hospital. Halftone after a photograph by W. Gregory and Co., London, 1899–1902. Wellcome Collection, CC BY.

the mud of the trenches still on their khaki puttees" to the extent that the receiving hospital "was as much a part of the war zone as any field dressing station" (P. Hoare 2001, 179).

The abstract railway lines and the diagrams of evacuation vibrated by linking war and country, away and home, together by technical but nonetheless expressive threads. They became a visual and aesthetic connection between distanced spaces, and one particular conduit was tourism (Lisle 2016). The evacuation networks developed during the Boer War were endlessly collated in a series of personal journeys for journalists, novelists, travel writers, and military personnel. Evacuation seemed something like a novelty, just as Kipling's Boer experience was characterized as a meandering good time with his family in Cape Town, "looking at hospitals and wounded men and guns and generals" (Kendall 2006, 29). The landscape and romantic watercolorist Edward Duncan's portrait of Netley in 1856 imagines a Venetian palace viewed from the lagoon like Southampton Water, as sailboats and rowboats pull by. In a much later postcard, a photograph sees khaki-clad soldiers at rest on the pier, lazing on the long boards, leaning nonchalantly on one arm with the great building stretching out behind them (figure 2.7). Patients would be encouraged to send messages and news to their families or loved ones via postcards such as these. Photographs from the hospital pier see it as a "therapeutic amenity" much like the piers adorning British seaside towns. The Netley pier was a sort of escape from the hospital and "an idyllic prospect, as though admission to the hospital were akin to going on holiday" (P. Hoare 2001, 173).

It was a place for viewing and for being seen. War had indeed been brought home by medical evacuation, but evacuation was also brought back. Evacuation by ambulance entered the civilian world (Goddard 2016; London School of Hygiene and Tropical Medicine 1909, 2). By 1909 London's Metropolitan Police took primary responsibility for evacuating the injured to hospitals for treatment rather than treating them on the street. The *New York Times* even noticed how eight of the surviving fallen who were taken by electric ambulance from the scene of the Triangle Shirtwaist fire in 1911 relied on a service that resulted from the "exigencies of war" (A. Jackson 1915, 12), led by Edward Dalton, who had been in charge of the field ambulance corps of the Army of the Potomac. Even Evatt complained that the human casualties of war had "really been nothing by comparison with that ever-constant, never ending pain endured by the mass of civilians," for whom an ambulance service was even more necessary (1884, 3).

The ambulance trains of the Great War went on tour, to be photographed and visited at local and mainline railway stations before they went abroad. For one shilling in Manchester Mayfield Station, or London's Victoria Station on platform 9 and 10—the train had to be split, it was so long—the trains became leisure attractions (*Manchester Guardian* 1916b). In 1915 in Brighton, 2,400 people visited a train, took in the compartments, and bought brochures and souvenirs. The brochure for the Lancashire and Yorkshire Railway Company shows a watercolor "general view" of the train concertinaing into a vanishing point. Almost a dream, the train floats on rails, abstracted from the realities of conflict or the wounds within it. In Liverpool people thronged, "cheerfully" paying their shilling "for the privilege of inspecting a new ambulance train" (*Manchester Guardian* 1915b, 6). As the evacuated came home, they "brought the horror of the conflict home to the public," and perhaps to assuage these feelings, exhibitions and publicity brochures emphasized the femininity and domesticity of the ambulance trains (National Railway Museum 2019). The popular representations around the trains were the female nurses, reinforcing "traditional values associated with domesticity" (Meyer 2019, 87; see also *Observer* 1915). The 1915 Victoria Station exhibition emphasized how the trains afforded comfort and convenience to the wounded and the officers and nurses caring for their patients with "ingenious little contrivances" that eased "the pain of the passengers" (*Manchester Guardian* 1915c, 3). Fresh-cut flowers in glass vases were placed on tableclothed perches in almost all of the interior photos, softening the white enameled surfaces; the steel, iron and copper fixtures of fans, hot water storage, and ceiling-hung cables; and the draping white linen. The exhibition goers could feel that the troops would be embraced (*Manchester Guardian* 1915a, 1915d).[21] In a brochure printed to commemorate the first British ambulance train designed for the American Red Cross, the train was characterized as "the first medical unit in which the wounded patients feel that they are really resting" (Gallie 1915, 185). They would be on the "road to recovery," a journey with therapeutic benefits (185). The returning wounded would perhaps then hitch a ride on domestic ambulance trains. A seeing, feeling space, the coaches of the train afforded a view of the British countryside from which the myths of landscape and nationalism could be reforged (Daniels 1993). A *Manchester Guardian* reporter wrote from one of these trains through misty eyes on its way up from Southampton. The train offered a romantic, kinesthetic encounter:

> The afternoon is hot but the sliding doors of the coaches are wide open, and the soft southern breeze is drawn pleasantly through the train. Groups of arm and head patients lean over the protecting bars or stand along the corridors of the extra coaches watching the fleeting countryside of Berkshire and Oxfordshire. The sunlight falls upon rich landscapes of meadow and woodland, on reapers in the cornfields, and on the sylvan Thames, where girls in punts and boats cheer and wave handkerchiefs as the train speeds by. The men drink in with delight the peacefulness and beauty of this wonderful old "Blighty" and contrast it with the ruin and desolation of the country they have fought over in France. (*Manchester Guardian* 1916a, 10)

EVACUATION AND EUPHEMISM

MEMORY, LEXICALITY, AND APHASIA—FROM THE HOLOCAUST
TO JAPANESE AMERICAN "INTERNMENT"

Evacuation finds its roots etymologically and conceptually in Galenic and Hippocratic-led medieval treatments, a manner of expelling deadly flows and accumulations from within the body outwardly. Evacuation was dispensed by royal physicians to medieval Catalan rulers as a regimen for how to live in a regime for organizing the body and its (e)motions in relation to its surroundings. The concern was how to govern the excessive buildup of the humors to dangerous levels within the body. Such advice, disseminated from the fourteenth to sixteenth centuries within the burgeoning vernacular medicine from the Iberian Peninsula, saw bloodletting become a common practice. Coitus was understood as a form of evacuating the buildup of semen. Evacuation was part of a broader regulation of superfluous bodily flows and affects. A plague treatise by Miguel Martínez de Leiva from 1597 argued that evacuations should be practiced with frugality or they would expend life, and that other solutions for "avoiding or eliminating superfluous bodily matter" should be sought out (Carrera 2013, 153). Robert Burton's *The Anatomy of Melancholy* (1628) even identifies "evacuations stopped" as quite a normal cause of melancholic spirits, but he warns not to overdo the remedying expulsions.

In glancing at early ideas of more bodily evacuations, we witness a normalizing spatiotemporal imagination at work. Internal accumulations need to be routinely expelled, what Julia Kristeva (1982) would identify in the movements of ingestion and evacuation. That which is taken in will require expulsion. Perhaps this extends Adi Ophir's characterization of the "catastrophic state formation" that externalizes through an "almost extraterritorial and devastating force that throws the inside outside" (2007, 118). Moving something to the external is a process of abjection—of making what is inside completely and utterly deplorable, unthinkable, superfluous, and excessive. As Kristeva puts it, "I expel myself, I spit myself out, I abject myself within the same motion through which 'I' claim to establish myself" (1982, 3). Abjection by evacuation, therefore, could be read as a constituting act (S. Moore 2012; Scanlan 2005). It protects not necessarily the evacuated—the ejected— but the container or the host. As Janice Delaney, Mary Jane Lupton, and Emily Toth (1988) explain in their cultural history of menstruation (see also Sobo 1993), however, this logic is applied differentially. Women's bodies were believed to hold an abundance of too much blood and were believed to "have a unique need for evacuation" as opposed to men's (Delaney, Lupton, and Toth 1988, 47). For Galen, it was other excesses: the product of an idle, domestic stasis, in which people heaped "up a great quantity of humours by living continuously at home, and not being used to hard labour" (Galen, quoted in Delaney, Lupton, and Toth 1988, 47). Continuing this denigration of women, attempts to regularize menstruation into a mechanically aided exercise blurred the distinctions between abortion practices and the cultural and political meaning of menstruation. Sigmund Freud's exploration of the sexual life of children suggests that their understanding of the reproductive system even confuses birth with excretion, imagining a baby *evacuated like a piece of excrement, like a stool*" (Freud 1908, 219). This might seem very far away from the concerns of this book so far, perhaps inappropriate.

From court cases to photographic archives, a number of different cultural critics and writers have explored what Ann Laura Stoler (2016) has called colonial *aphasia*—the neurophysiological condition that affects the ability to understand and speak words. Stoler uses aphasia as a kind of critique of the disremembering of colonial legacies and the durability of its power structures and inequalities. Aphasia, Stoler writes, is a "dismembering, a difficulty speaking, a difficulty generating a vocabulary that associates appropriate words and concepts to appropriate things" (125). The discussion of more clonic, excremental evacuations of the body seems slightly aphasic, and in some ways, those meanings and senses have been disremembered too and seem out of place. Following Stoler, the theorist and philosopher Ariella Azoulay has examined a range of images in an archive of photos taken by the Red Cross. Azoulay works through the collection to interrogate the process of the Nakba of Palestinian expulsion during the 1948 war following the formation of the State of Israel. Azoulay (2011) reconsiders the dominant imaginations and narratives of state and official photographic archives. Azoulay's focus is a critical exploration and imaginative reconstruction of the annotations and captions that accompanied the imagery. She calls these part of the "constituent violences" in uprooting 750,000 Palestinians from their homes between 1948 and 1950. Publishing the Red Cross images as redrawn sketches from the photos, *Different Ways Not to Say Deportation* (Azoulay 2012) examines the pictures' labels, captions, and language, such as *repatriation* or *transborderment*, but especially *evacuation*. The words neutralize the events. This formed part of what she calls a kind of "political jargon" in the "systematic relocation of populations in Europe after the end of the Second World War" (3).

It is worth pointing out that Azoulay translates the French *transfert* into the English *evacuation*, as opposed to *transfer*, which is revealing of the confusion of evacuation with other terminologies and modalities of displacement and of what she calls the neutralizing and negating language of evacuation. Azoulay translates the captions as "evacuation of their own free will," or "evacuation from a 'Jewish zone' to an 'Arab zone.'" In a photo showing deportations during the November ceasefire allowing an exchange of prisoners of war in 1948, Azoulay notes how the Red Cross caption describing the ten thousand who were deported uses the terms *transfer* and *evacuation* of the "wounded or ailing." Questioning the photographer's narrative, she asks, "Is the Red Cross photographer aware of the gaps between the calm evacuation he sees with his own eyes and the horror stories he is likely to have heard from the refugees" (2), and do they realize "the meaning of the picture of an apparently calm and orderly evacuation of the population?" (3). At stake are forms of mobility and the way they are named, drawn (photographically), and affectively redrawn—"calmed" (remembering the 9/11 Memorial and Museum's similar use of the term)—and given meaning, neutralizing the displacement at the hands of the Jewish state seeking to clear the way for settlers. The focus is on autonomy of movement as opposed to the conditions of the conflict. Azoulay's concern is for the "laundered language" that aesthetically and lexically confuses. It is a way of calling something what it is not, "evacuation, not expulsion; flight, not deportation; distribution of property, not looting; fair allocation, not dispossession" (13).

This chapter is about a way of thinking evacuation in these aphasic forms as a cover, cloak, or confusion for a version of evacuation that we could describe as expulsive and therefore having more in common with the more clonic, abject senses of evacuation with which we started the chapter. Evacuation holds this in potential. In the way that Stoler (2016, 18) reminds us that "concepts are not 'tranquil,' stable configurations in resting mode but in resistive agitation," as a word of meaning, with aesthetic and affective resonance, *evacuation* can come into contact with others that may "congeal, collide, and rearrange themselves around it" in the material infrastructures, institutional apparatus, and other practices that facilitate mobility. *Evacuation* becomes more than a misnomer or euphemism but indeed *is* these things and more. In the previous chapter, we saw vehicles as viapolitical arbiters of evacuation mobility; in this chapter, words become vehicles to transport the mechanism of murder, forced displacement, and imprisonment by concealing or changing the tonalities of meaning. The chapter shuttles—somewhat aphasically—between Nazi Germany, the occupied and administered territories in Europe, and the wartime controversy of "internment" in the United States. First, however, the chapter contextualizes the tautological problems of sense and meaning, taking two court cases as moments that reveal these issues in ways that escape the legal rigidities of language. Next, it teases out the technical and contextual evolution of evacuation in language. It then goes on to explore the infrastructural confusions and processes that multiple forms or versions of evacuation were performed through, where both language and infrastructure sought or served to conceal horrific violences and exclusions. Finally, the chapter concludes with the different ways the lexical structures of evacuation have been "detoured."

THE POSSIBILITY AND IMPOSSIBILITY OF TELLING

Hannah Arendt (1963, 61) described Adolf Eichmann as "the master who knew how to make people move," being the bureaucrat at the head of the Nazi deportations and transports. In Shoshana Felman's (2001b) reinterpretation of Eichmann's 1961 trial in a Jerusalem courtroom, evacuation seems to express the "impossibility of telling" the Holocaust through a clinical set of traces, sterile terms, categories, and valences. *Evacuation*, notices Arendt—frequently placing it in scare quotes—was the name, euphemism, and to some extent mechanism of deportation and mass killing. Even as she writes about these semantic games, Arendt's use of language and her own translation of the German *Aussiedlung* (resettlement) and *Evakuierung* (evacuation) in different editions of *Eichmann in Jerusalem* are inconsistent, sometimes conflating those words and translating them differently—a little like Azoulay.[1] Historians have sought to secure the actual conceptual and practical meaning of these terms, even within the court case *David Irving v. Penguin Books and Deborah Lipstadt* (2000), where the lexical disorientation of the Nazis continued in Irving's Holocaust denial. Holocaust denial has relied on confusion, attempting to take the term *evacuation* at its face or "literal" value. One of the issues Irving's case exposed was whether it was possible to hold schemes like the Madagascar Project—a ludicrously ambitious and probably fake plan to send four million Jews to Madagascar—to be actual evacuation plans that would not involve mass killing. Expert witnesses Richard Evans and Peter Longerich argued that references to evacuation were predominantly "euphemistic"—euphemism was a key technique of Nazi concealment *and* Holocaust denial (Vidal-Naquet 1992)—to the extent that the evacuation, deportation, and execution of the Jewish population were so "intimately connected that is impossible to draw a distinction between them."[2]

Commensurable with the layer of fictions, Nazi jargon, and testimony that interweaved deportation and mass killing was, Arendt noticed, a kind of suspension of reality. The courtroom dramatics and confusion of Eichmann's and others' testimony was "an exception" for Arendt (1963, 224), obfuscating speech and discrediting the trial. It is to this end that Arendt famously argued that Eichmann could not and did not speak except through the banalities and clichés of the Nazi coded language and officialese, in which *evacuation* was crucial—an "autistic ventriloquism of technocratic Nazi language" devoid of meaning (Felman 2001b, 204). Eichmann's testimony "amounted to a mild case of aphasia—he apologized for it was his 'only language'" (205n4). The speechlessness evoked by other writers (Bailey 2011), such as an author who went under the pseudonym K. Zetnik and collapsed from a stroke in the process of giving testimony, opens up something else. K. Zetnik took his pseudonym from the slang word for an Auschwitz inmate, but he was asked for and named by his legal name for the purposes of the court's legal norms. Felman explores this scene as a rupture between the legal accuracies and precisions required by the court and K. Zetnik's difficulty of being asked to oscillate between his name(s) and the experiences that his names would disclose. For Felman (2001a, 254), this reveals strange and difficult proximities suspended in an apparently clinical juridical setting that the law could not decipher. The courtroom setting can pull us into a different proximity to the labyrinthine names and practices bound up in evacuation and the abstract and affective incongruities with the practices they describe.

Irving's libel trial reveals a different set of slippages of language but is far less dramatic or intimate. Irving had repeated the Nazi tricks of evacuation

by deliberately misreading and misrepresenting evidence within his arguments, mistranslating words and phrases, transposing quotes and evidence
of Adolf Hitler's guilt and complicity so as to conceal it, and decontextual-
izing the use of certain terms in order to misdirect their intent (Evans 2002,
191). Within the trial Evans argued that evacuation could not be called what
it was. Irving had argued in his own glossary, which included *liquidieren*,
evakuieren, and *umsiedeln*, that such terms should be understood in a non-
genocidal manner and contested the broadly "euphemistic reading" of the
defendants and their expert witnesses. In a sense Irving and the witnesses
agreed that different meanings could inhabit single words, but what mattered
to Longerich's rebuttal was that the context within which the words could
be used and interpreted should shape the historian's understanding of events
to the extent that the German *Umsiedlung* could have meant resettlement
but "more often embrace[d] a homicidal meaning as well."[3] Nazi evacuation
meant a form of concealment simultaneously indistinct from the actions
aimed at the protection and expansion of the German state and its ethnic
community through the methods of displacing and killing.

Within the very different context of the "relocation," "internment," or
more accurately forced transfer and sometimes violent detention of Japanese
Americans, also during World War II, the courts were equally interesting
places to understand the use of *evacuation* as a way to describe the process
and purpose of the removal of 120,000 people from the western coastal states
of the United States. It followed presidential Executive Order 9066, of 1942,
which had been planned by General John L. DeWitt following the Japanese
attack on Pearl Harbor. In *Korematsu v. United States* (1944), a case taken
to the Supreme Court on the basis of the conviction of Fred Korematsu for
not complying with the military order to relocate to an assembly center,
from which he would have been "evacuated" to an "internment camp," the
Supreme Court's majority opinions found the "military necessity" given
for the movement and detention to be broadly inviolable given the circumstances of war,[4] even if racism was the outright motive for the military
decision. Justice Owen Roberts's famous dissenting view pinpoints other
manipulations of language and law, especially in the use of terms. Roberts
sees the label *relocation centers* as "a euphemism for concentration camps" and
indeed calls the system "a cleverly devised trap" to accomplish incarceration.[5]
The terms belied not only what the process was but what it was for. Roberts
regards the use of *evacuation* to describe the movements that Korematsu
could, or could not, have performed "voluntarily" as a "disingenuous attempt
to camouflage the compulsion which was to be applied," while holding to

evacuation's protective meaning as a possible approach in times of disaster or emergency, such as fire cordons or "the removal of persons from the area where a pestilence has broken out."[6] Necessary and unavoidable detention is Roberts's point of dissent. Evacuation for protection, voluntary or forced, is possible and part of the state's responsibility and usual practice of response in emergency, but it will not do "to say that the detention was a necessary part of the process of evacuation."[7] Roberts holds the protective possibilities of evacuation intact and in potential.

There are some interesting links between the events in Germany and the United States and the following periods of legal discussion. Lipstadt's (1994) book that sparked the libel case made some of her own uncomfortable loops and reassociations between the Japanese American mass removal and the Nazi version of "evacuation." Holocaust deniers, Lipstadt showed, would try to make "immoral" equivalences between the justification of the American detention of Japanese Americans and the Nazi "internment" of European Jews, a kind of relativism that exculpated the Nazi regime from prosecution because the United States had also incarcerated on the basis of race. Lipstadt demonstrates the problematic use of language, context, and distortions of fact in different deniers' prose but makes a strange mismatch around the culpability of the Japanese American citizens the government had decided to transfer and lock up, because "the Jews had not bombed Nazi cities or attacked German forces in 1939" (Lipstadt, quoted in Rosenbaum 2019). Lipstadt (1994) is of course right to find the use of the word *internment* to be a whitewashing of what the Nazis did to the Jews and others they persecuted, yet simultaneously she lacks critical reflection on its use in North America.[8] While the trials reveal the linguistic and conceptual complexity of evacuation, detention, and mass murder across several contexts, as well as approaches to conceal, cloak, and deny the existence of these practices, there is also something about them that is misleading, that continues the confusions the trials sought to undo. In the expressions of aphasia that continued in the spaces of the courts, they help develop a conception of evacuation that cannot be separated from slippages, intimacies, and distances of words and practices.

Perhaps one way of describing this might be expressed in the novelistic and forensic documentary pairing of W. G. Sebald's (2001) novel *Austerlitz* and the Theresienstadt transit-camp internee and historian and Holocaust survivor H. G. Adler, whose work appears in Sebald's astonishing novel. Adler's historical and biographical excavation of the Theresienstadt ghetto finds its way into Sebald's prose, especially around the complex architecture of the site and the engineering and structure of the "fascist jargon." Adler explained some of the problems at the start of his book, lumping these together as what he called the Nazis' "general deterioration of language in the age of mechanical materialism" (2017, xxiii). These games of confusion must, Adler suggested, be "conquered linguistically" (xxiii). Processes of mobility were key to this amorphous lexicon where "transport" implied "mechanical materialism," and "travel," "departed," and "travelled," explained Adler, euphemistically clouded the horrors mobility made possible (19). Adler surfaces what was occluded. Sebald's Jacques Austerlitz (whose mother's papers were also stamped "Evacuation" when she was assembled at an agricultural meeting hall in Prague to be deported) is undone by the labyrinthine complexity of this obscuration. By looking ever more closely at it, Austerlitz finds that the detail becomes overwhelmingly vertiginous, and he cannot cope with the endless "list of nouns" (Kohl 2014, 101). For Ruth Vogel-Klein, "it is exactly in this drawnout reading process that he [Austerlitz] loses orientation and meaning" (2014, 187). This is an aesthetic judgment and feeling of too much, an excessiveness brought by looking so closely at the (even abstract) details that it becomes overwhelming.

LEXICALITIES AND EUPHEMISM

In Cornelia Schmitz-Berning's collation of Nazi terminology, the entry "Evacuate/Evacuation" describes two opposing meanings: "a) to clear regions vulnerable to war or bombardment of women and children; b) to deport Jews with the object of destroying them" (1998, 5; translated in Torrie 2010, 225). Language was crucially important to the Nazis' apparatus of power. For Sara Horowitz, it was its most deadly weapon, which encoded "morally reprehensible acts in a vague idiom" (1997, 157), a "simulated innocence" (157, quoting Esh 1963, 134). What makes such a doubling or dissonance possible? How can evacuation be both for protecting space and people for and killing? How can *evacuation* be used for moving the vulnerable (defined as women and children) out of the way and also for moving another group of the populace precisely in order to expose them *to* danger—to make them vulnerable, to evacuate them to death? Part of the reason for this is the coupling or indistinction of terms Paola Giaccaria and Claudia Minca (2011, 74) explicate as the "semiotic indeterminacy" manufactured by the Nazis in what Felman (2001b, 202) might call a "space of slippage" of language and law.

   Evacuation, like some infrastructures, has meant the "laying down" of words—like train tracks—in a language (Larkin 2018, 179), which in other cases might be "lush" or florid (Mrázek 2002) but in evacuation has tended to be cold, rational, and neutralizing, operating within a different technical lexicon. For many historians, the Nazi terminology of *evacuation* is synonymous with destruction. But while evacuation primarily meant death in the Nazi context (Adey 2020), it is too reductive to suggest that that is all evacuation was in the broader Nazi imagination and in other measures and processes of mass population transfers. As Tim Cole (2011) has argued, these too were in motion. Evacuation was a protective measure provided for ethnic German mothers and children to evacuate outside of target areas. This version of evacuation excluded Jews. Evacuation in both versions was invented for the racial and territorial security of the state and its peoples, its discourses and practices produced through a spatio-regional notion of a Germany and its territories emptied of Jews, yet open for "ethnic" German settlement. Ian Klinke and Mark Bassin (2018) have explicated the Nazi geopolitical imaginary of the nation competing for survival, a physiological but expulsive body politic evacuating internal and polluting substances from within its boundaries, an "ontological ordering characterized by the systematic expulsion of its own internally engendered waste" (Clarke, Doel, and McDonough 1996, 476). Evacuation subsisted within a wider imagination of Nazism that articulated a "deterritorialization" of Jews into an archipelago of prisons and concentration and death camps and a "reterritorialization" (Barnes and Minca 2012; Clarke, Doel, and McDonough 1996) to nurture and protect German racial purity in settling the land. Evacuation could maintain and legitimize a spatial threshold of the state and its territories that justified the evacuation of that space. In some ways, it reinforced the psychological "doubling," or "double consciousness" (Yahil 2000), that others have noticed in the Nazi biomedical model of the state (Lifton 1986), expressing what Sara Horowitz (1997, 165) characterizes as "the coexistence of extreme cruelty and extreme decency," where murder, in the context of an ideology of anti-Semitism, could be murderous *and* appear "life-affirming, saving the body politic."

Evacuation evolved during the Nazis' time in power, gaining in indistinction from other terms and concepts. Even the civil protective measures (Torrie 2010) evolved in complicated ways in Germany. A series of articles published in the civil protection journal *Gasschutz und Luftschutz* in 1935 by a Munich police officer (Nagel 1935a, 1935b) argued for the value and benefit of evacuation (discussed as *Räumung*) for around 20–22 percent of the total population of a city with "medium vulnerability." It suggested plans to disperse urban populations by truck, railway, automobile, and pedestrian marches (see figure 3.1). Evacuation, the officer suggested, should be

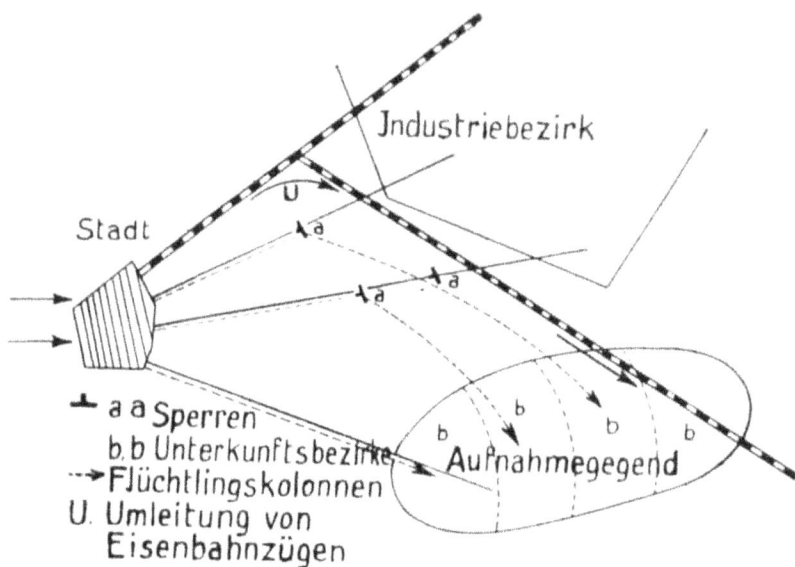

Ablenkung von Flüchtlingskolonnen.

FIGURE 3.1. "Das Räumungsproblem im zivilen Luftschutz" (The evacuation problem in civil air protection) (Nagel 1935c, 253). The diagram shows civilian populations leaving toward reception areas (*Aufnahmegegend*) and accommodation (*Untekunftsbezirke*) along arrowed pathways for columns of refugees (*Fluchtlingskolonnen*).

voluntary, but—in order to prevent panic—it demanded a level of control and coercion. Two years later in the same journal, a military figure would advocate erasing *evacuation* from the national vocabulary of civil defense altogether. Retired General Hugo Grimme argued that the word *evacuation* "as a civil defence measure" "should disappear from the air raid protection vocabulary of the German people" (quoted in Torrie 2010, 18). The German people should remain in place, prepared unto death. By 1940 evacuation was more regularly used as both an acceptable term and a practice. As the Allies' strategic bombing forced a reconceptualization of civil evacuation policy, Joseph Goebbels would begin to consider that the "'evacuation problem' was now critical and required emergency action" (Steneck 2005, 58).

The civilian "protective" and explicitly non-Jewish evacuations involved a softer German state unwilling to risk popular unrest at the measures, which proved not only unpopular but difficult to enforce. Hitler was averse to making civilian evacuation mandatory and was careful not to appear too coercive (Kock 1997; Stargardt 2011). At the beginning of the continuation of the *Kinderlandverschickung* (KLV) program of vacations for children, both Goebbels and Hitler determined that it should not be forceful so as to "drive down the mood quite hard" (Stoltzhus 2016, 217), especially for mothers. The evacuations were to appear "from the outside not as an evacuation but as a strengthened welfare program for recovery," and as Hitler's secretary Martin Bormann circulated it, "the Leader rather conceptualized the evacuations as a shining example of social welfare precisely for the poor" (Stoltzfus 2016, 217). As with the evacuations in Britain and France, many evacuees simply returned from the reception centers. Things came to a head in Witten in 1943 when over three hundred mothers protested an order attempting to deter returning children by withholding ration cards from those who had returned (Torrie 2006).

There are continuities with other state expulsions wherein abjection becomes an ideological condition underpinning the state's security, territorial expansion, and logics of social reproduction. For example, in the "evacuations" of the Khmer Rouge from April 1975, between 1.5 and 2.5 million people were forcibly displaced from Cambodia's cities by an anti-urbanist ideology intended to "evacuate" the urban population to agricultural collectives in the countryside. The evacuations were justified by the threat of American bombing, the "construction" of a food security problem by the Khmer Communist Party (CPK) regime, and a broader suspicion of Phnom Penh as a "debauched" and "parasitical" metropolis. Evacuation was a means to clear the bloated cities of this "detritus" (McFann and Hinton 2018, 225). The evacuations utilized metaphors of the abject, imagined as a cleanup of the "garbage and filth" of the social, material, and biological—streets overwhelmed with "heaps of garbage . . . rotting and dead cats and dogs" (quoted in McFann and Hinton 2018, 226). Survivors would recall feeling placed into the flow of social and human wreckage being "swept" from the cities. Many went on foot, by truck, and by train on the tracks first built during the French colonial protectorate. As death was the (teleo)logical conclusion of the Nazi transports and deportations, the evacuation was a condition of disorganized genocidal revolution and a continuity in a violent process of reproductive reorganization to an agrarian collectivist economy (Tyner 2017; Tyner and Rice 2016; Tyner et al. 2018, 166).

In the contexts of Japanese American wartime detention, however, the unraveling of these slippages and euphemisms has occurred through a longer process of legal and political redress. One of the results has been a reclaiming of terms and a new set of glossaries and dictionaries that have tried to recode

FIGURE 3.2. The Manzanar diorama, Manzanar Visitor Center, Lone Pine, California, 2013. Photo by the author.

and make visible the lexical and aesthetic categories Japanese Americans endured, particularly in the work of official photographers, well-known artists and landscape photographers whose aesthetic choices and composi-tions naturalized a visual lexicon around "evacuation." Ansel Adams's and Dorothea Lange's collections of photos for the War Relocation Authority (WRA) have been explored as two critical but very different collections. Lange's followed her interest in documenting migrant labor movements in California. As Catherine Gudis (2013, 51) suggests, her censured images of the "internment" demonstrate the failed promises of the "Californian dream" of settlement, citizenship, and security with forced mobility and precarity.

Adams was admonished by Lange for doing too little to criticize the removal and detention while his photography tended to repeat some of the political and aesthetic structures the "relocation" drew on (L. Gordon 2006). Adams celebrated the visual narratives of the frontier in the juxtaposition of modernist grid-like structures and cultivated gardens of the camp, with the snow-capped mountains behind them. This struck me when I saw the extraordinary camp diorama in the Manzanar visitor center (figure 3.2). The visitor's-eye view of the model appears akin to a few of Adams's images that were taken at altitude from a guard tower, suggesting for Thy Phu (2008) a kind of "uplift" or transcendence (Morshed 2015). The camp's vanishing point is directed by the paths and lines of telegraph poles straight into the mountainous Sierra Nevada backdrop, creating what Colleen Lye (2009) has called a "secret harmony" among the residents, the camp, and their setting. Adams would comment on these practices as a kind of response to the "harsh soil," to extract "fine crops," glowing gardens, "a democratic internal society and a praiseworthy personal adjustment to conditions beyond their control," the landscape providing "a vital reassurance following the experiences of enforced exodus" (Adams, quoted in Phu 2008, 350). For Phu, Adams's view exposes some of the contradictions of (im)mobility within liberalism that we have discussed. By contrast, the continuities of dispossession that Lange's photographs explore perhaps fetishize the fact of mobility, leaving one place to go to another. Lange's effort to explore the banalities of the mobility of the "evacuation" used a technique of normalization within a wider experience of Americanized mobilities (Cresswell 2001). The use of bus companies and moving vans from companies like Bekins offers aphasically juxtaposed bags of luggage with the sunshine, palm trees, and cruise-liner imagery of international tourism (see figure 3.3).

To reconcile the words of Japanese American detention, perhaps, rather than the visual imagery, the Japanese American Citizens League (2020) has published a guide to the use of words in their booklet *Power of Words*. The entry for *evacuate/evacuation* states, "The dictionary defines this verb/noun as 'the process of temporarily moving people away from an immediate and real danger, such as a fire, flood, shoot-out, or bomb threat.' . . . They were not 'evacuated' to protect them from a disastrous environment. By using these words, the government only made it seem that these individuals were being 'helped'" (9).

The guide recommends that the words *forced removal* should be used instead—which more accurately describes the lack of choice provided to Japanese Americans who were ordered to leave their homes, often by intimidation and force. It makes clear how evacuation encapsulated a political struggle over the meaning of the word and the determination of the events. The guide suggests that the use of euphemisms, such as "evacuation and assembly centers, made the government actions seem benign and acceptable in the context of wartime" (1). What's more, the use of evacuation impresses the belief of "being rescued from some kind of disaster (like an earthquake)," imputing the natural causality that often goes hand in hand with emergency

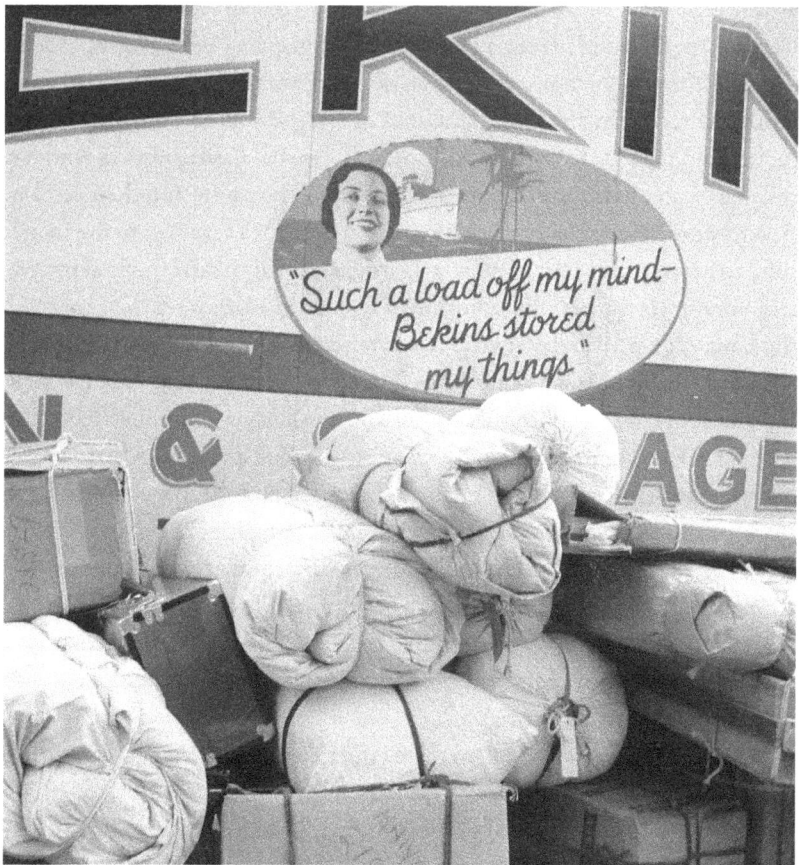

FIGURE 3.3. Baggage of evacuees of Japanese ancestry ready to be loaded onto the moving van, Hayward, California, 1942. Photo by Dorothea Lange. National Archives and Records Repository.

and disaster (Hata and Hata 2011, 68; see also N. Smith 2006). Central to redress, therefore, has been a reclamation of terms.

Evacuation wasn't even usable as a label to account for a narrative of "sacrifice" some Japanese Americans might have turned to. For Paul Shinodo, "there's no glory being in being evacuated. You can't say 'I'm an evacuee—veteran of the evacuation'" (interviewed in Tateishi and Daniels 2012, 58). Compared to *camp* and *centers* within the officialese, *evacuation* remained more underdefined. The massive six-hundred-page report the US Army compiled to summarize the "internment" doesn't include *evacuation* within its glossary. The evacuee is defined thus: "person of Japanese ancestry excluded

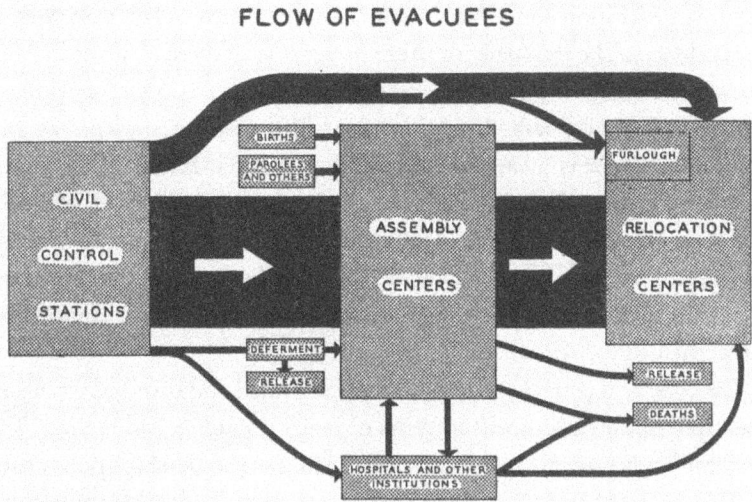

FLOW OF EVACUEES

BIRTHS

PAROLEES
AND OTHERS

CIVIL

CONTROL

STATIONS

FURLOUGH

ASSEMBLY

CENTERS

RELOCATION

CENTERS

DEFERMENT

RELEASE

RELEASE

DEATHS

HOSPITALS AND OTHER
INSTITUTIONS

FIGURE 3.4. The military's imagination of the Japanese American "Flow of Evacuees" (DeWitt 1943, 93).

from Military Area No. 1 and the California portion of Military Area No. 2, by proclamation of the Commanding General Western Defense Command" (DeWitt 1943, 513). "Internee," the glossary explains, is deliberately "NOT to be confused with evacuee" (513). When evacuation shows up, it is defined in the neutral, a kind of blank process of movement, a phase that will end in a "center" (another euphemism). Meanwhile, the "evacuee" is a subject of technical designation—the abstract outcome of a process rather than a person. Perhaps the evacuation flowchart found in John L. DeWitt's final report expresses this best, portraying it like any other mechanistic emergency action, reasoning away the motion of people as inputs and outputs in a flow diagram (see figure 3.4). Dillon S. Meyer, director of the WRA, called it a "large-scale migration."[9]

Once more here the protection of evacuation is so at odds with itself. The evacuation was an apparent means "provided to protect the persons, the property and the health of evacuees" (DeWitt 1943, viii). DeWitt claimed that the evacuation of the citizens would add a layer of protection to America's vulnerable zones, even though many Japanese Americans were landowners farming sugar beet. The protection the evacuation offered needed to be balanced against destabilizing agricultural production—working practices that the evacuees would be brought back to do as cheap labor. Justification for the evacuation relied on the principle of constructing a negative space of military exclusion zones that tallied with high levels of Japanese American residences. The rationale for the construction of these zones correlated Japanese American communities and farms—and the report gave evidence of farms in Santa Barbara—with "utilities, airfields, bridges, telephone and powerlines," whereas "nearby areas, equally fertile but lacking these installations, were virtually uninhabited by the Japanese" (US War Relocation Authority 1946, 149). DeWitt's argument was that while it may not display a conspiracy, it "was not mere coincidence" (DeWitt 1943, 9), the suspicions drawing on prewar racial prejudices and jealousies over Japanese landownership in the Californian inland (Carpio 2019).

The technical language of evacuation concealed and diverted attention away from eddies and undercurrents of racial prejudice and feeling. The WRA, explicitly critical of DeWitt's and the military leadership's policy during the war, went lengths to distance themselves from the flawed assumptions of the program and, in some ways, the real meaning of *evacuation* in the camps Japanese Americans were moved to. They referred to "preserving" and "protecting" the Japanese American community, noting that the camps could act as a "refuge" and a "haven" from the "storms of racial prejudice and the disruptions of total war" (184). As a refuge the camps could be characterized as place to give up the "vexing responsibilities" of the residents' former lives and "'ride out' the wartime period without too much personal discomfort and with a maximum of compensating personal security" (US War Relocation Authority 1946, 184). The WRA's last official report was oddly subtitled *A Story of Human Conservation*, proud of how the WRA had protected the Japanese American citizens from themselves, from the wider public, and from the poorly conceived government and military policies, suggesting that the evacuees themselves saw the camps as "the Government's 'compensation' to the people of Japanese descent for the 'monstrous injustice' of the evacuation" (184). The camps were meant to save the American citizens of Japanese ancestry from the evacuations that took them there, dislocating Roberts's dissent over *Korematsu v. United States*. In this sense, the camps were not carceral, but the evacuation was. The camps offered protection *and* redemption.

The movements to reclaim the language of the "internment" detentions became infinitely complex because of the incomparability of the Japanese American experience of incarceration in what were effectively concentration camps and the durability of meanings associated with the Nazi Holocaust. The WRA used the terms *relocation* and *assembly centers* and *camps*. High government officials and even President Franklin Delano Roosevelt used *evacuation camp* and *concentration camp* interchangeably (Herzig-Yoshinaga 2009). Official jargon slipped in and out of the euphemisms. However, it became problematic to align the Japanese American experience with "concentration camps" because of the latter's association with the ethnic cleansing and murder of millions of Jews. The consideration of the Manzanar detention center as a National Historical Site in the 1980s continued the struggle over terminology by avoiding the term *concentration* altogether. The Service for Asian American Youth, writing to the director of California Department of Parks and Recreation in response, bemoaned the "semantic game" that they saw the Historical Landmarks Committee was "playing with the Asian community" (Catton and Krahe 2018, 47; see also Bahr 2007). The Japanese American National Museum would call these instances of lexical disagreement battles within a "semantics of suppression" working to shroud "the gross injustice of the incarceration," distancing the "reality of the concentration camp experience from honest scrutiny" (Ishizuka 2005, 104).

The visual, word, and aesthetic struggles here are comparable to the difficulty of telling experienced by survivors of the Holocaust, examined by Felman, Arendt, Jean-François Lyotard, and others, where the language for describing the events becomes "the oppressor's language" (Felman 2001b, 229). For Felman, quoting Thomas Szasz, "the oppressor succeeds not only in subduing his victim but also in robbing him of a vocabulary for articulating his victimization, thus making him a captive deprived of all means of escape" (quoted in Felman 2001b, 228n46; see also Lyotard 1988, 13). Words can become realities, and there is a sense that the language of the detention became part of the furniture of Japanese American life, which "led more than one Sansei to believe that 'camp' stood for some kind of summer vacation that their parents used to go on" (Daniels 2005, 202), perhaps drawing on the WRA's conceit, not to mention wider racialized rumors around the "preferential" treatment the camp inhabitants were receiving on the government's dollar. The scholarship aimed at uncovering these euphemisms is equally confusing. Lynn Thiesmeyer's (1995, 346) analysis of the "official language" of the removal highlights "the euphemism of calling an internment camp an 'Assembly Center.'" For other advocates of redress, this is equally problematic: in fact, the Japanese American citizens could not be "internees" as they were US citizens, and *internment* is a technical term for the detention of enemy citizens. *Evacuation* is used as a technical phrase throughout the piece somewhat unproblematically too. However, they are surely right to suppose that the system of occlusion and aphasia is effectively an "endless succession of mirrors" (Thiesmeyer 1995, 328), in the same way that the Nazi jargon achieved some semblance of "elliptical" form (Kallis 2008) through the endless "conjoined nominalizations" such as "evacuation and relocation" and "relocation and internment" (Schiffrin 2001, 526) that occurred across both contexts.

DEBRIS: APHASIC INFRASTRUCTURES

In Germany the office responsible for the Jewish removals was Referat IVB4, separated into subsections a and b within the Reichssicherheitshauptamt (RSHA), the main security office of the Reich. These differed in emphasis before and after 1941 (Lozowick 1999, 2005). The department worked with the Reichsbahn on planning international train transports and managing them. This meant allocating available space on the track to ensure there was capacity and ensuring there were sufficient train cars to carry out the operations. The "evacuations" relied on existing expertise and physical and social infrastructures that had evolved over the duration of the war. The development of the human transports involved learned practices brought on by the evacuations, clearances, and settlements so that, "with precise timetables, cost calculations and contingency plans," the practices were "already routine long before the systematic murder of the Jews began" (Steinbacher 2005, 81). In this section I want to explore how these component parts reused and in some instances recombined and remade existing mobility systems.

The complex apparatus of a railway network, trains, and equipment could function as a machinery of care in parallel with the reused and repurposed machines within the wider social and discursive lexicality of suppression, incarceration, and mass murder. One way to think about the pairing of such extremes is to quickly recall one of the entangled vehicles of Indian Partition: the railway, with the unbelievable violence that accompanied it during the massacres of Hindus and Muslims on the so-called death trains. The train was the visible, modern colonial technology of mobility that moved people through violence and blood across sectarian lines during the partition of India. During Partition, evacuation became part of the largest migration in history, that of some 10–12 million people, many by train (Khan 2007). The technological rationalities of both the train and evacuation are somehow at odds with, yet facilitated, the extreme violence of the train massacres. The trains carried out, argues Marian Aguiar, a technocratic and colonial abstraction of passengers, enabling trains and train carriages to become a metonym for modernity, a marker of national identity, and, counterintuitively, an abstraction representing religious and national "communal identities that were abstracted from bodies" (Aguiar 2011, 85). The coexistence of mechanical rationality *with* but also *as* the means for inhuman violence sees the train as the perfect symbol for the contradictions and ambivalences of evacuation.

In Germany railway mobilities were essential to the camp system itself (see Sofsky and Templer 2013), where mobility could be "accurately routinized, and turned into additional occasions for torture and murder" (Giaccaria and Minca 2016, 152). The lines of the railways—the "railway stations, platforms, carriages, and tracks"—that permitted large-scale armament and munitions movements across Europe were equally the infrastructure for the Nazi killing machine itself (Gigliotti 2009, 65). Similarly, the Nazis' "evacuations" of the Jews to killing centers ran in parallel to the civil evacuation measures. The first German civil evacuation order for the general populace was given on August 29, 1939. This meant moving "nonessential" civilians living between the western border of Germany and the line of defenses and fortifications known as the *Westwall*, including towns, villages, and cities like Karlsruhe (Torrie 2010, 34).

These different versions of evacuation were linked "in innumerable concrete and abstract ways," as Torrie (2010, 131) details. The "same *Reichsbahn* officials who scheduled passenger trains to take children of the *Volksgemeinschaft* to safe areas in occupied Poland, arranged cattle cars to take other children to Auschwitz" (132). Some of the Jewish evacuations were illusions of the civil processes: the "evacuees" were advised to carry "travel rations" so as to avoid resistance to the real purpose of the evacuation. As Sebald narrates Austerlitz's investigation of his mother's deportation to the garrisoned town of Terezín, the Theresienstadt ghetto in Czechoslovakia, mentioned earlier, they would have thought they were being sent to a pseudo–holiday camp, euphemistically referred to as a "spa town." This was how the Nazis assuaged the international community (Cesarani 2016; Troller, Shatzky, and Cernyak-Spatz 2004) and misdirected the Jewish community. Theresienstadt served as a waypoint for transports to extermination camps.

Across the Atlantic, the camps intended for the detention of Japanese Americans were directly located in federally owned or managed wilderness areas intended for "rehabilitation" for settlement. Assembly centers were quite different. These sites tended to be positioned in locations such as Fresno's country fairground, or the euphemistically nicknamed Camp Harmony, an assembly center in the Puyallup, Washington, fairgrounds, where rollercoasters arced through the skyline (figure 3.5). The fair was built around a racetrack and its animal stalls, which formed the manure-smelling homes for the "internees." Tanforan, in San Bruno, and Santa Anita, north of

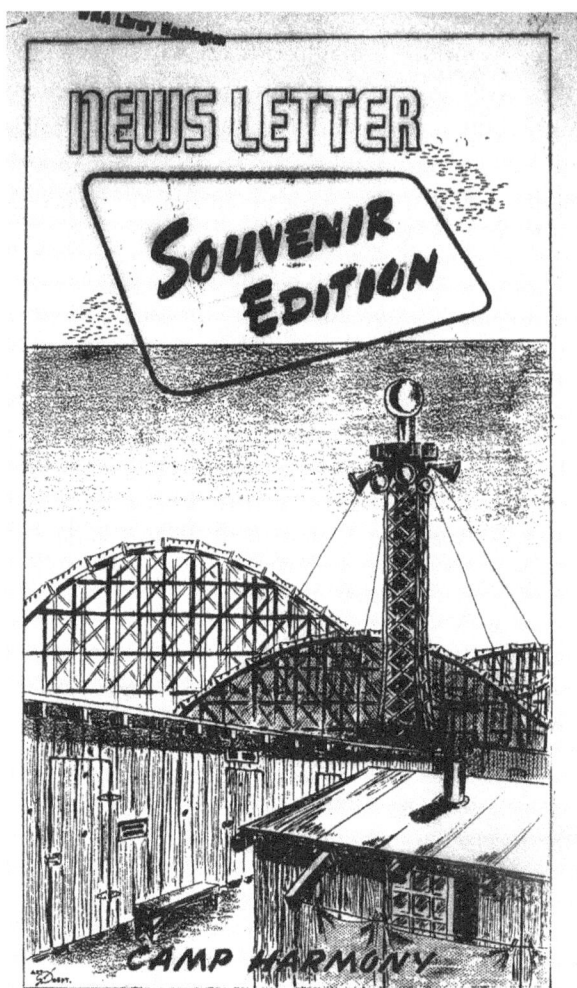

FIGURE 3.5. Camp Harmony, 1942. Drawn by Eddie Sato. Records of the War Relocation Authority, 1942–1945: field basic documentation. Located at the National Archives, Washington, DC, Alexandria, Virginia, Chadwyck-Healey, microfiche.

downtown Los Angeles—both Depression-era racetracks—became assembly centers too. Like in some of the German ghettos—David Cesarani's (2016) account of families arriving in Theresienstadt describes them being installed in a brick-walled stable and instructed to sleep on the ground—horse stalls were remade as temporary "apartments." The Camp Harmony spectator grandstand became a key site from which to observe and people-watch the endless lineups.

The rooms stank of manure from their former occupants and lacked proper floors. The "evacuees" slept on the concrete or bare soil that had been scraped of feces, although at Tanforan the stables had been divided into rooms with a swinging half door. Toilets that lacked dividing partitions between seats led to widespread constipation as people avoided defecating. The official report described the hastily constructed and remade assembly centers as a "caged-in government made ghetto."[10] The conditions must have broken the pretense. The citizens were treated as a threatening waste to be excluded to these centers, only deserving of the standards of shelter and hygiene previously afforded to the animals that lived there. Moreover, the personnel, purpose, architecture, and function of the "relocation" camps were equally embroiled within the wider institutional direction of the WRA. The WRA had been made up of agents and administrative structures from federal land conservation departments, including the Bureau of Indian Affairs. Placed in this context, there is much more to the notion of "conservation" that the WRA used in their final report. As Lye explains, the portrayal of the incarceration as a "conservationist act discloses more than an arbitrary use of euphemism" (2009, 159). The camps were built on federal reclamation projects, some purchased by the Farm Security Administration for "rehabilitation."

Poston, the largest of the relocation camps, was built by the Bureau of Indian Affairs under the supervision of head John Collier on the Colorado River Indian Reservation lands and was in part managed by the WRA. Collier was convinced to take on the project given the long experience of the Indian Service "in handling a minority group," which would "eminently" equip it "to direct this program" (Parker 1994, 89). The Interior Department's role could be expanded to protect the public and aid "the salvaging of dignity and self respect" of the Japanese American citizens who had become a dangerous surplus (Parker 1994, 89). Collier saw the camp as a continuation of the treatment and rehabilitation of not only the land but the Native Americans. In his welcoming address at the opening of the camp, Collier described the camp as a model community setting, "a truly happy place where individuals and families will be giving themselves utterly to the community and winning a reward of inward power and inward joy"; the WRA's function was as a "facilitator" and the protector of "liberties so far as we have the power."[11]

The "rehabilitating" and "salvaging" even went further than this. Someone detained in the Poston center (see the "evacuees" arriving by bus in figure 3.6) called it an "internment camp within an internment camp" (Mastropolo 2009), and Collier pursued a wider plan to use the Japanese American labor—"with their agricultural expertise"—as a means to irrigate and bring water to the region and encourage Sioux and Hopi Indian tribes to relocate there (Parker 1994, 89). This was what the bureau called "colonization," building on

FIGURE 3.6. "Evacuees" arrive at Poston on specially chartered coaches of the local Las Vegas–Needles–Phoenix Stages bus lines, 1942. Central Photographic File of the War Relocation Authority, Department of the Interior, National Archives ID 536311.

the existing Colorado River Indian project to resettle tribes from other reservations that were deemed "overpopulated" and where they could not "adequately support themselves." The water and agricultural development of the site was intended as a form of "offset" or compensation to the Indian communities for the government's use of the land.[12] Poston was a continuation of the "economic and ecological reconstruction purposes in response to the Depression" (Lye 2009, 170), tying together Indigenous and migrant diasporas in a strange matrix of exclusion and care.

Poston was run with the support of social scientists in the form of a sociological bureau that saw the camp in the same way Collier did, as a kind of colony, recognizing the problems of trying to turn the community into a willing workforce, or getting them to accept that it was for their own "protection and well-being."[13] In a memo, the bureau wrote, "The first step of setting up the colonies of Japanese is to make them as self-sufficient and self-supporting as possible. This, however, is to be considered merely as a preliminary objective. The ultimate goal is to put the colonies in a position to participate fully in the total war effort through the production of agricultural and manufactured products."[14]

This would apparently give them purpose and morale and would help to prepare them for redistribution back into society at the end of the war. The desert and arid environments into which the citizens were relocated had seemed an expression of their treatment as enemy subjects, incarcerated in abject, apparently empty settings. Yet Poston signals how the evacuation and camp system would be drawn into other narratives, aesthetic maneuvers, and institutional directions located within prewar New Deal structures.

Emphasis was given to several different kinds of (re)productivity, in contradistinction to the nihilistic and destructive focus of the Nazi camp system and more like the use of imprisoned labor for agricultural production in the United States (Lillquist 2010), mixed in with foundational myths of the American frontier. The camp authorities tried to dissolve the enclave tendencies they perceived within Japanese and Asian communities, including language and cultural identity. The idea was to play "a productive role in postwar reconstruction" and simultaneously to "prepare them for their ultimate reintegration into American life."[15] Evacuation is, once again, a kind of circulation. This could be achieved through communal living and cooperative enterprise. The original ambitions for the "colonies" were very high. Twenty thousand acres of land were to be used to farm fifty thousand chicks a year, five thousand hogs, and a herd of eight hundred dairy cattle. Within the correspondence and administrative documents of the Poston camp are peculiarly Jeffersonian ideas of American agricultural expansionism. This has been noted at the other relocation centers too. Ansel Adams's photography of Manzanar portrayed a longer transitional recuperation of the Californian landscape, "the internees as reincarnated pioneers of the old frontier" (Lye 2009, 202). In the Manzanar National Historic Site's visitor center, a display quotes from an anonymous poem published in the *Manzanar Free Press* of March 20, 1943: "Out of the desert's bosom—a new town is born."

Robert H. Rupkey, the Poston camp engineer, described the work of the "evacuee" residents in enclosing the land, finishing off a fence, and encouraging Hopi Native American reservation horses out of the acculturated territory. A Poston County Fair was organized for 1942, reimagining Japanese American communities and their labor within the terrain of the yeoman farmer of agrarian rurality—a position that they had previously been excluded from outside of their detention (figure 3.7).[16] And one poem, signed Danny Iwanaga, "So Interned," compared them to pilgrims: Yet "braver were

FIGURE 3.7. The Japanese American evacuee imagined as Jeffersonian yeoman farmer, 1942. Poston Community Council, *Press Bulletin*, Public Relations Committee. Online Archive of California, BANC MSS 67/14 C, folder J2.94, Bancroft Library.

we, when at Parker we docked, On desert waste without a rock."[17] In the exhibition of the new buildings; the extension, improvement, and widening of the canal and irrigation works; and the demonstrations of industrial machinery, the Poston project nurtured a kind of desert rebirth. Messages described the creation of a "civilization in this section of the Western Desert," the camp facing the "blue horizon of the future, glimpsing the prospect of vast agricultural and industrial projects to be planned and undertaken by our evacuees."[18] The language of others was just as hyperbolic and bucolic. Poston's public relations director suggested that the fair demonstrated "high promise for . . . agricultural development," and saw Poston in a transitional line of colonial settlement in the region from the conquistadors to the frontiersmen as a "last frontier."[19] For Rupkey, the irrigation project was but a "step in the progress of an old idea," recalling the explorer Charles D. Poston, the camp's namesake, who had imagined the valley as an Indian reservation to make "livable homes" in the Colorado Valley.[20] Others emphasized the fertility of the land, boasting that "the evacuee population" were "the ablest farmers in the entire country and we have soil and the water."[21] Nothing could stop them. The chairman of the camp's agricultural committee wrote a short piece on what he called "Poston Agrarians" by binding the settler and frontier narratives with the ethical possibilities of the "evacuee" experience. Postwar rehabilitation and reconstruction would "depend upon Postonians who are young today for the energy and the determination to restore humaneness to mankind."[22] The "horticultural rhetoric" is powerful, evoking the "uprooting, clearing, and replanting of a population, or nurturing the seeds of interracial understanding" (Lye 2009, 162). Once the camp was closed in 1945, the Hopi tribe were encouraged to settle there, captured by the Japanese American photographer Hikaru "Carl" Iwasaki (Hirabayashi and Shimada 2009) embarking on new colonization and "rehabilitational" practices, from building chicken coops to receiving instructional guidance on "home economics" (figure 3.8).

In sharp contrast to the German context, where "evacuation" was part of a broader process of violent expulsion and eradication in order to make way for imperial settlement but relied on the administrative and logistical infrastructure of mass mobility, the Japanese American forced removals and detentions worked within an already existing infrastructure of colonial practices, moralities, visualities, and words, which Japanese American detention would even later make way for.

DETOURS OF RESISTANCE

In the United States, the struggle over words, language, meaning, and narrative was suppressed and rigorously managed in the camps and various centers that housed Japanese Americans, and not only after the fact. The assembly centers were particularly vulnerable to authoritarian control, being run by the Wartime Civil Control Administration, a branch of the Western Defense Command and Fourth Army. Under the head of a military director, Colonel Karl R. Bendetsen, center regulations prohibited the Japanese language in pamphlets, "papers and periodicals" (Mizuno 2003, 852), such as the Santa Anita camp publication *Racemaker*.[23] The controls (Okamura 1982) created special problems for forms of dissent, yet solidarities and resistance emerged through language too, and even euphemism. As discussed earlier,

FIGURE 3.8. The Hopi tribe, settling in Poston, are joined by a superintendent of home economics, September 1945. Photo by Hikaru Iwasaki. Central Photographic File of the War Relocation Authority, Department of the Interior, National Archives ID 539890.

Sianne Ngai (2012) finds in the "interesting" a euphemism for a judgment deferred, a "forward reference," perhaps from feelings of the incomplete. Some of the euphemisms of evacuation have so far, however, been more about strategic disguise, to make something "uninteresting" by making multiple violences appear benign. And yet new euphemisms evolved within the camps through a series of slang words and nicknames—an "evacuese"—in order to go unremarked and avoid the WRA censors. It was a kind of critique of the depletion, displacement, and detention of the community by another set of euphemisms to generate critique and "interest" but under the radar of the authorities. Adams's use of the term *detour* saw Manzanar as a blip or "detour on the road of American citizenship" within a wider social-geoecological narrative of American values (A. Adams 1944). Citizens within the American detention camps performed "detours" in other ways, as playful critiques by distorting words and making others up in forms of resistance that Gilles Deleuze and others have made sense of in the molecular fluxes of the "detour"—little acts of resistance (Bissell 2016). Nancy Rose Hunt calls them "the strange, groping and effective, indirect procedures of the stricken" (2019, 449).

In one issue of the *Manzanar Free Press*, the reuse of a building block as an education center was reported as yet another evacuation. On the same page was news that the Southern Californian farmers had confirmed that they could produce "fine" vegetables without Japanese American labor. This was remarked upon by the author, who comments, "That's fine," but notes that what they really wanted was the right to "reside anywhere in our own country, a natural privilege accorded to all citizens and law abiding aliens."[24] These verbal protests mixed obedience with the euphemistic language in order to mock it and directly moved the meaning of government or exclusionary wording. Archaeologists would even find slang words in Japanese-language "graffiti" drawn with fingers and in pebbles in the wet concrete of some of the building works (Burton and Farrell 2013).

A satirical piece was printed in the first issue of the magazine *Trek*, a literary journal published in the Topaz camp that was edited by the artist and writer Mine Okubo. The piece came out in December 1943 under the pen name Globularius Schraubi. It is a kind of fake genealogy and etymology of camp language. For some commentators, the piece was "lightly coded," for within it lay strong contextual criticisms of the displacement, incarceration, and WRA policies and processes. Susan Schweik suggests that the inclusion of the word "'concentrated' into the first sentence" is a way of "writing without writing 'concentration camp'" (1989, 98). The piece presents the evolution of forms of Japanese American language use as a kind of "Evacuese language" that uses metaphors of the promiscuous and licentious behavior that occurred within the shadows cast by the camp floodlights: "Verbs never conjugate. If they do accidentally; they conjugate in any old way, without the slightest embarrassment or consideration for number, person, tense or pretense, and always end in *na, ne, no,* or *batten*. This last comes from the English word 'batten' which means 'to fatten' or 'to prosper at another's expense' a well-known pastime among the better class of Evacuese speakers" (Schraubi 1942, 13).

The piece undermined the camp authority's policing of relationships, from which flowed rumor and gossip, as well as love and group solidarities. Given their incarceration, the mention of "tongue locomotion" was highly ironic. So too was the animalization of the lexicon, which mocked and admonished the camp infrastructures, especially those of the assembly centers built within racing-horse stalls "and other racetracks in which the Evacuese language was born" (Schraubi 1942, 15). The piece savors the olfactory atmosphere by seeing the language as "aromatic" and hints at an almost zoonotic contamination through language and smells between humans and animals. The author also mocks the apparent ease and convenience of the "Evacuese language," hinting at the violence and inconvenience of the "evacuation." "Logic," Globularius wrote, "is characteristic of the Evacuese language" (16), mocking the contradictory, irrational, and illogical nature of displacing, dispossessing, and incarcerating American citizens under the exact guise of an emergency rationality.

Deleuze's understanding of creative poetic prose as a kind of stutter is helpful for how we might read such creative resistances. Deleuze sees writers using a kind of "minorization" in which "they make language take flight . . . ceaselessly placing it in a state of disequilibrium, making it bifurcate, and vary" (1998, 109). Globularius mocks while reusing the official language of the evacuation and relocation program against itself in what Deleuze might call a wandering but "affective and intensive language" (1998, 23). The technical jargon of evacuation struggles to accord sympathy among its users even if it becomes habit, yet the words vibrate in another resonance in how they are uttered and given different meaning. These creative appropriations shape the potential for how the words might be read and become productive of their own "atmospheric quality, a milieu that acts as the conductor of words," where alternative meanings and passions "reverberate through the words" (Deleuze 1998, 107–8).

Perry Miyake's *21st Century Manzanar* (2002) offers a more recent detouring of the Japanese American evacuations and detentions—very much in the shadow of the racialized suspicions, divisions, and security practices seen in the wake of 9/11—in a rereading of the Japanese American wartime experience as a new fictional imagining. In the novel, *revac* is one of a number of wordplays used to describe and upset the re-vacuation—an "evacuation: the sequel"—of Japanese Americans under a new twenty-first-century cold war between the United States and Japan, when Manzanar is opened once again. In one crucial moment the camp's own vindictive and hostile director, Lillian, contemplates an order to evacuate the camp: "'Evacuate an evacuation center?' Lillian looked at Jenny quizzically, then smiled. 'Temporarily relocate a temporary relocation center?' Lillian walked down the hallway to prepare her next move" (Miyake 2002, 248).

The aesthetics of the sanitary facilities was a matter for disgust and complaint in the real Manzanar, yet collective experiences of going to the toilet facilitated solidarities among the camp dwellers—especially women—who expressed themselves in poetry and camp communications under the oversight of the camp authorities. In contrast, in Miyake's novel, the sight and the stench of the toilets work in slightly different ways. The toilet and shower blocks become moments of humorous encounters that afford privacy in the intimacy of homosocial relations, similarly away from the intrusive surveillance of the guards, while the toilets too become a crucial site of resistance when a riot emerges in the protest around toilet hygiene. In the chapter "Evacuation," Miyake even juxtaposes the clonic evacuations of a soldier emptying their bowels with the evacuation of the camp and the escape of a group of the incarcerated.

Camp poetry went with and against the longer narratives of evacuation, "conservation," and rehabilitation we have considered. The poems that Okubo commissioned, and those that we saw in Poston's agricultural fair, tended toward images of agricultural longing and (feminine) reproduction, aping the colonial frontier narratives that had overdetermined the evacuations and the function of the colonies/camps. Read metonymically, they did something else, revealing the community's alienation from precisely these national narratives "of an American promised land which kept revealing itself to immigrants from Japan as a wasteland" (Schweik 1989, 188). At the end of Miyake's novel, the escapees find safety and refuge on an Indian reservation, a "refuge in a land of exile and discard that had been reclaimed by its original inhabitants" (2002, 381).

CONCLUSION

This chapter has explored more expulsive meanings and versions of evacuation across two apparently very different events and contexts. These examples help us see how evacuation's meanings and its relationship to notions of protection can be highly distorted and uneven. These moments give us windows into different formations of (post)colonial and imperial aphasia (Stoler 2016), wherein evacuation was mobilized as a way to occlude other kinds of violent and murderous acts and create strange and difficult associations between incoherent and incompatible acts, as well as infrastructures—materials, practices, visualities, language—to the extent that it becomes difficult, but not impossible, to make sense of things amid confusing and dissonant aesthetics and aesthetic judgments. These uses of evacuation are disturbing, and they can be derived from older, perhaps even more common-sense and "clonic" ways of understanding evacuation: evacuation as expulsion. The recurrence of evacuation as an expulsive move did not, however, separate it from a protective one. Instead, deployed in this way, evacuation could displace from it those who might be protected by it. Displacement through evacuation, or whatever else evacuation would name—it could be argued—could be used to protect someone else or something else: the state, a military zone, a status quo, a more valued ethnic community. What's more, evacuation's recursions saw other aphasic assemblies.

By the end of the war, the Nazis were "evacuating" or "liquidating" concentration camps, death camps, and prisoners of war. Thousands of prisoners were killed through the camps' own "evacuations," what Daniel Blatman has called a "voyage of murder and horror" (2011, 52). If this kind of murder through moving was a very different form of systematized and yet decentralized killing compared to the Nazis' usual means of genocide, it was a product of confusion and inconsistency, and for Blatman, aimed at not merely the protection of the Reich but, as the Nazis saw it, "the survival of hearth and home" (423) through a nihilistic kind of drive. The railway became a kind of moniker for this affective release. As some of the camp evacuation trains tried to make their way back into Germany, they would come to the "end of the line," as it were, halted by encroaching Allied forces or destroyed bridges and railway lines. Blatman records yet more euphemisms used by a guard discussing the inability of a train to travel any further, who referred to it as a possible *Himmelfahrtstransport*, a "transport to heaven" (300). Despite this, the camp "evacuations" saw some of the most intimate solidarities of survival in motion, as bodies commingled into "intertwined micro-communities on the move" (Cole 2016, 182), to the extent that "it was hard to say who was doing what and where one body ended and another began" (182). A survivor, Hania Laks, writes Cole, was carried by her sisters in a manner akin to the embodied stretcher-like configurations we explored in the previous chapter: "The three sisters with their limbs intertwined appear as a kind of three-person unit pulling and sliding along the snow-covered road, such that it was unclear to each of them exactly who was doing what" (182).

"THE CITY IS TO BE EVACUATED"

ROADS, RACE, AND AUTOMOBILITY DURING
THE EARLY COLD WAR

That's when I began toying with this idea of evacuation, which is simply the utilization of space.
— VAL PETERSON, INTERVIEW IN *U.S. NEWS & WORLD REPORT*, 1955

If atomic weapons and mutually assured destruction would create some kind of human limit on survivable life in nuclear Armageddon, evacuation meant taking that limit into an experimental and anticipatory terrain of planning, scenario building, role-play, and simulation. In 1949 the architect and futurist Buckminster Fuller asked his students at the Chicago Institute of Design to respond: "The city is to be evacuated," he began, outlining a scenario calling for the widespread dispersal of the city; "everything not decentralized will be destroyed" (quoted in Díaz 2014, 145–46).

From Norbert Wiener's exurban "life-belt" evacuation mechanism for the future metropolis portrayed in *Life Magazine* in 1950, to the stochastic models that could seek to anticipate the characteristics of mass evacuations from urban centers, the chapter explores the abstraction of urban life inside these new cultures of simulation in the 1950s, and from within a wider process of urban dispersal and decentralization that saw evacuation penetrate a whole plethora of disciplines and concerns in North America. Those abstractions work on a broadly different scale and through different techniques from those explored so far. Where buildings from high rises to camps and fairgrounds and trains have taken most focus, we shift attention to cities and their infrastructures, especially the road and the car. In this context the urban, understood almost metabolically through its arteries, roadways, and communities, is considered as a "problem" space. People are further abstracted into classed, gendered, and racialized categories tracing similar and different potentialities to those explored so far, especially through the aesthetic and pseudoscientific concept of panic. The chapter explores the intertwined origins of evacuation thinking primarily in 1950s America. While existing studies have demonstrated the duplicity of Cold War nuclear planning with urban design, science and sociology, national and local layers of administration and governance, and race relations and civil rights, none has put evacuation center stage.

First, the chapter examines how evacuation was considered at the same time as road problems emerged in the aftermath of World War II and fears of invasion, as well as other urban "problems." Second, we trace the evolution of nuclear emergency exercises that sought to stage and practice evacuation but that additionally institutionalized evacuation science within the academy, for whom the city became a laboratory of evacuation. Finally, the entwinement of race and automobility appears key to how evacuation was negotiated.

ROADS AND RATIONALITIES IN COLD WAR CIVIL DEFENSE

In the postwar period, the United States' leaders quickly realized that the civilian population required protection from the threat of an atomic bomb. Civil defense planners had tended to mistrust the population during World War II, seeing them as irritable, erratic, affectively panicky subjects, liable to regress into a mob, in not-dissimilar terms to the characterization of the evacuating subjects of chapter 1. The automobile was the ideal foil, the structuring object that would frame so much thinking around evacuation planning. It was a thing of mechanistic efficiency and power, a symbol of American modernity, while it tugged at the passions as the principal object of desire in postwar American consumerism. But the car was not universal, even if universally desired. As a site of racial antagonism over Black ownership, driving routes, and dispossession from urban expressway and highway development, it sang as a metaphor of liberal individualism with all of its contradictions (Seiler 2009).

The road and the car capture something of the oscillations of what became nuclear evacuation policy in the United States: the atomic event was so exceptional that it demanded response, even as it was made so ordinary that there was nothing to fear from it. As Americans were considering how to evacuate by driving, Richard Gerstell told them how to say it in his popular *How to Survive an Atomic Bomb* pamphlet. Say "ee-VAK-u-ate" (quoted in Boyer 2005, 325). The industries of advice and guidance would follow the governmental management of the peril and terror of gas and air attacks during World War II as a practical problem. Gerstell suggests packing "one change of clothing, including shoes," putting everything in a suitcase. Wear a hat, and, if possible, wrap rubber and rags around your shoes. Gerstell warned against "rushing out of the city to escape attack." Consider, he suggested, "Where are you going to end up? Who's going to take care of you? . . . All decisions should be left to the authorities. . . . [T]hey'll do it according to a plan" (quoted in Boyer 2005, 324–25).

Evacuation posed a serious problem. For one commentator to the *Bulletin of Atomic Scientists*, it meant weighing up the "serious social disturbances" of leaving one's home and neighborhood, along with the "dislocation of production schedules." War production, they argued, "takes precedence over the risk of civilian causalities," while removal of "surplus populations" not essential to the war effort would be "desirable" (Peter 1950, 255). Val Peterson, the director of the Federal Civil Defense Administration (FCDA), in an extended interview on civil defense argued how "this idea of evacuation" meant that the "remain-where-you-are concept, keep-working, stay-at-your lathe concept" was over (Peterson 1955, 74). While the desire might have been to present evacuation really as "simply the utilization of space" (Peterson 1955, 74), this may have been because the research studies, policy considerations, scenario planning, and exercises that were carried out would show it was so much more than this. During the period before the formation of the FCDA by President Harry S. Truman in 1950, various evacuation studies and reports were immediately commissioned. The consultants Dean R. Brimmhall and L. Dewey Anderson for the National Security Resource Board, set up in 1947, predicted violent social breakdown, especially in complex multiethnic cities such as Chicago and New York.

One of the key touch points for evacuation was the road. Dwight D. Eisenhower appointed the military general and engineer Lucius D. Clay to form a President's Advisory Committee on a National Highway Program, which reported to a congressional subcommittee on public roads in 1955. Clay's report was indicative of the shifting logics around which an "interstate" network was conceived as a way to wrap connected roads around the country. The committee revolved around the priorities of civil defense and population evacuation. Clay argued that the interstate would be "extremely urgent to the civil-defense program for the evacuation of our cities," to economic life, and, last, to what Clay called the "social aspect of the automobile . . . in the daily life of almost every American citizen." Ghosting the eventual centrality of the automobile in many evacuations, the interstate highways could undergird evacuation and reduce highway accidents.[1] Eisenhower would make similar claims in his statement to Congress, noting the thirty-six thousand people who died yearly on the highways, seeing the enormous emotional and social costs of those accidents as a "gap in the family circle," and noting the $4.3 billion toll on the economy. The so-called "road net must permit quick evacuation of target areas," and the current system was portrayed as a licentious "breeder of deadly congestion."[2] In affinity to Gregg Culver's (2018) notion of everyday lethal mobility, evacuation and civil defense measures were cast in the same breath as the personal tragedies of road accidents and the snarling up of US highways. Eisenhower and Clay's plan was eventually known as the National System of Interstate and Defense Highways (Kaszynski 2000) under the Federal-Aid Highways Act of 1956. The interstate was effortlessly associated with an exceptional kind of American freedom, such that constraining it would ring of the evils of the anticapitalist ideological threat of communism.

New York's Robert Moses contributed to the Clay Committee (Raymonds 1947). He was venerated as a doer, linking the chronic problems of the city "deep in traffic problems" with urban evacuation, reminding the panel that cities "must not be forgotten or neglected" and that, for the interstate system and cities themselves, the "strategic, military and evacuation aspects of arterial construction are vital."[3] Moses had been asked by the mayor of New York to prepare a plan, and he contrasted the experience of "sane" and "rational" men and the "stunts and impulses" of officials trying to meet the "unrecognizable" (1951, 8). Moses was somewhat critical of the future gazing of the civil defense imperative and far more enamored by massive, authoritarian, and militaristic road development—such as Baron von Haussmann's modernization of Paris and the German Autobahn. He would try to conjoin these to civil defense (Moses 1942, 61; *New York Times* 1940). The railway lobby was far more critical in total, claiming the interstate plan seriously distorted competition and would "'handicap' the railroads with the possibility of eventual nationalization."[4] The nation's "backbone," they argued, was "utterly indispensable in time of national emergency."[5] The trucking industry expectedly emphasized national defense, suggesting that in an evacuation the "highways of the nation" should be the "principal avenue for movement," and while a damaged railway could not be quickly fixed, "a detour for a bombed-out highway is nearly always feasible." The general counsel to the Advisory Committee to the Trucking Industry even suggested that "the threat of devastation due to the lack of mobility imposed by inadequate roads will continue to constitute the most serious impediment to our nation's defense until the road situation is remedied."[6] By 1957, however, Moses was fed up. His civil defense "Fiasco" piece in *Harpers* pulled no punches. He argued that evacuation could be more dangerous than the bomb itself: "You cannot get the inhabitants quickly out of the town without killing and maiming more in the process than would be lost or wounded in the explosions, fires, concussions, and fallout" (1957, 34).

The debate converged into a chemistry of urban political, architectural, engineering, and urban policy goals. For Jennifer Light (2003), national defense became the pied piper of not only militaries and national government but academics, civic and urban authorities, architects and professional planners, consultants, political commentators, and business strategists. This "intellectual exchange" became most acutely focused around the notion of dispersing America's cities. In 1950 the *Bulletin of Atomic Scientists* announced Moses—bizarrely—as the New York coordinator of "Evacuation and Rehousing" to approach civil defense. The *New York Times* illustrator's depiction of Moses as the many-limbed Hindu god Kali was somehow apt, given Moses's countless other state and city roles that involved movement and displacement. For Moses, atomic threat, panic, and congestion were all the same problem, and evacuation was really the strategic capital with which he could continue his renewal of the city, including "slum clearances" (Moses 1940, 151).

Preinterstate campaigns worked with an understanding of the city as threatened not only by the nuclear bomb but by decline and decay. Road building was a means to eliminate blighted communities, a knife to carve out slums. Thomas MacDonald headed up the US Bureau of Public Roads, an influential prewar department intent on creating a network of highways and using "highway construction to eliminate 'blighted' neighborhoods and to redevelop valuable inner-city land" (M. Rose and Mohl 2012, 97). Traffic and congestion gave rise to urban social decay. The city was a "patient," "not fully aware of his condition," the expressways a kind of surgery to address "New York's malignant tumours" (Tochterman 2017, 70). Moses's "traffic menace" was deliberately equated with the "Red Menace." "Bombing or no Bombing," Moses argued in a *New York Times* piece, the civil defense solutions could also solve "traffic abuses—arrogant, lawless, almost out of hand," which could "threaten our safety health and business" (1951, 15). For

Moses, "civilian defense looks to us like defense against overcrowding in all its aspects—against fright, hysteria and mob psychology" (15). Traffic and crowd panic in this formulation become one and the same thing in America's fight with the evils of congested automobility: "We need not wait for a long range Russian bomber to teach us dramatically what street congestion, multiple parking, overbuilding and lack of open express arteries can do to inspire fear, panic and unceasing public fury as distinguished from mere inconvenience, delay . . . retrogress and slow rot. . . . Our own attack must be on congestion as such" (1951, 15).

Moses was vastly critical of the FCDA's shelter program (Ingraham 1950a, 1). Through the city's mayor, Vincent R. Impellitteri, Moses gave testimony to the congressional committee hearings on the Federal Civil Defense Act in 1950.[7] Impellitteri's view was as clear as Moses's: "We reject the 'take to the hills' and 'make for the caves' philosophy.' . . . We shall not adopt a policy of dispersion and decentralization" (Hinted 1950, 1). No one could take seriously "that we prepare to evacuate 8,000,000 people. . . . New York will not be abandoned" (1). Even in Moses's unbridled skepticism toward civil defense, he saw that the means to evacuate citizens out of a city on short notice could have a dual or even triple purpose in bringing forward longer-term plans for reconstructing the city along more efficient lines of movement (Wallander and Moses 1951). As Light argued in the context of the dispersal debate, "urban professionals capitalized upon concerns about urban security in order to advance arguments for their own pre-war goals . . . such as reducing traffic and congestion, and slum clearance" (2003, 17). This could mean plans for one-way street traffic, anticipating on a permanent basis the use of interstate contraflows during emergency as both a "defense measure and [a way of] tackling traffic congestion" (Ingraham 1950b). Moses was really admitting that they could do little for evacuation, hinging his conclusion on the unknowability and the irrationality of the car and the driver, so that "any thought that you can evacuate a large population in a short time from any large city, even if you have a place to move them to, is so much moonshine. No experienced, responsible official will advocate it" (1957, 32).

Interstates, expressways, and multipurpose road improvements were often synonymous with urban segregation, as "engines of homelessness and destruction" (Graham 2018, 528). Moses remarked on the interstate that the "greatest obstacle is the one which has been stressed the least—moving people off the rights-of-way so as to get the land cleared and construction started. People don't like to move" (quoted in Pierce 1948, 37). Similarly, the FCDA's Val Peterson would make the comparison between broader patterns of urban dispersal and slum clearances, a "difficult and sensitive field . . . when we tear out slum areas whereas they should become parks, we should then, too, thin out these centres of population." Wider streets were necessary for people to get "in and out," "if we are going to survive . . . regardless of the emergency."[8] Peterson built on a subcommittee's review of the Project East River, which reported to the government in 1952 on how to address the atomic threat through urban dispersal. Slum clearance, other environmental projects, and circumferential urban roads were the "indirect means" to make evacuation concrete. Even shopping centers were perceived as a form of defensive weaponry. In 1951 the annual convention of the American Institute of Architects saw "shopping centre designers" publicizing "their value as evacuation centres" (Tobin 2002, 21). Victor Gruen, guru and chief designer of the American shopping center ideal, saw shopping malls as "suburban crystallization points" to meet a broad variety of social needs, including temporary housing for evacuees (Mennel 2004, 121; Hardwick 2015). Yet the building of the interstates and urban expressways "often meant black removal from the central-city area" (M. Rose and Mohl 2012, 100). While the interstates could be used deliberately or inadvertently to erase Black communities from the urban center, in other contexts it was a policy of segregation "protecting" so-called sound neighborhoods of the white middle class (Connerly 2002). In Atlanta the interstates continued a policy to use roads and highways to enforce segregation and restrict Black mobilities (Bayor 1988). The interstate and expressway building projects created an itinerant population of what Samuel Zipp called "clearance site evacuees, cast out by the destructive energies of progress," who were "said to resemble the displaced persons of postwar Europe" (2010, 10).

CITY AS LABORATORY: PANIC AND
THE EVACUATION ACADEMICS

Evacuation gained in resolution through research projects, literature reviews, and various tests, exercises, or simulations designed to play out evacuation. For Tracy Davis (2007, 90), the emergent evacuation drills were a "rehearsal for a performance." They were dramaturgical. Public information films tended to be "unexceptionally white and middle class," and their performance and demeanor expressed nothing but "flat affect" (Oakes 1995, 101). The plans and exercises aesthetically aligned role-play and drama with ways of meeting an uncertain and ineffable future (O'Grady 2018). Whiteness was mixed with rationality, and anything other than that was a dangerous problem. New York City organized a program of civil defense exercises. On September 31, 1952, the city endured its fourth atomic bomb simulation. The event was staged by dropping incendiaries on Staten Island, and in a foreshadowing of urban renewal, the bomber was supposed to continue on to the Bronx, Brooklyn, and Queens (Feinberg 1952, 1). The main audience was at Broadway and Times Square, where some estimated 100,000 people gathered to watch. The *New York Times* fostered the looter myth, publishing an odd image of police holding at gunpoint apparent looters attempting to rob a bank. It was part of the drill. Over a megaphone at the edge of Bryant Park, the police commissioner declared the drill a great success, although he may not have taken kindly to Moses's declaration that these were "silly city drills" that people paid no attention to (1957, 34).

Eager to serve urban evacuation practices was a university and urban planning sector. Northwestern University's traffic group would advise the FCDA and urban authorities on traffic handling, creating a course on supervising and regulating street and highway traffic during an evacuation for police and civil defense officers (Federal Civil Defense Administration 1957, 123). In 1954 the group met with police, city, and civil defense authorities and the FCDA at city hall in Milwaukee, Wisconsin, to discuss measures to evacuate the city given its prior interest in evacuation, having set up a civil defense committee in 1951. It would also develop a general plan for its expressway system in 1955 following a consultative report that considered mass transit, parking, and "mass evacuation" combined (Parsons et al. 1955).

Elsewhere, evacuation drills became commonplace. In Robert Oppenheimer's Los Alamos atomic community, the first evacuation drill of its kind was practiced in November 1952. Eight thousand people were loaded into three thousand cars and driven out of the proximity of Los Alamos to a civil defense parking area (Hunner 2014). By the mid-1950s the FCDA had carried out several ride-outs and walkouts. Peterson's mark of success—focusing on the ride-out by automobile from Mobile, Alabama, in 1954—was that, he claimed, forty-nine thousand people had left the city in nineteen minutes, and there was not one "scratched fender."[9] In the same year, Peterson, along with Barent Landstreet, who headed up evacuation for the FCDA, engaged a conference of the National Women's Advisory Committee to the FCDA, who were especially active in the focus on the home and on preparing it for shelter and evacuation. As Davis suggests, the performative aspects of these schemes tended to rely on "systematic desensitization" and "reciprocal inhibition" (2007, 42), in which the family mother played a directing role. The theme of the 1954 conference What the Family and the Community Will Do for

Survival in the Age of Peril cemented the traditional gendered associations of femininity with motherhood, family, and community preparedness. It involved several presentations by advocates from the Women's Voluntary Services of Great Britain, as well as Marjorie Child Husted, the home economist behind the General Mills Betty Crocker cookery and homemaking brand (Federal Civil Defense Administration 1954). Landstreet began by admitting that "there is no problem in civil defense that concerns more of our defense . . . than this whole problem" (1954, 13). There had been no more "emotion, more uncertainty, more misunderstanding than [over] the 'whole question' of evacuation." The object of their concern was the "family and the individual in the family," but it was a "qualitative job," to which they were adding "a quantitative factor" requiring certain "data."[10] Of the many factors the FCDA (in collaboration with the National Academy of Sciences) would try to examine in evacuation, they wanted to know more about humor and eliminate it. Evacuation needed to be taken seriously; to be controllable it needed to be less emotional.

Landstreet hinted at the other problems the FCDA was concerned with: that "things happen" when evacuating industrial areas with a "polyglut of nationalities, of religions, of different income levels into a quiet, middle class, mostly agricultural town." Studies of urban analysis meant exploring "the racial breakdown, the religious breakdown," the "parts that go to make up a city," to the extent that when evacuation was begun, "people can be fitted as closely as possible into the kind of community structure to which they are accustomed."[11] Landstreet was building on a technical rationalism born of defense intellectuals and other academics focused on personal and social behavior in disaster that tended to treat racialized communities as "problems" with quantitative solutions. One study compared the housing densities and ownership of dwellings of New York to explore the possibility of "matching" segregated evacuees. "Usually less hostility develops if hosts and evacuees are matched with regard to social characteristics; however, other considerations may make non-segregated billeting preferable" (Iklé and Kincaid 1956, 12), the report suggested. The disproportionate lack of housing share within the Black community made this especially contentious. The report's recommendation meant deliberately placing Black communities in housing more likely to be "dilapidated" or without key infrastructure, producing drastically different billeting densities within the reception communities.

Irving Janis, who had taken part in a RAND study on the psychological effects of war on the population, was an important voice. Janis reported from the Morale Division of the US Strategic Bombing Survey on "fairly severe terror states," hysteria, disorganized "flight." He even spoke of ways to offer punishments to those who might not evacuate should an organized evacuation be called in advance of a bomb, "against nonconformists in the event of an evacuation crisis" (Janis 1950, 261), although he admitted, "There is likely to be a fairly high degree of conformity to the evacuation plan" (260). The growing research on disasters relied on, for Sharon Ghamari-Tabrizi, a "grand analogy," expressed in an "atmosphere of doomed credulousness. . . . What did they think they could know by looking at the one world, the peacetime present, and the other, the world of atomic, then soon after, thermonuclear war?" (2013, 336–37). Most who have traced the lineage of academic scholarship in this area look at the work of the National Opinion Research Center at the University of Chicago, whose disaster studies program developed a plan in 1949 to find "common elements" from which to transfer experience and knowledge, from tornados, industrial fires, and train crashes. Charles Fritz, Enrico Quarantelli, and Harry Williams—who was the National Academy of Sciences technical director—all carried out research through the National Opinion Research Center. The Committee on Disaster Studies, led by Williams, was formed under the auspices of the National Academy of Sciences. The Stanford professor Bertrand Klass managed the social science projects (Ghamari-Tabrizi 2013, 350). Walking to the committee's archives in Washington, DC, on a drizzly March afternoon, I realized I was following one of the many blue evacuation route signs directing escape out of the city.

Evacuation seemed pitted against or with the concept of panic. "With a few exceptions," Charles Fritz and Eli Marks claimed, "the literature on disaster places heavy emphasis on panic behavior—so much so that one gains the impression that this is the most important (and most common) type of disaster behavior." "Panic behavior can, in some cases, cause more damage than the disaster itself," which is the most common claim about why keeping calm and cool is crucial to evacuation, yet they suggested that "it is an *unusual*, rather than a common, reaction to disaster" (1954, 28). Panic was an erroneous label, a name given to disaster behavior that expressed less a lack of individual control than "deficiencies of coordination and organization" (33). Panic was commonly expressed in an emotional, affective "flight" behavior and in actions "inappropriate to the situation" (Janis 1950, 260). The vague practice of "Panic stopping" was a way the government leaders sought to suppress both the reaction and the susceptibility of citizens to it. As Joseph Masco (2014) observes, other metaphors and aesthetic categories were reached for. Val Peterson, the FCDA director, in 1953 even compared

FIGURE 4.1.
Preparation
Prevents Panic
(Georgia, Civil
Defense Division
1952). Georgia
Archives, Defense,
Civil Defense,
Georgia Alert
and Responses
in Georgia, RG
022-05-042.

FIGURE 4.2.
The National
Defense Pattern
(Georgia, Civil
Defense Division
1952). Georgia
Archives, Defense,
Civil Defense,
Georgia Alert
and Responses
in Georgia, RG
022-05-042.

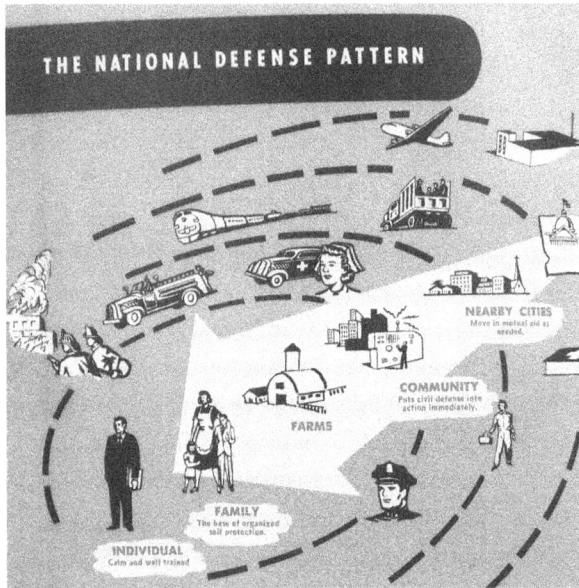

the properties of panic to nuclear fission itself, both chain reactions of volatile
energies and destructive forces: *"Panic is fissionable"* (Masco 2014, 49). In
Georgia, as I will explore later, material was produced for schools with the
slogan "Preparation Prevents Panic" (see figures 4.1 and 4.2). A national defense pattern envisioned protecting Georgia's "peace," portrayed as two boys
fishing, by a radial scaled diagram of protection with the smallest possible unit
being the individual, who should be "calm and well trained" and whose "jit-
ters," "fears," and "tensions" could be soothed by the communicable affects of
purpose, "calmness and quiet handling of a potential danger" (Georgia 1952, 7).

Later Fritz and Williams characterized panic as "trampling each other
and losing all sense of concern for their fellow human beings," reminiscent
of the high-rise garment factory fires discussed in chapter 1, before "they

turn to looting and exploitation, while the community is rent with conflict" (1957, 42). Flight, or what the FCDA was trying to control as evacuation, did not mean automatic "panic"; most of the evidence they found pointed to "people continuing to think of others and continuing to use critical judgment," often the "the only rational choice" (44). Evacuation was not cowardice or lack of bravery but "a rational, adaptive form of behavior" (44n2), in similar terms to Gilbert White's assumptions of behavior in disaster. They aligned resisted accusations of panic in the way Rebecca Solnit (2010) describes "elite panic," as an "erroneous impression of wide spread fault-finding and search for a scapegoat," usually by people in positions of authority, who would "use the disaster to secure power, status, prestige or other rewards for themselves or for special interest groups" (Fritz and Williams 1957, 49, 50).

Despite the gravitation toward behavioral psychology, urban sociology, and human geography, which were, at the time, wrestling with new quantitative and positivist approaches in a time of scientific relevancy, panic persisted. These experts became closely involved in a cadre of evacuation studies commissioned by the FCDA and the Disaster Studies Committee, and I want to focus on two that were carried out in 1954 in Alabama in the city of Mobile, and another in Bremerton, Washington, near Seattle. Harry Williams turned to the Washington State University geographer Edward Ullman first when preparing a team of academic observers to take part in Operation Rideout in Bremerton. Ullman could not go and recommended his colleague William Garrison. Garrison's work was especially tailored to exploring mobility and connectivity between cities and regions. It was a close disciplinary match (Barnes 2001, 420) that aligned with geography's pretentions to scientific models, quantitative methods, and "laboratory-like interventions" (Barnes 2012, 9)—almost perfectly expressing the penchant for simulation tests. Williams included John Mathewson, a transport and traffic engineer from the Institute of Transportation and Traffic Engineering at the University of California. Operation Rideout was to involve mainly naval shipyard personnel in Bremerton, across Elliot Bay from Seattle in the Puget Sound lowlands. Following an earlier exercise operation "walkout" in Spokane in April 1954—where B-36 bombers had simulated bombing runs over the city to time with a mock explosion and actually dropped leaflets with diagrams of the bombing area (O'Connor 2010)—the intention was to compare the evacuation of an urban area carried out on foot and the one in Bremerton to be conducted by car: to "ride out."

Some municipal leaders had assumed the "complete abandonment of all privately owned passenger automobiles in case of attack." Bremerton could provide "an actual laboratory for such an examination."[12] Would evacuation mean confusion, where "traffic, complicated by panic, will be snarled, and will result in increasing rather than decreasing people's exposure" to all of the risks of a nuclear bomb?[13] Before the exercise, a questionnaire was sent to all of the civil defense workers to consider during the exercise. Question n was, "Did everyone drive safely and sanely or were there those who crowded, pushed, became panicky, etc.; and who failed to observe the common courtesies of driving?"[14] Bremerton's physical and urban geography of waterways and transfer routes, features of a low-lying metropolitan peninsular area, would give it a "congested character" yet were "typical" of most American cities.[15] Of the twenty-eight thousand residences occupying the central business district, around half were assumed to be destroyed in the exercise. The observers noticed the rhythm of population movements in and out of the city each day. They identified approximately twenty thousand workers across five hundred acres, with a population density of twenty-six thousand per square mile. The advantage of Bremerton was its apparent generalizability, having the "features of population dispersion, population concentration, and congested population movement that approximates the 'American City'"; they suggested that "the observer can approach laboratory-like conditions."[16]

Shuttling between different perspectives, several Civil Air Patrol aircraft (CAP) were used to record film of the evacuation over the two days, resulting in detailed observations of "streams" of traffic leaving Bremerton. Seven CAP aircraft perhaps most approximated the abstractions of the burgeoning spatial science Garrison had developed by offering a synoptic view of traffic flow. The exercise reflected a multitude of visualities or aesthetic registers, similar to but substantially different from the targeting gaze of the aerial reconnaissance or bomber's-eye view (Saint-Amour 2011). Plans for the CAP aircraft to be adorned with different cameras took account of poor or foul weather, devising a system of coverage so that the aircraft took circuits over the city. They focused on shaded areas over particular traffic interchanges, timing the spacing of each shot in relation to the pre-alert and post-alert phases. The cars were to have white numbers or letters painted onto their roofs so as to be easily visible from overhead. The committee continually experimented with different approaches to observing the exercises. In one letter inviting the San Francisco planner Lawrence Livingston Jr. to chair the Spokane exercise, its author discussed the involvement of "outside observers" like RAND Corporation, academics, and consultants in order to "learn something of the natural history of disasters from a distance great enough to ensure freedom from direct participation and yet close enough to permit some detailed examination." The process was compared to a documentary film, planning shots from a hovering-observer perspective as things may not be picked up again from the "cutting room." They imagined "a good traffic engineer, or city planner, or behavioral expert sitting in a helicopter, observing tactical errors, bad planning, panic and what have you."[17]

In Bremerton a "roving" fire department vehicle brought the observer's perspective to a more grounded position. The team observed the evacuation, which began from the naval repair station, of 9,500 pedestrians and 1,000 cars. Traffic movement in the city was "smooth and rapid."[18] They observed family units coalescing, believing fathers were driving to their homes to pick up their wives and families before leaving. Those awaiting public vehicles were older-age people, teenagers, or mothers with children. The dedicated evacuation routes were considerably underutilized; they acknowledged that the one-industry nature of the city made it a particularly simple evacuation, avoiding multiple streams of traffic. By the end of the test, two thousand cars were believed to have left the city. The approach to realism was mixed. Klass noted that "the lack of public participation could be interpreted to indicate apathy and disinterest. On the other hand it could be equally indicative of nothing more than an unwillingness to expose oneself to some inconvenience."[19] Because of the widespread communication, he made the unrealistic assumption "that everyone was fully aware of the exercise, of escape routes etc."[20] "Nonparticipants" closed their shops early or kept away from the major streets. Some were lining up for public transport, waiting for the alert signal. Interviews with "informants" indicated good knowledge of the exercise and instructions, yet many had already evacuated before the exercise alert began, "an unfortunate deficit in the test."[21] Another method of observation, enthusiastically proposed by Garrison and prepared by Mathewson, was a "time-space tracking" plan where "control vehicles" were inserted into the traffic flow and released at particular starting points. The researchers were concerned at the exercises' smoothness, it being "unlikely that any informed person, however optimistic, would have predicted the phenomenal freedom of movement of traffic." Only "car A" was forced to stop because of traffic congestion, and the cars retained high average speeds, between 26.5 and 37.4 miles per hour.[22]

A "most promising civil defense technique," the automobile became the primary vehicle for evacuation. That there was plenty of space available in the cars and buses for more people was a takeaway from the exercise, although they had been encouraged to take "full loads."[23] The researchers concluded that "many more participants could have been evacuated without increasing the numbers of cars."[24] This sense of empty capacity flowed into the kinds of abstract, utilitarian, and budgetary thinking that occupied their approaches and tended to decontextualize automobility from the car and the mobility cultures and politics of the time. A "logical" follow-up question would have been, "To what extent would people make space in their cars available to others?" What proved more troubling to the observers, and a "larger question," was that the "freedom of movement" identified in the traffic streams and recorded speeds in the artificial "closed circle type of operation"[25] would be vulnerable to car breakdowns and traffic accidents—how could they be adequately controlled and regulated? What was the maximum volume of cars possible? The test "represented a rational ideal . . . the refined unreality of abstraction" (Barnes and Farish 2006, 820), yet was less attuned to the felt and social meanings, the habits and differences, that pattern the cultures of automobility, especially in the United States.

Another emergency exercise study enrolled John Rohrer, a social psychologist and director of the Urban Life Research Institute at Tulane University, New Orleans, who was invited to the city of Mobile, Alabama, for Operation Scat. The emphasis of the study was entirely different. Using "participant observers," the study sought to understand the community's perspective on the exercise and the FCDA's preparations more generally. The "role and effectiveness of mass communications" were to be their main objects of concern.[26] The relationship between evacuation and race was not explicitly mentioned in any of the planning documents I examined, which was strange given the context. It was in Alabama, and the *Brown v. Board of Education* ruling, on the controversy over school desegregation, had been issued just a month before Operation Scat. Rohrer was invited by Williams as a "specialist in urban problems"—perhaps a loose euphemism for Rohrer's interest in racial and social structures.[27]

Rohrer's report reveals something of his wider interests in social conformity (Rohrer et al. 1954), Robert Parks's influence on a human-ecological urban sociology, and Rohrer's longer-term concerns for Black social identity in America (Hammack 2018). The report's opening discussion sets the scene with the human "ecological patterning" of Mobile as a "transitional zone" of industry with immediately adjacent residential areas in some "decay," occupied by "lower class and lower middle class people." Estimating that forty thousand members of Mobile's Black community were evacuated, the most important generalization he concluded from the research was that the areas most likely to be bombed were going to be populated by "minority groups," "who have value and belief systems distinct from those of the 'community leaders'" and who would not use traditional routes of communication, such as the radio or newspapers. Rohrer recommended looking at "informal means of mass communication," supposing that a "strong urge to conform" to "white authority," "probably typical of most Negroes in southern white urban areas," may have mitigated the problems of communication.[28]

Rohrer's participant observers looked closely for racial antagonism and signs of humor—an interesting aesthetic and affective companion to evacuation. They wondered whether evacuation was being taken seriously or not, once more assessing the aesthetic and affective life of evacuation through aesthetic judgments in which play and humor were a concern. Of course, in part, the use of Cold War simulations was a way of reducing worry, even panic, through the minor aesthetic categories of the "interesting" in Sianne Ngai's (2012) and others' interpretations of evacuation plans, calculations, and war games. Humor could suggest a kind of interestingness gone too far.

The Scat observers reported a rumor that the exercise was a cover for the government to drop a bomb on Mobile so as to eradicate the Black population and avoid school desegregation. Interestingly, the schoolchildren in Mobile had been given "dog tags" by the local civil defense office, presumably so that they would be identifiable should an evacuation occur. One of the most telling anecdotes recorded a Black woman walking down the road, holding two suitcases, before a Black couple picked her up in their car: "She said 'she didn't want to get hit by that bomb.'"[29] The researchers confirmed the presence of these rumors amid a more ephemeral "vague unidentified fear," but their line of questioning seemed at odds with their conclusions of almost total submissiveness. They even suggested to a waitress in a café that they could just hide in the back and the authorities would be "none the wiser."[30] Lexically distanced by the observers' language, the Black citizens repeated a rumor that they would be fined $50 should they not comply.

The observers expressed strikingly classed and racialized concerns and prejudices themselves. Observer W approved of the opinions of "good, solid middle class people" who were "real involved with the evacuation" as opposed to the Black population, who conformed but were somehow not as invested in its outcomes in the observer's estimation.[31] The observer noticed that some people in a store were making fun of the exercise, saying "Scat" when they said goodbye. Much later, the emergency controller of the Mobile exercise acknowledged Rohrer's findings and suggested that they were fortunate to have used the word *scat* because it made sense to *even* the lower-class and Black population, some of whom had problems understanding a word like *evacuation*. Public transport for returning was very poor. The buses had to pass by an estimated 1,500 people waiting at the stops as they were so full. The report echoes the evacuation failures of today: "There was ample evidence that public transportation facilities were inadequate. . . . Apparently it was not realised that a majority of the people living in the area did not own cars but were completely dependent upon public transportation."[32] This fact required "more attention than was given to it," the report summarizes. For those who could get on the buses, it "was not a cheerful kind of experience," yet high conformity was observed. The suggestion that drivers pick up people walking was not followed. Black and white Americans left the city with only "partially filled cars," it being "the exception rather than the rule to see cars picking up pedestrians."[33] Far from a "utopian scenario of interracial carpools," social and spatial segregation continued (McEnaney 2000, 139).

It is questionable whether the FCDA really did imagine interracial carpooling. While Observer B did get picked up by a white driver, the driver complained that "the Negroes were not being picked up and that she probably should have waited and picked up Negroes instead of us."[34] A policeman and segregationist, detailed in the account of Observer S, gives a startling and embarrassed response when his opinions of multiracial car sharing in the evacuation are challenged. Under the bomb, "racial barriers" could break down, and he believed that "Negroes would be able to ride out of town with Whites." So could evacuation be a leveler? Not so, thinks the observer, who suggests preferential treatment would be given to evacuate whites from the area first before the Black community would be given a thought.[35] Observer W was picked up by an office worker and their secretary, who did not "hesitate" to drive past two Black males. When questioned about this, the driver responded, "Oh, those boys live just a few blocks over there. They are on their way home, will be there in just a few minutes." Here the observer asks the driver to "tell me a little bit more about the Negro problem. . . . There were a lot of automobiles in this little parade of which I was a part and I did not see any Negroes at all, even driving their own automobiles." When the driver responded that they were coming from another area, the observer reflected, "There was an 'apartness,' geographically or in some way between themselves and the Negroes in Mobile."[36] Another observer saw a great many people of color walking along the Virginia Avenue route in the spirit or atmosphere of a "holiday or picnic type of feeling."[37] The "remainers," an observer noticed, had "neatly segregated" on either side of the street.[38] An FCDA-produced film used multiple recordings of the event to idealize the mobility and social divisions, whitewashing racialized automobility. The cars move in calm, orderly lines—in contrast to observer B's observation of "irritability," expressed in the use of the horn. They snake almost side by side, gleaming in the sunshine, shiny new models of American manufacturing progress, the nation moving in concert and together.[39] All the while, concludes Gretchen Sullivan Sorin, "the city's more than 14,000 black residents never appear" (2009, 39).

RURALITY, RACE, AND VEHICLES FOR SURVIVAL

On September 28, 1955, Operation Lifesaver was intended to evacuate forty thousand people from the Canadian city of Calgary. Some five thousand people took part. Calgary was an important military site; a key location for oil and gas production, wheat distribution, and meatpacking; and an important Canadian railway gateway (see Davis 2007). The car could remove people along designated routes marked by colored signs to towns where they would be billeted. Individual automobility would save the nation. The authorities were hugely concerned, worrying about a "mad rush . . . chocked with people fleeing in all directions" (Sopko 2015, 22–23). Like in the United States, the car represented a form of liberalism in motion, expressing the "hunger for self control" even if evacuation seemed to threaten it within "communal assembly points and public transportation," which was "reminiscent of communism" (Reilly 2008, 60). In practice, though, "to consume was to survive; a family car and a full tank of gas could spell safety from attack and there was no need to depend on anyone else" (Reilly 2008, 60–61).

In Operation Lifesaver the liberal freedoms and free-market capitalism, which were distilled in the metaphor of the automobile (Seiler 2009), had to live with the regulated and coordinated control and the frequent diagramming of action that imbued evacuation planning. Residents were written to directly with instructions for how to comply. First, they should expect citywide sirens on a sustained blast. Sequences of action were prescribed, from assembling to packing, loading the car, collecting passengers, and driving out. Only one instruction was given to those who were not relying on an automobile. Procedure demanded automobility discipline, reassurance, and easy communication. The authorities insisted that traffic signs be made of plywood, eight by four feet in size, with black lettering on the top half and route coloring on the bottom, to create easily comprehensible directions (Reilly 2008, 66). Codes of conduct strengthened the oxymoronic tensions of the liberal "open road" that held together (Seiler 2009) powers of individualist autonomy with doctrinary disciplination. Guidance advised that the "instructions (which may not fit exactly with your personal wishes), are intended to ensure that any evacuation in which you have to take part

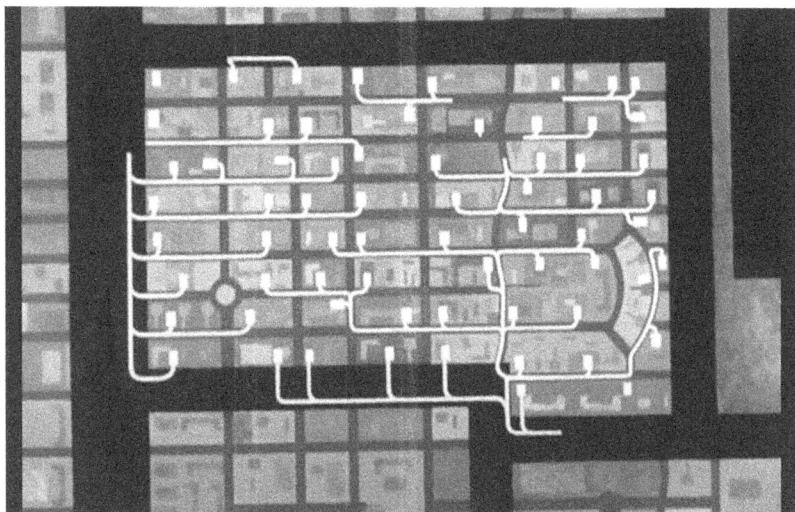

FIGURE 4.3. The household/evacuation network, Operation Lifesaver (Kehoe 1957).

is an ORDERLY WITHDRAWAL." Any attempt, the guidance suggested, "to disobey these instructions, and to substitute for them rules of your own making, can only result in the reproduction of mob law with all its attendant evils." And going in the other direction: "The majority will not be sacrificed to the whim of any individualist."[40]

Operation Lifesaver (Kehoe 1957) became a short fifteen-minute film by the Canadian Broadcasting Corporation (CBC). Ground zero was a traffic intersection in the Calgary city center. Cars and pedestrians from the 1950s create a drizzly black-and-white agora. A narrator declares, "From this point on they've been instructed on every move they'll make. . . . Carelessness and delay can cause panic." The film centers around a family whose children quickly make their way home from school. Mr. George La Pierre has driven home from work. The neighborhood is now referred to as sector B3, the new zoning of the evacuation drill. The film displays a graphic showing every household plugged into the evacuation network by white interfacing line (figure 4.3). He loads his Oldsmobile 88 Holiday, a family four-door hardtop sedan with front bumper "bombs" and a V-8 engine, with his family and their belongings (figure 4.4). Oldsmobile marketed the car as faster than a rocket.

The film shows the civil defense control room taking in percentages of the residents who had evacuated from city neighborhoods. At xhr (exercise hour, or the beginning of the exercise) + 60 minutes, the film moves to a helicopter

FIGURE 4.4. Mr. George La Pierre loads his family and luggage into the car during Operation Lifesaver (Kehoe 1957).

shot over the neighborhood of deserted streets, the voiceover proudly explaining that the city has been emptied. Farther north, another "spotter" helicopter crew follows cars moving to the reception towns. These fulfill the logical promise of the evacuation diagrams and a federal metropolitan imagination of rural North America. Andrew Burtch reminds us that the evacuation highlighted a transition and "interdependency" between city and country (2012, 102). In the United States, the FCDA's pamphlet *Rural Family Defense* (Federal Civil Defense Administration 1956) demonstrated that urban evacuation policies relied on rural areas where evacuees could be temporarily sheltered, fed, and potentially rehoused. For the rural and agricultural historian Jenny Barker-Devine, "Rural people were typically treated as a homogenous group; ready and willing to help refugees and rarely in need of direct assistance" (2006, 421). Colorado had led the way under William L. Shaffer, whom Peterson described colloquially as a "farmer" ambling into his office one day to surprise him with the civil defense capacities of America's rural areas. Shaffer was recruited by Peterson to the FCDA as a rural consultant in 1956.[41] The FCDA was encouraged by Shaffer's optimistic plans to feed and accommodate Denver's evacuees using the agricultural wealth of the state's farms and what Shaffer would describe, in testimony to a US Senate subcommittee, as the simple hospitality of rural dwellers.[42] Colorado practiced just this in

Operation Welcome, although it involved a comparatively small number of evacuees, some six hundred, compared to the 200,000 imaginary ones organized within the exercise from its control center.

Lionizing rural Canadians and Americans into welcoming but passive recipients tended to demonize city dwellers and elicit concerns over both control and fear, as revealed by the FCDA studies: "How could white suburbanites or rural white families deal with the hordes of hungry, thirsty, and frightened urban poor streaming into their communities from the target cities?" (Garrison 2006, 40). The *Rural Family Defense* pamphlet came across as controlling, demanding that "rural people—you, your friends, and neighbors—will be in charge of reception centers for evacuees, have positions of responsibility and authority" (quoted in Barker-Devine 2006, 422). Peterson was widely quoted, in an interview for the *Bulletin of Atomic Scientists*, demanding a "higher responsibility" from the nation's homes. He could not dispute his "responsibility with the evacuees as they came into my front yard" (quoted in Barker-Devine 2006, 422).

The *Operation Lifesaver* film ends in a curious way. The La Pierre family are registered at the Innisfail reception town and, despite the weather conditions, consume hot meals using china plates and tea cups, cooked on gas stoves outside. The family balance their meals precariously on the bonnet of the white family car. The scene is so obviously staged, the white car standing out among the throng of other evacuees who make do without the automobile-cum-table. Mother and daughter are in the shot, looking perturbed, while the voiceover says they "didn't have to worry about the dishes!" (Kehoe 1957) as the duties of gendered reproductive labor are apparently suspended and reconfigured on a car bonnet. Mr. La Pierre grins. This moment of Cold War evacuation consumption by private automobility ends with the car becoming the platform for their own bodily sustenance and the fulfilling of the nuclear family's shared meal. News reports described the event as having a holiday atmosphere, with residents "packing picnic baskets and loading up the car with family to go on an excursion to the countryside" (Burtch 2012, 102). One of the family's children was quoted in the press as saying, "Gee, we had a swell time and saw a free movie!" (Kehoe 1957). Back in Calgary and in front of a "movement chart," the civil defense coordinators congratulate one another. Uniformed male defense workers bodily diagram the circuits of evacuation coordination as they shake hands. Smiles are passed on. Women, perhaps the administrative support for the coordination team, are just visible through a window into another office room. The Canadian Federal Civil Defence coordinator reminds viewers of the importance of evacuation drills such as these, before commenting, "The swift movement to the countryside is your best protection against mass death. Civil defence provides that organization." But he then signs off: "It is up to you to use it for your own sake."

At the same time as Operation Lifesaver, US urban evacuation practices—similarly relying on the car—were translated into guidance. Leaflets circulated to addresses in the Savannah-Chatham and Atlanta areas in Georgia stand out (see figure 4.5).

Public communications focused almost solely on the car as the vehicle of escape, and on the family as the key unit for survival by this "route." The mobility possible for evacuation was closely tied to automobile access and ownership, "which in turn was tied to broader concerns for privacy and social order" (Farish 2010, 232). In Atlanta the first evacuation plans were drawn up by the Atlanta Metropolitan Area of Civil Defense in September 1955 and considered the same spot where Scarlett O'Hara had observed the wounded evacuees from the Civil War. The densely populated Five Points was its ground zero, with an estimated 7.4-mile damage radius. A "timed" and "supervised" evacuation plan, imagined through an abstract plan of "escape ways," was drawn up (figure 4.6). The plan, with some features reminiscent of Robert Park and Ernest Burgess's (1925) famous radial concentric ring model of urban development modeled on Chicago—whose sociology had informed Tulane's John Rohrer's understanding of Mobile in Operation Scat—contained a simplified urban geography of the city with a centralized system of organization and coordination, along with instructions and assumptions of a disciplined public compliant with instructions. The centrality of the automobile to this system was paradoxical, oscillating between the centralized dictates of the civil defense hierarchies and the car as a reliable engine of individual and familial mobility, as shelter, as conveyor of comforts—food and warmth, clothing and autonomy.

Your Route to Survival spelled out the use of evacuation routes by car away from the Five Points and target areas such as the Lockheed factory. Another leaflet, *4 Wheels to Survival*, emphasized the car's properties.[43] The automobile was democratic, moving a person many times farther than the "strongest" might on foot, provided that the car was kept in readiness—the battery should be in "tip-top shape," the gas tank "kept half-full at all times." With economical driving habits, which meant driving at speeds between thirty and forty miles per hour and coasting when possible, the car would get a family out. It was a shelter that could endure fallout, as well as an "information center" to tune in to the federal CONELRAD emergency radio stations. Why not swap the properties of the home to the car? Shelter from

EVACUATION GUIDE FOR A CIVIL DEFENSE EMERGENCY

Your

ROUTE to
survival

ATLANTA METROPOLITAN AREA OF CIVIL DEFENSE
Elliott R. Jackson, DIRECTOR

FIGURE 4.5. *Your Route to Survival* pamphlet, 1955. Atlanta Metropolitan Area of Civil Defense, Georgia Archives, MARTA (Metropolitan Atlanta Transit Authority), Atlanta Transit Authority, Administrative Records, RG 087-31-066.

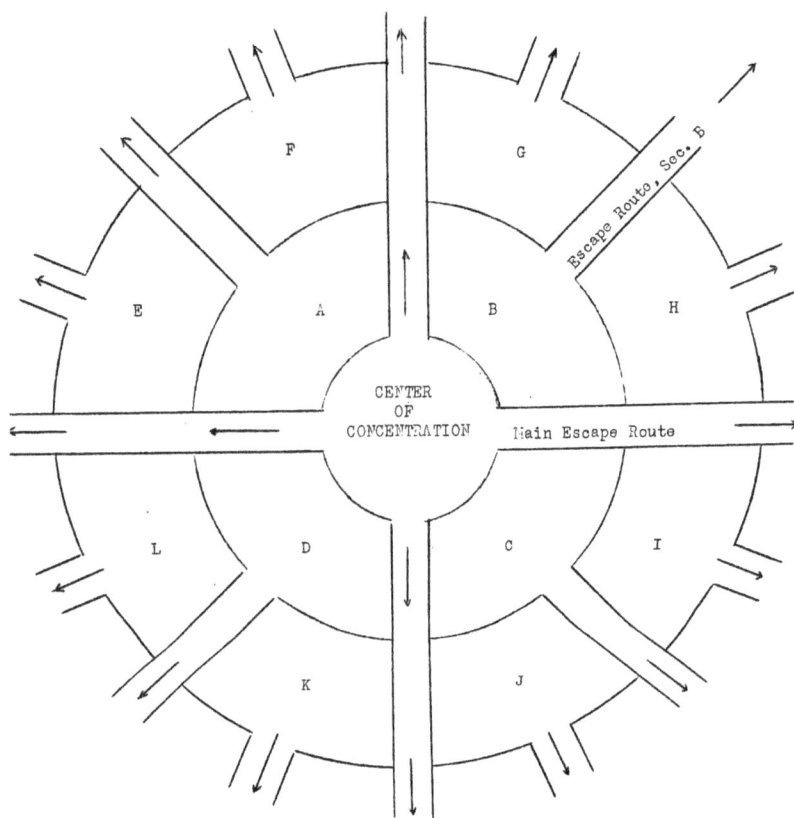

FIGURE 4.6. The "oversimplified" and "hypothetical" evacuation plan to prepare "escape routes" within Atlanta and neighboring cities, 1955. Atlanta Metropolitan Area of Civil Defense, Georgia Archives, MARTA (Metropolitan Atlanta Transit Authority), Atlanta Transit Authority, Administrative Records, RG 087-31-066.

a blast and much more, the car could be "a small movable house. You can get away in it—then live, eat, and sleep in it in almost any climatic conditions, if necessary, until a civil defense emergency is ended." From a house to a "shopping center," the car could be an effective storage container for the stockpiled seven-day supply recommended by the FCDA under its "Grandma's Pantry" program (Garrison 2006).

Combined with good automobile civilities—"courtesy, cooperation and careful driving"—the car meant individual isolation and mobile protection for the family, somehow squeezed into the cramped conditions. It could be a defensible membrane separating them from the outside and other road users. "Don't crowd or try to beat the other fellow," guidance suggested. If broken down, try not to block traffic. And if the traffic was slow or stalled, do not "lean on the horn. Your impatience may become someone else's panic. That can cost lives!" If you don't have a car, the guidance advises that you begin walking "towards an evacuation route. Someone will pick you up."[44]

In Alabama and Georgia, other plans were determined for children. In the context of Jim Crow, within a de facto segregated American South, Jonathan Leib and Thomas Chapman point out that the evacuation plans reinforced segregationist practices just as Mobile's ride-out had demonstrated. The plan *Escape from the H-Bomb* (Chatham-Savannah Defense Council 1955), which was delivered to all homes within Savannah and Chatham County, placed most of its emphasis on the car. People were advised to drive to designated "escape ways" that would lead them out of the area (Leib and Chapman 2011, 583) and to head to reception centers. There is little critical discussion of the implicit transport injustices embedded within a plan such as this, in terms of limited Black ownership of vehicles, access to vehicles, and the cultural disciplination and alienation of Black drivers—issues revealed in Mobile. The city developed an "innovative" approach to schools. Planning bus services and chartered trains were utilized within the plan, in which 32,240 children were to be taken to one of forty-two communities throughout southeastern Georgia and one in South Carolina. As Leib and Chapman show, "The vast majority of these evacuation destinations were to be racially segregated" (2011, 583). Destination areas were to be segregated by white and Black students, in some effort to match up the evacuee and receiving communities, as the FCDA studies had predicted and recommended. Train cars were most likely to have been "racially sorted" by assigning specific railroad cars to specific schools. The segregation of evacuation, however, also meant the absence of people of color within the records and photographic images kept on the civil defense committee and its membership. Racial partition was aesthetic, writing out Black lives from the civil defense narratives. When an evacuation drill was held in Operation Boxcar in 1955, two of the three "African American" schools were singled out as having "problems" with their evacuation drills that "'marred' the otherwise outstanding evacuation drills in other schools" (Leib and Chapman 2011, 586). The local press and the *Georgia Alert* civil defense newsletter regularly reported on these drills, congratulating local civil defense coordinators, many of whom were women, for their "splendid" work.[45]

By 1957 the United States had changed position, questioning the separation of civil and military defense administrations and running hearings exploring national survival rather than evacuation. In Congress, Peterson was berated for leaving possible evacuees stranded, "alerted and scared . . . to death," telling "5 million people out at Los Angeles to get out of that city on the 4 roads that go into the desert, and you haven't told them what to do when they get there." Peterson's response was to direct his detractors to the studies and simulations: "Every one has been successful. All of the evidence indicates that evacuation is a valid concept," he argued, yet perhaps Los Angeles was a "special situation" (US House Subcommittee 1956, 1216–17). The FCDA was not always so sure, and "special situation" was probably yet another euphemism for expected social and racial tensions.[46]

The issues anticipated in this chapter expose quite starkly the failings of evacuation planning that we have seen today. They demonstrate the complicated folding of evacuation through the car. Even fifty years before Hurricane Katrina, evacuation policies built on the assumption of car ownership were clearly problematic. Prehistories of Hurricane Katrina have explored the continuities of race and mobility in disasters elsewhere in the American South, such as during Hurricane Camille, when segregated evacuation buses were used, and Black people were denied access to an evacuee center—evacuation once again a microcosm of evolving and recurring race relations in the United States.

A 1980 study by the Federal Emergency Management Agency (FEMA) on the evacuation of New York stated that the "half million Hispanic and African-American Bronx citizens" could "experience special problems" (quoted in Garrison 2006, 162; quoted in Preston 2007, 150) as FEMA planners began to take evacuation seriously again as "crisis relocation planning" (CRP) (Blanchard 1985). Ronald Reagan's appointment of Louis O. Giuffrida as the director of FEMA underpinned these approaches given Giuffrida's prior interest in "evacuating" dissenting Black Americans to "relocation camps and assembly" (quoted in Roberts 2013, 152) centers in scenarios of severe social unrest—a subject of his 1970 thesis and a direct continuation of the methods and euphemisms of the Japanese American incarceration discussed in the previous chapter. Another recursion, a "revacuation," as Perry Miyake (2002) names it, CRP was met with ridicule. For the political scientist and commentator Louis René Beres, CRP's imagination of transfusing "normalcy" within the "veins of a nuclearly devastated body politic" should be responded to with "side splitting laughter . . . not the stuff of intelligent public policy" (1983, 8). It was another kind of fiasco that continued the reliance on the car as the primary vehicle of escape yet again. What about the millions "without access to automobiles?" Beres asks. The plans depended on high car occupancy plus belongings; otherwise, car owners should carpool or pick up nondrivers (8–9). Despite the sincerity and humorless ways they were distributed, and the logical ways they sought to impose order, for Beres, the evacuation plans were once again a performance, a "desolate canon of nuclear theology" (1).

And yet the interstate highways and Moses's evacuation-inspired or evacuation-facilitating expressways of course *did* become evacuation infrastructures in a manner. Moses's 1946 plan for New Orleans's waterfront, partly resurrected under the interstate federal funds in the 1950s and 1960s, is well known for its failure to blight the white and rich Vieux Carré riverfront, owing to the strong resistance to the elevated expressway and the "Frankenstein" monster of the automobile, as one activist called it (Kelman 2003, 198). However, an elevated expressway and Interstate 10 spur was constructed along Claiborne Avenue. The development has been criticized for destroying the life and community of the tree-lined avenue in one of the most important Black neighborhoods in the United States (Kelman 2003), as "another group with far less power, the city's African-Americans, watched the destruction of a promenade that they deemed as valuable as the waterfront" (216). Since Hurricane Katrina it has been earmarked for potential remediation and removal but to significant disquiet. The Claiborne portion of Interstate 10 was one of the primary contraflow lanes used by those who could leave on the interstate by car in advance of Katrina—its multilanes were turned one direction to allow greater volume of traffic. The Claiborne Avenue on-ramp, often the scene of many traffic collisions, was advised by police and city officials—and repeated through rumor—as the primary pedestrian escape route for those with no other means to leave and reach the relative safety of the interstate and the authorities. Some who did attempt to walk the ramp, however, were left without shelter or food provision and were treated as potential looters and criminals on the inhospitable elevated roads.

It was as if the road's evacuative infrastructural promise was able to reach through from the past. The road was another of evacuation's recursions, still uneven and duplicitous, rising above the surface of history and floodwaters and damaged levees. It was a beckoning expressway for evacuation, but all the while some authorities regarded those acts and actors as transgressive and threatening.

COMPANION EVACUATIONS
AT THE BOUNDARIES OF LIFE

As the farmer got out of the truck and headed toward the train, the dog knew what was going on and the dog began to howl with a super-human, ungodly just horrible screaming howl that was just the most heart-broken—it was just an expression of the whole evacuation.

—LINDA PAGE, IN L. GORDON, "DOROTHEA LANGE PHOTOGRAPHS THE JAPANESE AMERICAN INTERNMENT"

One might or might not accept the authenticity of the well known Biblical story of Noah, The Ark, and the Great Flood. But whatever one's view on that matter, the account illustrates well that organized efforts to move away from an identified danger is as old as human history. In the alleged incident there was an acceptable warning and collective evacuation from a threatened locality.

—ENRICO LOUIS QUARANTELLI, *EVACUATION BEHAVIOR AND PROBLEMS*

In the State Library Victoria, in Melbourne, hangs a painting by William Strutt. The painting is an intriguing image of a kind of evacuation. It is titled *Black Thursday, February 6th, 1851*, painted ten years later when Strutt was back in Britain (figure 5.1). The painting, which looks akin to a scene of war, is valuable for it was one of the only images available of the bushfire created by an eyewitness, even if it was refracted through Strutt's impressions and experiences many years after. It shows what the newspaper

FIGURE 5.1. William Strutt, *Black Thursday, February 6th, 1851*, 1861. Oil on canvas, 106.5 × 343.0 cm. State Library Victoria, Melbourne.

the *Argus* (1883) described as "the most thrilling episodes of the terrible event," by depicting "men, women, and children, horses, oxen, and sheep, wild and domestic fowl, dogs and kangaroos" fleeing and dying. There are of course different readings of Strutt's hellish painting of dark and fiery tones. An "apocalyptic sublime," eschatological interpretation (Paley and Frye 1986) is possible, given energy by religious end-of-the-world imaginaries of Armageddon that were common in the nineteenth century but are becoming increasingly durable with the increased regularity of wildfires in the twenty-first century.

Strutt acknowledged that the image refers to the hardships of settler life endured as a kind of comeuppance of the colonial project: "This is the country the empire has claimed as its own. This land of opportunity and gold. We have been settlers and explorers in hell, that's all, and this Black Thursday should remind all such that the devil will have his due" (quoted in Dahlenburg and Curry 2011).

The painting demonstrates some of the hallmarks of a social and contextual hierarchical reading of disaster. A man is seen on a white horse, wearing a very rounded straw hat–like halo and rescuing a woman from the fire. Everywhere in the image there is religious and gendered primacy. The presence of animals signals other things (Dahlenburg and Curry 2011). For some, the fire itself moved at the "speed of a racehorse, carrying all before it clean as a chimney newly swept." Others compared it to a "monkey" leaping from tree to tree. Or it was a "dreadful volume, rolling along like some supernatural monster" (Wivell 1883, 6). More than the representational inscription of meaning onto the animals, the painting confuses the morass of lives leaving together in the face of danger with an ethical plea of mutual vulnerability and distress, distributing sympathy across life.

Animals have often been ignored in how we think of evacuations as "organized efforts" to move away from danger, in Enrico Louis Quarantelli's (1980, 2) definition quoted in one of this chapter's epigraphs. Yet the Christian ark story he reminds us of was about the protection of human and animal life through an evacuation of sorts in a mobile shelter to weather out the storm. To look at Strutt's imagery is to see animal lives as caught up in our own trying to escape and vice versa, recalling the table discussion on sea creatures I had with my son, described in the introduction to this book. It is a flattening scene of evacuation. Life fills a narrow horizontal segment, fixing the eye on the mass of bodies—human and nonhuman—moving into the foreground and out of the distance. In this chapter I want to signal evacuation's relationship to the human and animal. In moments of emergency, these relations can revert or resettle in difficult and alarming ways. Emergency pushes at the limits not only of where animal and human stop but also of where life is cared for to such an extent that it may be taken away; when life is rendered within inhuman states of disposal, salvage, and abandonment.

Lisa Stevenson (2014) has described at length the medical evacuation of Inuit and Indigenous populations from the Canadian Arctic to metropolitan hospitals in the south from the 1950s. The removals often went hand in hand with the killing of Inuit sled dogs, known as *qimmiit*. Stevenson connects the medical evacuations, and the longer duress of Inuit life under settler colonialism, to a violence held in parallel to the evacuations. The Qikiqtani Truth Commission, set up to investigate dog killings between 1950 and 1975, saw some participants recalling the shootings as a similar kind of traumatic memory to the tuberculosis evacuations. Leah Natinine, one of the participants to give testimony, enfolds her childhood experiences of evacuation and *qimmiit* killing as one. As she was to go aboard the infamous *C. D. Howe*, she remembers when her and her sister's wooden dolls made out of driftwood were thrown away. She recalls being tied to her bed in Hamilton, "down south," as part of her convalescence and treatment, and she remembers RCMP officers shooting her father's dogs and threatening to kill more of the dogs if Leah did not go to the school. The nonlinearity of these memories is instructive for the transversal and overlapping violences and traumas of colonialism, in which evacuation, displacement, and animal violence become confused and complicit with one another. The bonds of mutual strength the Inuit had with the *qimmiit* became a "point of vulnerability when these groups were relocated" (McHugh 2013, 165). The dog shootings were part of a longer process of force and coercion to make the Inuit assimilate to Western practices of settlement and waged livelihoods where they might be dependent on motorized sleds. The Qikiqtani Truth Commission inquiry was intended to explore the widely repeated Inuit belief that the dog killings were a deliberate method to colonize and subjugate, "to deprive Inuit of their mobility and autonomy and to tie them to settlements" (Qikiqtani Inuit Association 2013, 13).

The coincidence of the dog shootings and the tuberculosis evacuations signals when the treatment of animals becomes a pathway to treat humans or some animals as killable, as disposable. Lisa Stevenson (2014) focuses on a moment within an interview conducted by anthropologist Toshio Yatsushiro, a Japanese American formerly "evacuated" under the World War II Japanese American "internment" program we explored in chapter 3.[1] Yatsushiro had carried out many interviews in Iqaluit and raised questions regarding the RCMP dog shootings, repeating the RCMP claims that the animals constituted "a menace to the community, especially the white residents" (Yatsushiro, in L. Stevenson 2014, 70). In this moment of an interview, the respondent Jamesie articulates a common Inuit anxiety, that the shooting of the dogs—their euthanization on the grounds of public health—would be a first step before the Inuit were turned on next. For Stevenson, this is evidence of the "psychic life of biopolitics" (7) in which the state's apparent care and twisted benevolence through the killing of the dogs is exceeded by a perceptible indifference of the act to the lives and livelihoods that communed with and depended on the life of their dogs. In this we see something of an affective *momentum* of evacuation and its relation to killing, running away with itself as the "dreadful volume" of the bushfire told of a rolling accumulation, as the failure to evacuate turns dreadful. What might evacuation mean for those who are not human? How might we better account for not only intraspecies mobilities but also what Timothy Hodgetts and Jamie Lorimer (2020) have recently called "animals' mobilities" in evacuation? And how do these mobilities mobilize reconfigurings of relations of care from human to animal, from animal to human, and from human to human?

The chapter explores these relations across two diverse contexts: first, the so-called pet massacres of World War II in Britain, focusing on the way pet owners were encouraged to euthanize or dispose of their companion species in advance of civilian evacuation. Second, we turn to Hurricane Katrina and the role of animals and specifically pets and the evolution of policies, primarily in the United States, to include pets (mainly) within evacuation policies and practices in times of disaster and emergency; we also explore, more broadly, the animalization of the Black and minority populations of New Orleans and the continuation of evacuation principles based on the assumption of private automobile ownership. Finally, the chapter draws together the pet massacres with the deadly choices made at Memorial Medical Center during Hurricane Katrina, when pets and humans were euthanized in advance of and during the evacuation, in the belief that they were unevacuable.

KILL THE CAT

Historians have pointed out how concerns over food shortages during World War II led companion animals and pets to be viewed as problematic life. In the midst of the evacuation of children from urban centers to the countryside and suburbs in Britain during September 1939—in the first wave of the evacuations—mass domestic animal culls were carried out. What was odd about this violence was how much animals had brought remarkable companionship during the moments of emotional or affective loss that had already anticipated the war (see the excellent Howell and Kean 2018). "It was one of things people had to do," writes historian Hilda Kean, "evacuate the children, put up the blackout curtains, kill the cat" (quoted in Feeny-Hart 2013). Figures estimated that as many as 400,000 animals were destroyed in London in one week of September 1939 alone. Though the National Air Raid Precautions Animal Committee (NARPAC) initially guided away from euthanasia, the committee eventually supported households with a series of booklets that gave the impression that animals were, a little bit like how children were imagined (Welshman 1999), a liability in a bombing war and best evacuated to the countryside. "Owners should make up their minds whether they can take away their dog or cat themselves," one Air Raid Precautions brochure advised. Those who could not do this, the brochure recommended, "should decide whether the animal is best destroyed or evacuated to the care of friends in the country or to one of the numerous dog and cat homes situated convenient to large cities" (Home Office 1939, 2). Evacuation is once again brought together with death. If evacuation could not happen, death was preferable, and it was described in odd but functional terms, such as *humane destruction*.

Well before the Blitz, the battlefield provided the backdrop for animal evacuation and convalescence, when animals were not the simple conveyor of human mobility and escape—although they were crucial for the mobilities that constituted the lines of communication we explored in chapter 2's examination of medical-military evacuations in World War I—nor burdens marked for destruction but themselves evacuable lives for treatment and potential recuperation away from the front lines. Writers such as Philip Hoare (2018), Hilda Kean (2015), and Derek Gregory (2019c) have begun to account for the lives of animals in war. The "more-than-human casualties" on the Western Front during World War I moved along pathways and through circuits of care, sometimes completely in parallel with the chains of human casualties being evacuated to clearing stations and to military hospitals. As Gregory (2019c) argues, they were underpinned by the same "instrumentalism" that had "guided the RAMC's casualty evacuation model." That instrumentalism was underpinned by a logical rationality of energy, labor, and materials (Corvi 1998). Animal evacuation contrasted with expectations that animal welfare was a luxury afforded only to the spaces of peace (F. Smith 1918). Both American and British forces emphasized Edwardian ideals of machinic efficiency, the desire to avoid wastage at all costs. If war was a "system of extermination," "the conservation of horses in war was" simply "an economic proposition" (Marshall 1917, 90–91). It meant the "disposal" of animals unable to perform their required duties by evacuating them away from the front lines either for second-line service or for rest and recuperation before returning to action, just like the troops. As Major-General Sir John Moore of the Royal Army Veterinary Corps put it bluntly, "For a unit to be efficient in the field, it must get rid of its ineffective animals, and have them replaced by fit animals" (1920, 119). Having "sick" animals in the field would "inhibit mobility," and in a system imagined as a series of inputs and outputs of time and energy, it would lead to the "absorption of the attention of a proportion of unit personnel who might otherwise be employed in combatant duties." He called this a "paradise" compared to the wasteful methods employed during the Boer War, where animals were destroyed, abandoned, or dumped. Barges provided "comfortable and easy circumstances." Special evacuation trains were possible. Most were moved overland by walking, tethered together by a rope. The Royal Society for the Prevention of Cruelty to Animals raised enough money for twenty-six motor horse ambulances (J. Moore 1920).

In life on the home front in Britain during the next war, the animal-loving sympathies of evacuated children were expressed in letters to their parents from their homes in the country about their beloved dog or cat. The cull in September 1939 came to be known as the "animal massacre," or for the National Canine Defence League, "the September Holocaust" (Kean 2015, 742). Animals were considered problems of ownership in a time of limited resources and the distinctive problems of air raid precautions for sheltering in the cities or evacuating from them. Panic—that vector of blame pursued in the previous chapter—formed around the differing narratives for how animals should be treated. For one commentator in the *Veterinary Journal*, the second most important reason for the air raid precautions guidance for animals was the "protection of human beings from panic-stricken or gas-contaminated animals" (Bywater 1940, 221). Equally, the initial culls, for some, were a product of apparently panic-stricken owners, which one animal league sought to ensure would not be "stampeded into premature action" (Kean 2015, 744). The warnings and guidance anticipated a sort of zoonotic transfer of passions that would make both humans and animals excitable to the other in a familiar affective loop.

The Ministry of Home Security, under some pressure from police constables, animal societies, organizations, and charities, did not want to plan for evacuations, at least initially. Animals would not be allowed into communal public air raid shelters either. By 1940, when the government began planning a compulsory evacuation of children and some mothers, it conceded that denying a child the ability to evacuate with their pet might have adverse effects. Because this might cause problems in reception areas, it was felt preferable to "allow them to bring their pets with them if they are sufficiently small, rather than that forcible measures should be taken" (Kean 2015, 749), for "it would make owners infinitely easier to deal with if at least some hope can be held out to them of seeing their pets again."[2] Materials from the government show indecision. Writing privately in response to a question in the House of Commons in the summer of 1940, John Anderson, head of Air Raid Precautions and minister of home security, wrote that he could not give any "positive assurances" that people could take animals with them on the evacuation trains. It all depended on the "speed at which the process of evacuation would be carried out, and the accommodation available on the trains."[3] The official circular "Wartime Aids for Animal Owners," produced by NARPAC, went into graphic and diagrammatic detail on exactly how an animal should be killed. Under "no circumstances leave them behind in the house or turn them into the street," it suggested. Death was preferable to nonevacuation. The guidance, and instructions for captive bolt pistols within the pamphlet, suggests protection by destruction. The directional arrow and point illustrate the best point of entry for the trajectory of the bullet or captive bolt driven into the animal's brain and drawn back out again.

Railway stations considered baskets and containers to temporarily hold "abandoned" animals. In July 1940 the Home Office wrote to chief constables of East Anglia and Kent coastal towns that "no animals or birds would be allowed on the evacuation trains" and that owners should make provision for them in some other part of the country or hand them over to NARPAC for "disposal."[4] *Disposal* was regularly used around these debates. If one could not evacuate with their animals, then they should be "disposed of properly." This meant making provision to supply NARPAC veterinary surgeons and animal societies with ammunition, chloroform, and hydrogen cyanide and to set up "stations" near entraining points "for the destruction of pets left behind." They also considered that the removal of dogs and cats could create an epidemic of rats and mice and potential disease, although this was later derided by the Ministry for Agriculture.[5] Other instructions imagined the use of pens for the animals so that, once the evacuees left, the pets could be killed and put into pits with lime.[6] By mid-1942 they had resolved a pet amnesty, where pets would be given to NARPAC by a certain date and arranged to be evacuated. After that date, any animals left in the town would be destroyed. They were not that worried about strays, who were considered "likely to look after themselves and keep the rat population down," while, on the other hand, it would be dangerous for "dogs to run wild in the evacuation areas."[7] Communities would be warned of the "cruelty of leaving animals, particularly dogs, uncared for."[8]

For owners of smallholdings with rabbits, pigs, or poultry, the government advised them to seek similar "disposal" in advance of compulsory evacuation, which meant either destroying them or, through slaughter, putting them into commercial exchange.[9] The spatial imagination at work here is complicated. Evacuation meant not simply the emptying of the coastal towns and villages but a clearing in advance out of fear of enemy invasion. In their valuation of the usefulness of meat carcasses versus live animals, the government considered it impractical to attempt the evacuation of most animals, especially pigs, but they would permit and encourage voluntary relocation of "certain valuable pedigree and breeding stock."[10] By late 1940 government was unwilling to rule on wholesale destruction of livestock and composed a draft circular to agricultural executive committees to suggest slaughter only "as a last resort."[11] By the end of 1942, NARPAC had recruited about 1,500 veterinary surgeons and 40,000 animal guards, which also meant they could relieve civil defense workers for "duties connected with human beings."[12] For animals that might be killed by enemy action, "salvage" could be possible so that the animal might be recycled for human consumption. The calculus was between convalescence and treatment for further work, or salvage for food production. A spoof letter was circulated within the Ministry of Home Security. It was supposedly written by an Adam S. Windle, based in Houndsditch, who cheekily proposed that a cat ranch be built outside

Brighton. Windle, acting as manager, would be given a million of the cats to be disposed of, which he predicted would produce twelve kittens a year, estimating this would give 12 million skins a year, producing 1s 6 pence each per cat, or £2,500 a day. They would require about a hundred men to skin the cats, costing £25 in their daily wages. They would feed the cats on rats and start a rat ranch. And feed the rats on the cat carcasses. So, he says, "The cats will eat the rats and the rats will eat the cats and we will get the skins." The anonymous author signs off: "P.S. Eventually we will cross the cats with the snakes and they will skin themselves, thus saving men's time and wages for skinning and also getting two skins from one cat."[13]

Outside of such levity, the government advice seems to be a way of sterilizing the language of killing as well as the monospecies imagination and provision of mobility. It revolves around an assumption that the life to be taken is to be taken from a life not worth living, or not of value to live. An animal becomes a thing to be disposed of, killed in name or concept if and when the connection with an owner or carer or companion is cut, severed by evacuation. This seems so nihilistic and yet unbelievably rational and instrumental. As soon as the relational thread between owner and animal is frayed—even in potential—their lives become unlivable and therefore expendable, even as a way to be humane to a life that cannot be retrieved. The animal remains were to be put into a different kind of circulation, perhaps for their economic value as a carcass for consumption. And yet something is still salvageable beyond the economic, perhaps an affective relation from seemingly ethical treatment, or an absence of worry that might have come from the owner's own evacuation.

TAKE YOUR PETS WITH YOU

As we have seen through this book, and a wider array of research on evacuation, Hurricane Katrina offers some of the best and the worst of evacuation and has distilled and exposed the complex configurations of race and automobility, which have not been learned from despite their visibility. One of the other reasons Katrina proved to be such a key moment is the relations it laid bare between emergencies and animals. Part of that story is dealt with here, and we will recognize how many of those who were displaced or left behind by evacuation experienced sharp social-familial ruptures between themselves and the animals they cohabited with, depended on, and owned. Although statistics vary, some suggest that as many as fifty thousand pets (Nolen 2005) were stranded in the city, and fifteen thousand (Eugenios 2015) were recorded as rescued. The legislation and apparatus that partly evacuated New Orleans during Katrina was mostly inadequate for individuals, households, and their pets. Pets were turned away from evacuation and rest centers or on transport vehicles. Some were taken in an ad hoc disorganized way. At Memorial Medical Center, in events picked up later in this chapter, some patients and hospital workers had taken their pets to the hospital for refuge during the storm. Many would not evacuate their homes without their animals.

During Katrina a small boy, in advance of being evacuated from the Superdome, was tearfully separated from his dog, Snowball. This story was repeatedly rehearsed by political commentators as an example of the callous nature of the authorities toward people evacuating with their animals (Grimm 2015). Some of the debates and amendments that followed Katrina introduced state and federal legislation that tried to cater to human-animal relations. Federal law in the PETS Act, or the Pets Evacuation and Transportation Standards Act, has since tried to produce more adequate provision to support evacuation with pets. For some emergency managers and city authorities, the new provisions became something of a headache. In 2007 *Time Magazine* reported how the then–New Orleans director of emergency preparedness, Jerry Sneed, shot their reporter an exasperated look when they asked about the new provisions, "a look that says that the pet issue has become something of a pet peeve" (McCulley 2007). In Louisiana today, the guidance and advice to pet owners reflects what Marsha Baum has identified as support for a "joint pet-owner evacuation whenever the evacuations can be accomplished without danger to human life." The "preservation of animals," writes Baum, "is to aid the animal owner or caretaker, not to preserve the animals. The animals are considered to have no interests of their own" (2016, 115). Even then, while the legal provisions require state and federally organized evacuation and emergency facilities to include pets, they do not demand private locations or hotels to do the same. The United States' Centers for Disease Control's guidance for pets in evacuation centers still characterizes the presence of pets as one of potential risk for their owners, other owners, and maybe other pets. Stress from close proximity appears possible; pets may be a source of minor injuries, a vector of germs and contagion, and improper sanitation; and pets may carry a range of potential animal-to-human diseases, from leptospirosis to diseases spread by ticks or other insects (Centers for Disease Control and Prevention 2018).

One of the bizarre forms of guidance to emerge was a music video by Johnette Downing titled *Take Your Pets with You* (2019), produced by Louisiana's Department of Agriculture and Forestry (LDAF), which has hosted the video on their website and offers guidance and brochures using the same title. The video shows Downing, a local musician, singing and dancing in New Orleans with the iconic Huey P. Long bridge in the background. Within the video and the parallel guidance that the LDAF has produced, we see how animals are portrayed as family members, setting the moralistic context for the song, which does not refer to the problems of evacuating animals or the issues involved in transporting or including them within public or private facilities of transportation and shelter, but states that people should not leave them behind. The rest of the instructions provide relatively detailed guidance about the paraphernalia that pet owners would require in order to bring the animals with them, yet it endorses the suspicion that those who might or did leave pets behind engaged in some form of negligence. Pets have the status of property, which was precisely how several custody battles over animals that were left behind were argued (McNabb 2007).

Some principles of human-animal relations in emergency evacuations rendered through New Orleans's experience of litigation, legislation, and now guidance in the wake of Katrina still resonate. Like the LDAF brochure titled *Take Your Pet with You* (figure 5.2), the singer exclaims, "Tornado, fire, or flood, take your pet with you. Any time you evacuate"—the diction suddenly becomes very clear, and she nods, because you know the drill by now, "take your pet with you" (Downing 2019). At the same time, the relative ease of pet evacuation shown in the video continues to contradict some of the conditions of possibility for doing this so easily. It is as if the song's returning refrain—"take your pet with you"—another of evacuation's recursions, keeps instantiating the returning assumptions and implications of mobility privilege over and over again, to the extent that we do know what's coming. While cuteness might describe the pull of the animals on feelings of empathy and care (discussed below), turning this guidance into a jovial but childlike song collides with other, dissonant aesthetic categories and feelings. The song could seem to domesticate and potentially trivialize the very things it seeks to take seriously, especially in the face of the "pet peeves" of some emergency planners, that pet evacuation was not a serious issue.

A clip embedded within the video sees a female pet owner packing a white SUV with food and pet accoutrements, a travel box, and the dog, before they reverse out. They are on the way to evacuative safety in as smooth a way as possible. But the video and the brochure seem strangely out of tune and even vindictive given that many people would not be able to evacuate themselves or their pets in this way. Many people simply couldn't take their pets with them during Katrina, especially if they were stuck without a car. The video silences

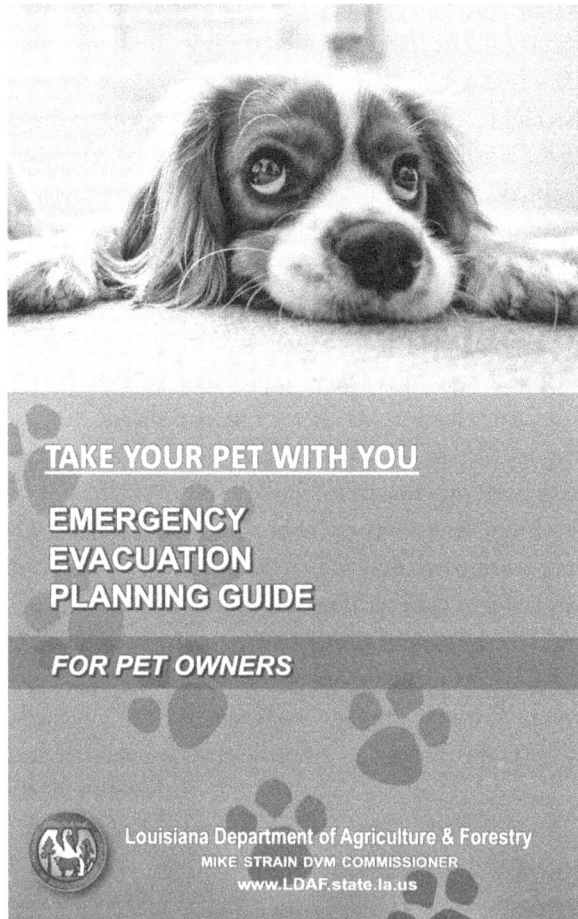

FIGURE 5.2. *Take Your Pet with You* pamphlet, 2019. Louisiana Department of Agriculture and Forestry. https://www.ldaf.state.la.us/wp-content/uploads/2019/05/Takeyourpet.bi-fold.pdf.

the automobility injustices so common within evacuation, especially during Katrina. The video showing smooth, easy, and fun evacuation—making sure pets are "comfy and cool" and absolutely not abandoned in the car on their own ("That's the rule!")—along with the "take your pets with you" campaign, forms part of the parallel game of blame so often encountered during evacuation and captured acutely in post-Katrina custody battles. In the video the owner has remembered the dog's leash and a certificate of vaccinations, and the dog is transported with a travel box and stowed.

Pet evacuation is portrayed as naturally mediated by automobility. Animals are strapped like passengers into the back seat and handily packaged up through different kinds of mobility apparel. The message is quite dissonant even if the guidance and instructions are warm, affective, and full of "cuteness" (Ngai 2012). The video shows a dog peeking out at the bottom of the frame between smiling children/siblings, again in a packed car—conforming to the familial imagination of the pet-companion rendered as a child and packaged up and conveyed within the heteronormative ideal of the automobile-family-unit assemblage—reminiscent of the car-centric Cold War evacuations of chapter 4.[14] The message is also bureaucratic, encouraging a host of apparel and adornments, including a combination of locative information, identification, disease and health status, and technologies required for the animal's care. These are to be ticked off with a checklist, as recommended by the Humane Society of America and other government organizations.[15] Particular emphasis in the different checklists is given to information that would help relocate pets should they be separated from their owners, such as basic biopolitical-level information around sex, diseases they have been inoculated for, and medication they need or have with them. The Humane Society warns of animal-society breakdown once human presence is removed, drawing on the notions of societal regression and colonial ideals that would see dogs without human owners or carers become "trash animals, as noxious Others" (Srinivasan 2019, 379), taken outside some of the more inclusive models of zoopolitical citizenship. They warn of animals' potential escape through "broken windows"; "turned loose," they are likely to "become victims" of a hostile environment, from "exposure, starvation, predators" (Humane Society 2008). To leave an animal chained outside "is a death sentence." Stray animals left behind in New Orleans's evacuation here become another narrative for blight and social decay; when "broken windows" are a vector of animal escape and social stigma (Harcourt 2009), dangerous reproductions appear to breed unchecked in abandoned housing, multiplying rapidly among drug dealing, sexual violence, and murder (Jonassen 2012).

The Take Your Pets with You campaign helps us explore a set of unresolved issues that point to the status of what a pet is as a political life and the politics of blame and responsibility that are so often distributed in harmful ways during and in the wake of evacuation and emergency. One of the sharp definitions used within custody debates over animals was whether a pet was deemed "lost" versus "abandoned." Each term has a starkly different meaning within US state law and changes the duration in which a pet can be quickly adopted by a new owner. Abandonment is morally coded as a signal of neglect, referring to an animal cast adrift by poor and inhospitable, uncaring, and negligent owners who "completely forsake and desert an animal" under Louisiana's State Laws.[16] In the courtroom custody battles that followed

FIGURE 5.3. Rascal the Chicken, 2018 (Ready.gov 2018).

Katrina, it was reported that many "rescued" pets from New Orleans ended up with rich, white families in other parts of the state or country who could "treat them as genuine 'pets'" (Zelinger 2018, 95). At the heart of these struggles were perhaps good intentions mixed up with classed and raced notions of care and ownership, alongside legal principles that equated the pet with property. These commingled with more traditional applications of culpability in and through evacuation and emergency.

From this subtext, the singer of the music video shakes, gyrates, and bops along in the background, lightening the issue through various formations of the category of the cute (Ngai 2012). We see pet mug shots of cute animals, puppy dogs with puppy-dog eyes, on the move or in a domestic setting, sometimes with smiling children cuddling, holding, or even sleeping on their pets, garnering sympathy from the viewer as if to say, "Don't forget me," or a more accusatory, "How could you?" (see figure 5.3), prompting a kind of ethical and affective sensibility that has flowed through the sections of this chapter; this has to do with the way the evacuation guidance addresses a reader or addressee with different kinds of animal faces. The instructions and checklists use a set of locations, ID information, to pin the animal down to their identity, to their owner, to their temporary accommodation, to any diseases they might carry. The constant display of animals' faces provides a different kind of address and addressee, aimed at cultivating an ethical sensibility toward the animal as a particular life, even an individual, which could be lost if evacuated, or harmed if not evacuated.

Emmanuel Levinas's (1991) philosophy of ethics and the face is relevant here, especially in his encounter with a dog that he and his comrades named Bobby during their time as prisoners in a Nazi labor camp. During that experience, Levinas recounts how Bobby recognized him and his campmates as human, in stark contrast to the Nazis, who stripped Levinas and his comrades "of their skin" as if a "gang of apes" (quoted in Plant 2011). It is self-evident that the animals' faces—of course a dog and a cat are photographed in the evacuation guidance—are used in much the same way as a human face is: to garner sympathetic responses. Showing the face of a pet, however, is not really about recognizing it as an animal but used to prompt the recognition of a particular animal. The front of a Humane Society brochure is interesting in this way, because it features four dogs, some looking at the camera. Others are turned away, but both groups appear amid flood debris and floodwaters, looking somewhere between a "pack" of animals and unworthy recipients of their former owners' presumed abandonment. Elsewhere in the brochure (like others of its kind) are singular animals, less threatening, but in the position of being saved from various situations in crates on rafts, being held or carried through floodwaters. In the other textual materials, they are individuals, looking directly at the camera with their glassy, watery eyes—looking cute once more. What is interesting about this is that the animals are in various processes of being saved, often by different emergency workers. So both the individual portraits of the animals' faces and these more scene-ly portrayals of pets being rescued cement quite a one-way relation between people and animals. The animal is the recipient of care or concern and sympathy, as well as practices of risky rescue in emergency and evacuation. Unlike the front cover of Lauren Berlant's (2011, 266) *Cruel Optimism*, which suggests the potential of "companionship, reciprocity, care, protection" in the dog's face, in the animals of evacuation the affirmative relations seem more conditional. Vulnerability is distributed from the animal, it seems—the same animal who might threaten its owner, bring rescue workers into increased risk, and hold its owner back from evacuation and help. People's bodies seem brought into vulnerability through their relation to their pets, rather than being made stronger by them.

This helps us to distill a crucial point about the nature of animality in the kind of abandonment Levinas and others explore and have applied in relation to Hurricane Katrina and the evacuations and nonevacuations of the city. Disproportionate evacuation policies and structural racism become the cause and effect of the failure to recognize the human in others, which saw the characterization of Black Americans as subhuman and disposable. Henry Giroux (2015, 176) has frequently referred to how a Floridian radio show host described those made homeless by Hurricane Katrina as less than human, as "parasites, like ticks on a dog." As Kevin Dowler argues, this was about the way in which distinctions over humans and animals as political entities were treated and conjoined with species treatment in the flood zone: "what will die, and what will be saved, and under what conditions such choices are made" (2019, 2).

Of the many artistic responses to Hurricane Katrina, Brad Benischek's (2007) graphic novel *Revacuation* is one of the most interesting for dehumanizing all of the characters in his family story of evacuation. Benischek imagines the evacuees as grounded birds taking flight from the city, being warned away from help and services, and kept from coming back by zombielike scarecrow officials, whose stitches show their patched-together and monstrously unemotive faces. The cover image pictures a hellish vertical scene of the urban geography of the overhead expressways, on which evacuees walked, sheltered, and were threatened (figure 5.4). The structure deforms and twists helter-skelter, to the extent that the urban infrastructure takes on an almost animal-like form. In one of the last images of the book, which changes perspective dramatically, Benischek imagines a US bomber flying over the city far below, perhaps tracing the infamous image of President George W. Bush's dispassionate aerial view in the days that followed Katrina's devastation. Benischek imagines the deliberation on board of whether the plane should drop a bomb, but they decide not to waste this "sacred object" on those "pathetic animals down there. . . . Besides I already set off the (Bureaucracy) bomb. It's a slow painful way to go but practically untraceable!" (Benischek 2007).

The Katrina evacuation was treated as "alien" in language and imaginaries that continued the depersonalization and dehumanization of evacuees. Their reception was even preceded by powerful watery metaphors of "waves of not-white people" who were "engulfing" and "swamping" cities in Texas. Evacuees became a "large unmanageable, and amorphous blob moving to

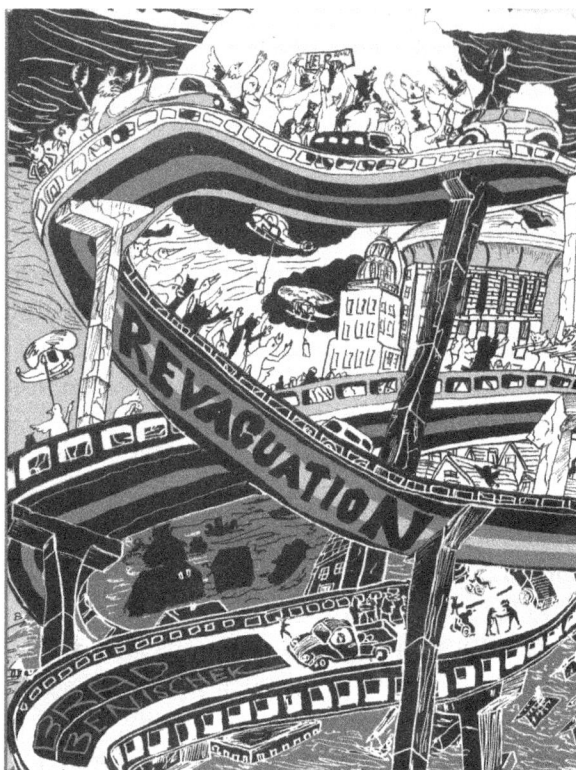

FIGURE 5.4. Brad Benischek, *Re-vacuation*, 2007.

distant locations, producing an alien threat" (Lacy and Haspel 2011, 28). For Rebecca Solnit, the fear of panic by those in power "unloosed even more savage attacks on the public because that public was portrayed as a monster out of control—a collective King Kong or Godzilla" (2010, 131). The authorities seemed to panic themselves; as Solnit states so cogently, "Imagining that the public is a danger, they endanger that public" (130). This also drew on wider raced and classed revanchist notions of the city as a "feral" space, which became the source of justifications for militarized and security-type, and even militia-type, interventions that characterized New Orleans and its law(lessness) (Solnit 2010).

Pets illuminated other hierarchies of treatment within the evacuation and the post-Katrina cleanup, "in the way companion species were differentiated from the generalized description of humans trapped in the zone as feral creatures" (Dowler 2019, 4). The use of some sort of hierarchical taxonomy between "some animals" and Black Americans stirred a different type of politics of comparison. Evacuating and caring for pets seemed to show that animals were prioritized over less privileged Black and minority citizens: "Herding a mob of refugees is difficult enough. To add on the duty to herd dogs, cats, hamsters, boa constrictors, llamas . . . I came away from the experience wondering whether the American public found it easier to feel sympathy for dogs and cats than for low-income black people. I'm not sure I want to know the answer" (Noah 2005).

Many worried whether the treatment of animals was actually better than that of Black Americans. The rescue images of stranded pets being evacuated from the city made possible a "sickening juxtaposition to the conditions faced by tens of thousands of black residents trapped by the storm" (Harris-Perry 2010). Claire Jean Kim worries about such a position of comparison, where the effect is to ignore how species-level differentiations almost always enable the "infliction of unremitting and profound suffering on animals" (2015, 287). As Irus Braverman (2017) has remarked on the zoometric techniques utilized by Israel's governance of Palestinian bodies and animal species, the distance and closeness of these comparisons and separations create problematic hierarchies of threat, and conflations of agency, which can prove highly dangerous, especially in contexts of settler colonialism and strong nationalism (Alloun 2018). What's more, the animal becomes another abstraction or index; "to say that animal rescue efforts show that whites love 'pets' more than they do Black people reduces nonhuman animals to instruments for measuring degrees of anti-Blackness" (Kim 2015, 287).

LIFE AND DEATH AT MEMORIAL

Evacuation efforts in New Orleans took place around some unlikely places too, such as the international airport (Vankawala 2005), which meant the sorting practices that airports all so readily perform (Lyon 2006; Salter 2007) were replaced by actual triage criteria. Some patients and hospital workers had taken their pets to Memorial Medical Center on Napoleon Avenue in Uptown New Orleans for refuge during the storm and were stranded there for five days. The story has been the focus of a Pulitzer Prize–winning investigation by Sheri Fink (2013). Fink's book concerns the forty-five deaths that happened at the hospital, five of them before Katrina struck. Floodwaters meant access to the hospital was possible only by boat. An aging helipad—built in the mid-1980s on top of the Magnolia Street parking garage—meant helicopter rescue was possible but difficult. With the power out, hospital staff carried patients down numerous stairs, up to an elevator that did work, through a hole in a wall by stretcher, up several more flights of parking-garage stairs, and up the remaining two flights to the helipad, where Coast Guard and private helicopters were sporadically airlifting patients away. Hospital doctors appear to have tagged and ordered patients according to who was most likely to survive. This meant that those with do-not-resuscitate orders and the sickest of patients who were held within the private hospital within a hospital known as LifeCare, which leased the seventh floor and was owned by a different parent company, were evacuated last, not first, in contrast to usual triage practice. Memorial doctors would later rationalize this choice as "battlefield triage" (Fink 2013).

The hospital's story is one of neglect and care, inadequate planning, racism, love, and abandonment. Intense fear of violence, exhaustion, and what hospital staff referred to as warlike conditions were common experiences. The hospital felt penned in, under siege. Many medical practitioners would recall the rumor that the city was under martial law. As Sheri Fink (2013) explains, the windowless rooms made the wards feel like an "army field hospital." Nurses and doctors and maintenance staff and their families had first hunkered down in the hospital, bringing their pets with them to wait out the storm. People came in to shelter at the hospital, as they had for other storms and hurricanes. The boats that came refused to take the animals. Like people, the animals suffered in the humidity and heat, and some lapped the unsanitary water that had flooded the lower floors and basement of the hospital. If they could not get out, if they were unevacuable, then they were perhaps also irretrievable as lives. This kind of ethical logic began to be applied to the animals in a way not dissimilar to the animal massacres we saw in Britain, and then eventually to the patients.

Things escalated around two of the male surgeons and later a cancer specialist who was eventually charged with four counts of second-degree homicide, as were the nurses who assisted her. Some of the animals in the hospital had been left behind by their owners, who asked that they not be abandoned or left to suffer. A doctor made out prescriptions of sodium thiopental and injected several dogs and cats, causing their deaths. He used injections of potassium chloride to euthanize several cats he believed would be abandoned. What happened at Memorial seemed to see the killing of the many pets begin to overshoot itself, to carry forward. A senior nurse and manager expressed frustration that they were euthanizing the pets but not talking about what they could do for the patients. A doctor brought her a sick cat to be looked after. "How could the doctor express more concern for a cat than the patients all around her? Pets were everywhere—everywhere!" is how Fink (2013, 322) explains the nurse's exasperation. Some patients were in severe pain, unable to be treated for their underlying illnesses. The hospital had shut off piped oxygen, and nurses were manually breathing air into people's lungs. Others seemed to be entering the Cheyne-Stokes pattern of breathing before death, characterized by rapid, rasping breaths that may be followed by apnea, distressing both staff and patients.

The manager suggested dulling the pain and discomfort of the patients to such an extent that they could withdraw from consciousness and "no longer care that they were smelling the feces they were lying in, that panting dogs were weaving past and licking their hands." Others took that as a cue to discuss euthanasia of the patients. A doctor, Fink explains, said he would "handle" how they dealt with the sickest patients. The doctors, according to Fink and the district attorney's investigators, began to organize the staff. Some were exhausted from struggling to move the frailest and could not imagine how they would tackle moving the heaviest patients (such as the three-hundred-pound Emmett Everett, sixty-one years old) up to the helicopters—perhaps in an upside-down version of the slower journeys that heavier people made down the stairways of the Twin Towers during 9/11, as we discussed earlier. And, remembering Berlant, Everett's size appears to have marked him as a person "of excess (already negated as both too much and too little for ordinary social membership)" (Berlant 2011, 113). Memorial staff rationalized that if and when the patients could be evacuated, the journey itself would almost certainly cause more suffering. Evacuation journeys did not seem survivable. The doctors appeared to use a combination of overdoses of morphine (for the pain) and midazolam (to subdue patients' breathing unto death). For enraged family members quoted in Fink, "euthanasia is something you do to a horse or an animal. When you do it to people, it's called murder." It is ironic given the repeats, refrains, and inversions that torment this chapter that the owners of the hospital held the palindromic name Tenet. In 2011 Tenet settled a class-action suit for $25 million and sold the hospital.

CONCLUSION: SALVAGE

This chapter has been, in part, about excess—the excesses that challenge evacuation planning assumptions, the excesses that push at the limits between humans and animals. We have seen how the nonhuman evacuants that Strutt's painting imagines are surely worthy of consideration: jumbled up with and in combinations of bodies both human and animal. Our accounts of evacuation are regularly much too human. But here different aesthetics are at work, which modulates that fact. Cuteness and humor humanize animals to a limit while reducing their agency, perhaps to the status of property. Meanwhile, emptiness and empty spaces appear to condition a dangerous animality that must be avoided.

This chapter has traced a manner of flow between human-animal relations and relations and decisions over life, and life and death, that play out differently from the other examples we have considered in this book, such as in chapter 3, even if it once more distorts evacuation's logics. When life becomes perceptively unevacuable—for both humans and animals—it seems to be made unlivable and therefore disposable, as if killing were the kindest thing. In the middle, we see continued struggles to retrieve rights and justice for those humans treated as animals, and animals treated as eminently evacuable property. In the process of rendering life disposable by or because of evacuation, killing becomes justifiable because it seems to be a way to recover something, to salvage a quality—perhaps a kind of redemption—a closure of immediate pain from terrible circumstances. Pet and livestock animals might be salvaged as food or exchangeable goods from their companions. At Memorial, doctors tried to recover patient and pet dignity in a painless and comfortable or unconscious death from inhumane conditions. They did this in light of the evacuations they feared would kill, or never come at all.

Novelist Jesmyn Ward (2011) suggests that there is also a different kind of animal ethic involved in the human-animal relations possible from Katrina. Ward's book *Salvage the Bones* tells the story of a poor Black family set against a hurricane in the midst of a New Orleans periphery. It challenges and exposes "the processes by which forms of life are positioned as disposable through the manipulation of the human–animal binary" (Brown 2017, 7). In Ward's story, in her use of unlikely metaphors, Black lives are not animalized as inhuman, even if abandoned by law and government provision that treats them as such. Ward's approach is to animalize the subjects of her book as susceptible, caring, sympathetic subjects, like those of Berlant's epilogue. A family struggle to move through the floodwaters, debris, and wind of the hurricane that enfolds them. The family hold their pets, a white fighting pit bull called China and three baby puppies. Ward juxtaposes this with her characters Skeetah and Randall, crouching at the edge of a roof, comparable perhaps to Benischek's *Revacuation* drawings, "hunched like birds, feathers ruffled against the bad wind, both of them holding their bundles closely" (Ward 2011, 232). The puppies are their babies, just as others compared their pets to infants: "They mewl still. I feel them with my hand, still downy, their coats just now turning to silk, and they squirm at my touch. The white, the brindle, the black and white. They lick for milk" (229). China is placed in a sling to carry her: "China's head and legs are smashed to his chest, pinned under the fabric. She is his baby in a sling, and she is shaking" (231).

Salvage the Bones is perhaps not redemptive or about recuperation but it articulates a very different affective economy and a different politics of emergency. The family's escape from their flooded home is another moment in a more constant emergency, a chronic one for the book's protagonists, who live in partial poverty, dealing with an unplanned pregnancy with an indifferent partner; an ill, injured, and abusive father; the recourse to petty theft; the loss of the dog's puppies. *Salvage the Bones* is less about the sharp shocks of emergency and evacuation, the terrible life-and-death decisions over animal and human life. It is more a constant picking up of the pieces, a way of creatively making use through human-animal survival and generosity of the rubble of racial, political, and economic divides to the extent that evacuation is an almost permanent condition.

A DISENGAGEMENT

EVACUATION TRAUMA, COLONIAL VERTIGO, AND NATIONAL REPRODUCTION

In late August 2005 a wide variety of public commentators, politicians, and religious leaders would begin to make a peculiar comparison to one of the settings we examined in the previous chapter. They compared a series of forced removals in the Gaza Strip and the stumbling evacuation of New Orleans during Hurricane Katrina.[1] Within a day of the completion of the removals, Katrina had made landfall, and mandatory evacuations were called a short time after. Rehearsing Adi Ophir's (2007) "catastrophic" version of emergency as resting on divine judgment, for some, the havoc wreaked on New Orleans and the temporary shelter of ten thousand residents evacuated to the city's Superdome stadium—about the same number of expelled people from the Gaza Strip—was God's vengeance on the United States. Several religious leaders in Israel, such as those reported on the conservative news site *WND*, suggested, "Katrina is a consequence of the destruction of [Gaza's] Gush Katif [slate of Jewish communities] with America's urging and encouragement." Rabbi Avraham Shmuel Lewin, the executive director of the Rabbinic Congress for Peace, explained that the "U.S. should have discouraged Israeli Prime Minister Ariel Sharon from implementing the Gaza evacuation rather than pushing for it and pressuring Israel into concessions" (quoted in A. Klein 2005).

Similarly, the controversial Rabbi Ovadia Yosef lashes was lambasted by Knesset leaders and the public for suggesting that Katrina was the product of American influence over the evacuation of the twenty-one Israeli settlements in the Gaza Strip, most from the area known as Gush Katif. Seeing President George W. Bush as "behind" the removals, the rabbi said, "We had 15,000 people expelled here, and there 150,000 (were expelled). It was God's retribution. . . . God does not short-change anyone" (quoted in Alush 2005). It is important to note that the rabbis' extreme positions did not reflect a consensus in Israel, but they were followed and paralleled by other commentators and contributors—some from within the United States—who saw the evacuation as a kind of divine retribution. Some blamed the disproportionate consequences of Katrina on its Black population because, they argued, Black Americans were unlikely to be studying the Torah and were, therefore, without God (Ashkenzi 2005). "A home for a home," said one New Orleans resident (M. Cooper 2015).

These senses of harm or sacrifice came from ideologically and theologically inspired connections made between apparently unrelated phenomena, connected by those espousing theologically inspired emergency politics and far-reaching conspiracy theories (Ophir 2007). An organizer of Columbia Christians for Life deemed Hurricane Katrina as retribution for the existence of abortion centers, bizarrely comparing a satellite image of the storm to a pro-life image of a six-week-old fetus: "Providence punishes national sins by national calamities. . . . Greater divine judgment is coming upon America unless we repent of the national sin of abortion" (Ashkenzi 2005). Even the sexual freedoms associated with New Orleans, as Gary Richards (2010, 522) showed, fitted a religious, conservative critique of New Orleans as sinful, befitting a retributive emergency of unequal evacuation.

The evacuations in Gaza were not actually the removals of Palestinians through the bulldozing of homes and property, or through other destructive methods, as seen in the removal of Palestinian homes and settlements in West Bank and Gaza, and in the longer history of Palestinian, Arab, and Bedouin removals. Many scholars have made visible these domicidal violences (Graham 2002; Segal, Tartakover, and Weizman 2003), where the Palestinian or Israeli home has huge political and cultural significance, especially within a logic of security. Under Prime Minster Ariel Sharon's influence and leadership, Israeli settler homes were the forefront strategic positions of Jewish territorial security. The Gaza evacuations formed Israel's "disengagement plan," which became operational in 2005 for what the Israeli government called the wider "evacuations" of Jewish settlers from the Gaza Strip. These removal operations were assigned to the Israel Defense Forces (IDF), who decided to use noncombat and noncommissioned officers for the purpose. The event performed an inverted form of what Elya Lucy Milner has called "the settler colonial logic" (2020, 269) by forcibly removing Israel's Jewish settlers, to deoccupy, in at least one dimension (Weizman 2007).

In law, the disengagement plan restricted Israeli freedom of movement within the Gaza Strip. Preceding the evacuation day, thousands of so-called infiltrators had moved into Gush Katif in solidarity with the settlers facing evacuation. The use of the term *evacuation* drew the evictions into some kind of consistency with the "humanitarian" measures designed to govern Gaza from Israel by abandonment (Weizman 2011), where concern for the population's mental health or biological sustenance in the face of an Israeli blockade saw a form of biopower enrolled in order to provide the Gazan Palestinians with the minimal amount of life (Li 2006). The Knesset bill papers that set out the disengagement in law in its glossary of terms could not but show that *evacuation* would mean a very circular thing, defined as "the evacuation of the evacuated areas, including the evacuated communities."[2] In other words, evacuation meant those who are evacuated, the areas they are evacuated from, and the process through which they were evacuated. Like the tortuous relations and definitions we have been following in this book, evacuation is a thing that seems to refer to itself, a process but a self-referential one. Thought aesthetically, it implies a making clear or vacant or being free of something, even if nothing is disclosed about the process. As an evacuation, the disengagement's evictions of Jewish settlers had been somewhat linguistically unloaded in such a way as to try to depoliticize those displacements.

The Israel State Commission of Inquiry investigating the disengagement recognized that the "use of the term 'evacuation' is controversial" (2010, 18). The report also refers to other studies of the "linguistic debate" that soon emerged.[3] For the linguist Pnina Shukrun-Nagar, applying a Bakhtinian study of linguistic polyphony to a series of writings on the disengagement plans written in advance of the removals, and evaluating the different standpoints and ideological positions involved, the disengagement evacuations were bound up in other ideologically, theologically, culturally, and both positive and negative emotionally loaded designations and terminologies.[4] Both *evacuation* and *disengagement* felt ambiguous as terms. Defense minister Shaul Mofaz considered, "I am not sure that the word disengagement exactly expresses what we want to do. We should look for a more successful word" (quoted in Shukrun-Nagar 2008, 353). One problem was that for some of the Israeli public the term *disengagement* (*hitnatkut*) could be considered as "a false term that represents a contemptible attempt to hide the fact that this is a withdrawal under fire from Gaza" (the linguist Ron Kuzar, quoted in Shukrun-Nagar 2008, 353). Israel's disengagement was actually an "exit," an act of being "deported" from Gaza or forced into the position of a "runaway state." For Shukrun-Nagar, words like *withdrawal* even came with what she calls a kind of "negative thrill," so that the use of *withdrawal* could anticipate or precipitate further violence in a longer cycle and reiterate the withdrawal from southern Lebanon in May 2000, which was perceived to have had led to the Second Intifada. A planned withdrawal from Gaza, if perceived as an escape or "flight"—perhaps an evacuation out of necessity rather than choice—might precipitate more aggression. For others, disengagement was an abstract and dispassionate act, an "exit," a painful act, a "deportation" (Shukrun-Nagar 2008, 352).

From the left, even the evacuations that preceded the 2005 disengagement were duplicitous and deplorable. Labeling them the "fake evacuations," activist and author Gideon Levy was highly critical of the earlier evacuations of small hilltop settlements in the northern West Bank in 2003. For Levy (2003), the evacuations were really a front: "His one hand appears to be evacuating while the other is assassinating." Evacuation is conjured as a kind of spectacle—a benign movement in order to disguise in plain sight the encouragement of illegal Israeli settlements in the West Bank, or to draw the intense scrutiny away from the targeted assassinations of Hamas leaders and operatives in Gaza. Even Sharon was not always quite sure of what terminology he should be using. In an interview the Israeli prime minister said, "I have given an instruction to carry out the evacuation," before correcting himself, "pardon me, the relocation, of seventeen localities from the Gaza Strip to Israel" (quoted in E. Lewin 2015, 24). For Lewin, whereas the term "evacuation implies the traumatic wresting of people from their beloved land, the alternative term 'relocation' is more neutral. It describes what people do when they go to study, find work or fulfill professional ambitions" (E. Lewin 2015, 24). We will explore the use of *relocation* in another context later in this book. The slippages are curiously akin to the wordplays around Japanese American incarceration that we explored in chapter 3 too. Shmuel Livne, a critic of the disengagement, took aim at "disengagement," explaining why the polysemy of the Hebrew term *hitnatkut* could imply a manner of disconnection (for example, from electricity). "What is it? Is it the disconnection of the electricity, water and other systems (I do not recommend it) from the Authority? No. It is about the cessation of the lives of thousands of Jews in their settlements built on migratory sand dunes, which no one wants or uses" (quoted in Shukrun-Nagar 2008, 357).

In this chapter we explore the entanglement of the Gaza disengagement plan with its controversial acts, practices, and terminologies of evacuation, which form an aesthetics of mobility governed, anticipated, represented, and resisted. Of course there are numerous examples of evacuations and otherwise forced displacements in this context. And Israel is no stranger to these different acts of (de)politicization of evacuation, mobility, and the lexical structures that name them. As discussed earlier, Ariella Azoulay's (2011) analysis of photography within the 1948 war navigated a neutralizing language that saw *evacuation* applied to dilute the expulsions and deportations of Palestinians from their land and homes. It is not the case, however, that the disengagement is simply another form of these removals; instead, the disengagement expresses the recursions and confusions of evacuation that were made so present and folded into or on top of one another. I pay particular attention to the words and translations of *evacuation* in Hebrew and English. These allow geographies and places to be quickly and aesthetically linked, as Gaza is connected, as we saw, to New Orleans, just as other versions of Jewish and Palestinian persecution, removal, and forced displacement are repeated and rehearsed within the disengagement's meanings and experiences and within the deeper cultural and political landscape of Israel's settler colonialism. The expressions and characterizations of disengagement as trauma saw the disengagement evacuations as repetitious, a refrain of persecution and the displacement of Jewish peoples, but this time by their own government. The disengagement also produced distinctive aesthetic experiences and judgments, what critical scholar Nicola Perugini (2014, 2019) identifies in the vertiginous.

First, the chapter examines the relationship among the articulation of different reproductivities, evacuation, and the disengagement plan. It moves from the theological and vertiginous registers of the disengagement to explore how trauma became an important narrative construction and felt experience, especially as the evacuation seemed to retraumatize as a recursion, a re-vacuation of the settlers and their memories and collective histories of previous displacement. Finally, the chapter concludes through the unusual sympathies and solidarities that cut across evacuated and evacuator.

HOME, EARTH, UNSETTLING

Paul Carter and Francis Maravillas have explored different versions of colonialism and its tendency to express a state of groundlessness. Carter evokes a kind of "imaginative transference" wherein colonizers appropriate land

and cultures and transform them in their own image. But by so doing, "in dissolving the ground of difference," the colonial settlers unseat their own position and realize that "they had nowhere to stand" (Carter 1996, 28). It is a tendency to forget that the ground of settler colonial appropriation may have been already inhabited—that settler colonialism is "haunted by the very ground upon which it performs its appropriative acts of dispossession" (Maravillas 2012, 20). For the Gush Katif settlers that the disengagement evacuations were meant to remove, the haunting of colonialism was already constant, less vertigo inducing but almost affirming through the routine threat of suicide and rocket attacks under the watchful protection of the IDF. The disengagement evacuations, as many of their critics argued, undermined Israel's future and claims to a past. It was widely seen as a reversal of the expansion of settlements in Gaza since the 1967 Six-Day War.

Evacuation became an enemy of settler reproductivities. The Gush Katif settlers were ideologically construed in ways that connected religious and national growth with the embodied colonization and agricultural transformation of an apparent empty and arid land (Berg 2014; Neumann 2011; Weizman 2017). The settlers had transformed Gush Katif from an arid area known locally as "the cursed land" of empty sand dunes, without even "insects or weeds" (Fendel 2010), into an "agricultural paradise" full of "beautiful hothouses, orchards, flowers, red roofs, streets, factories, crops, schools and synagogues," whose only crime—continued one op-ed within the repeated language of the Zionist settler acculturation of the land from wasteland—was to "have made that wasteland bloom" (Meotti 2012). Gush Katif was known as a profitable and productive agricultural oasis. Even in the face of imminent evacuation, as the soldiers approached, reports lamented that "some youths were still busy laying concrete foundations for a new building next to the synagogue and the sprinklers were on in the gardens" (Hasson 2005). Gush Katif was frequently and widely praised as miraculously fertile. A regional council member, Yigal Kirzenshaft, eulogizing the heroic efforts of the settlers' tomato production, suggested that "one acre produced 20 tons of tomatoes. It was a supernatural blessing, despite the fact that before we came it was a desert, totally desolate" (quoted in A. Lewin 2019). The official report on the evacuation, critical of the government's management of the event, began with precisely the same sentiment of spiritual amplification: "The settlers are the salt of the earth. With hard work, enormous sacrifice, capability and innocent faith they built magnificent settlements. They overcame difficult physical conditions and the land responded to them and yielded a hundredfold" (Israel State Commission of Inquiry 2010, 35).

The evacuation reversed this desert blooming, to leave a "desolation" or "barrenness" and a different kind of fertility, not of crops or the agricultural nirvana the Gush seemed to be praised as, but for rocket strikes and terrorists. Left on its own or in the wrong hands, it could only revert "to its former state of total desolation," serving "as a safe haven for terrorists" (A. Lewin 2019). Even Sharon would anticipate the move—metaphoricized as a transition from hothouses to a hotbed—blaming the evacuation on both the growth of the Palestinian population and the intensity of their apparent malice: "We can't hold on to Gaza forever. Over one million Palestinians live there, with their numbers multiplying every generation. They live in incomparable crowdedness in refugee camps, in poverty and distress, in hotbeds of ever-increasing hatred, and with no hope in sight" (Sharon, quoted in Kra-Oz 2013).

The settlers, the aid workers, and even the military personnel tasked with moving them later reported, recorded, and were documented to have suffered human rights abuses and lasting trauma from the experience. Children have borne much of the brunt of these studies. As symbols of innocence and futural reproduction, children were mobilized in images and video footage of them being carried by parents out of their homes, or photographed and videoed being carried away by IDF soldiers. Children were the addressees where the true trauma of the evacuation took place in some narratives. In August one group of surfers, reportedly between sixteen and twenty-one years old, even claimed that they would undertake a group suicide. Their intentions were framed by the report as influenced by a Palestinian suicide bomber killing a family member. A local security officer suggested that the "boys" were in "very deep emotional distress" and that, going further, he would do everything to prevent it, "including shooting them in the legs" (Shragai 2005; see also Lord 2015; McGreal 2005). Perhaps the most infamous confrontation between the IDF and Gush settlers was with a number of female minors, notably Chaya Belogorodsky, whose mother lived in Gush Katif. Belogorodsky was arrested during a protest after defying a banning order following a previous arrest. Belogorodsky was effectively detained by an Israeli judge, who argued that she would incite others to protest on her release (Fendel 2005). For an IDF general, the young were even susceptible to an early death given the disengagement's particular kind of trauma that could produce "heartbreak," causing them to "die prematurely" (quoted in A. Lewin 2019).

The IDF had simulated many different scenarios in preparation for the evacuations: drills for the evacuators, not the evacuees. Simulations that took place in the Kibbutz Kerem Shalom included female soldiers dressing up in pregnancy suits, their blue, purple, and orange clothing made from

FIGURE 6.1. The IDF's simulated protest in Kerem Shalom. Sebastian Scheiner / AFP via Getty Images, 2005.

cheap fabric and IDF unit shirts and sweatshirts—possibly to mimic settler dress—worn over the top of their green khaki uniforms and military boots (figure 6.1). The IDF were worried that the settlers might use the apparent vulnerability of pregnant women and women with babies to front the protest and embarrass the soldiers, or even goad them into rough treatment.

The protesters held plastic dolls as if they were holding newborn babies. Footage shows the women chanting as the camera looks back at them. Some hold chains; some clutch their swollen fake bellies—many gesticulate, with their hands open, trying to grasp something; others hold their hands over their eyes; some fold their babies into their bodies. In another, the women stand up to a line of soldiers; several of them carry babies in slings. The women shout at the line of troops, and one tugs at the shoulder of another. The male soldiers look down or away. A clearer image has been published by Getty: an IDF soldier holds a smartphone or camera to record the event in the background. The caption explains that these are "extremist Jewish settlers" protesting troops "trying to forcibly evacuate them."[5] This may have been another form of ungrounding. The IDF had criticized Hamas for using women and children as human shields, narrativizing this as a potent signal of Palestinian inhumanity, where "(children, women, and the elderly) are the human weapon Hamas uses against Israel" (N. Gordon and Perugini

2016, 179). For Neve Gordon and Nicola Perugini, certain political logics follow from the distinctions as to "why only certain subjects can become human shields while others are excluded" (179). The IDF seemed to assume that the settlers would do what they imagined the Palestinians would do, weaponizing their children and mothers in order to frustrate and humiliate the evacuation.

Some of the evacuees from Neve Dekalim, one of the settlements comprising Gush Katif, built a memorial museum in an apartment in central Jerusalem in 2008. The museum is a kind of rehoming (Ahmed et al. 2003)—an evacuation and reproduction of a version of Gush Katif in a new location—and a reminder of the absence of that home owing to the displacement. It is, in Paul Carter's terms, a reperformance of the vertigo of colonialism, perhaps both an admission of "existential anxiety" and a "phoenix like reproduction" (1996, 30). The museum uses terms like *displacement* and *evacuation* rather than *disengagement* to describe the process (Kra-Oz 2013). A large aerial photograph of Neve Dekalim is displayed. Visitors, including a family mentioned on the museum's website, are able to identify their own homes on the map; a publicity photograph shows them pointing to it. The gesture of pointing is a kind of rehoming, an identifying with a place of the past, while the "room of expulsion" focuses on the forced mobilities of children, "looking through the window of the buses that transported them from the settlements to the hotels, containers, and pre-fabricated houses where they were relocated by the Israeli government," in order to point toward their lack of a future (Perugini 2014, 50).

As a visiting US South Carolina representative, Alan Clemmons, emphasized during a visit to the museum in 2011, "there are memories here of good times, memories here of a beautiful place that was carved out of nothing, of a beautiful green oasis, that God fertilized and watered through the people of Gush Katif and turned a wasteland, a desert, into a beautiful garden blossoming like a rose" (Arutz Sheva / Israel National News TV 2011). Such forms of homing or rehoming are virtual. Gush Katif has a web page, "Israelis at Home—Gush Katif Summer 2005," where sun-kissed beaches, greenhouses, and community spaces are represented.[6] Amid hothouses, playgrounds, camouflaged IDF security towers, tomatoes, and palm trees, children look happy. In nostalgia for something evacuated and now mythologized, Gush Katif is portrayed as "a place where life was good and happy and the sun shone all the time until the moment the government came and turned off the light" (L. Hoare 2013).

For critics, the museum writes out many of the problems that the Gush Katif settlements symbolized in a kind of nostalgic and amnesic inversion of settler colonial relations. Perugini argues that the museum's memory flips the Israeli settlers into the position of a dispossessed indigene, possessing a paradoxical space threatened by the Palestinians, who portend to overtake their position within the Gazan soil, an "inversion" aimed "at erasing settler colonial dispossession; transforming the colonizer into the colonized, and vice versa" (2019, 44). Disengagement is portrayed as profoundly unsettling and indeed suspending—temporally and vertically. The official committee characterizes many of the evacuees in 2010 as "still living in temporary caravan sites; the construction of most of the permanent housing has not yet commenced," while "the decisive majority of the public structures in the evacuees' new settlements have not yet been built" (Israel State Commission of Inquiry 2010, 8). The evacuated settlers are in permanent suspension.

EVACUATION TRAUMA

In the aftermath of the disengagement, the Disengagement Administration responsible for handling and resettling the evacuees and known by its Hebrew acronym, SELA, gave families twelve therapy sessions to work through their experiences of the disengagement and evacuation process and to adjust to life afterward. In some instances, creative art therapists were given funding to work with children and young people who were identified as suffering from both "trauma and adjustment issues" (Czamanski-Cohen 2010, 409). The language of trauma was a frequent trope and measure of treatment for the disengagement, especially as the evacuation was experienced as a recurring, repetitious event that appeared to dislodge experiences and memories and narrations of prior displacements. Therapists, some using drawings and music, reported that the settlers endured recurrent dreams and fantasies where they lived the disengagement out again. Some would wake up from nightmares of the Nazis. Others fought the evacuation in their dreams (Ben 2006).

One of the group was encouraged to draw an expressive image from their nightmares. They rehearsed the common figure of the wandering European Jew, famously expressed in the work of Marc Chagall (Berg 2014). The disengagement seemed to re-uproot the community to an airy, inconsequential, impossible place—up in the clouds. Peculiarly, not only was this imagery a common imaginative trope to draw from in the evacuees' therapy, but the vertiginous almost etherealization of Gush Katif was a way to deal with the evacuation and even combat its trauma during the disengagement itself. One of the IDF commanders, Major General Gershon Hacohen, who was placed in charge of carrying out the IDF's mandate, has been particularly significant in the years since the disengagement as an active critic of the policy. His quite open public statements and interviews reveal how a particular kind of political theology (Goldberg 2018) guided and inspired his approach to the evacuation (Ginsberg 2014). Hacohen describes encouraging his soldiers to imagine that they were evacuating only a "lower Gush Katif," the physical bodies, homes, and objects that encumber spiritual life. Even if the soldiers evacuated the physical Gush, the "spiritual and ideological Gush Katif" (Hacohen 2019) would be sustained. In the context of several Jewish spiritual leaders claiming a kind of miraculous exceptionalism to the Gush Katif settlers, Michael Feige wrote, "Gush Katif was slipping away from physical existence, it was already entering the realm of folklore and the popular imagination" (2009, 266). It was place, as he quotes from another critic, "that cannot be detached because it is already not within the natural order of things" (266).

Ironically, fears of rootlessness continued within the medicalization of the disengagement, which focused on the postevacuation health of the Gush Katif residents in terms of obesity and morbidity, somehow making solid the imaginative, ethereal, and spiritual versions of Gush Katif into weighty corporeal matters, akin to Lauren Berlant's (2011) examination of "slow death." This might mark a very different sense of what Perugini (2014, 51) identifies in the making-real of curious special effects of coloniality through haunting by the "potential return of the colonized Palestinian." The logics of this continued to oscillate and twist even after the disengagement by other hauntings of an erased other, as if "to be ungrounded was to lose touch with one's human and physical surroundings" (Carter 1996, 30). The Palestinians had to consider how and whether to appropriate the Gush houses and buildings that had not been defaced or bombed by the Israeli air force. If architecture was a direct means of the occupation, argues Eyal Weizman (2012, 227), inhabitation and reinhabitation could be thought of as a kind of "co- or trans inhabitation" with those who had been there before. Where the figure of the "ruined Arab village" was erased by Israel's cultural and destructive appropriation during the 1948 war, the destruction and ruining of the Gush settlements into deserted buildings and rubble saw Israel performing what Gil Hochberg (2015) has called another "psychic bomb."

To apply the understanding, treatment, and representation of the settler evacuations with medicalized notions of trauma and ill health, the settler body appeared eroded and degraded into a state of lethargy and temporary stasis. One medical report even praised the active lifestyles that settlers assumed in Gush Katif, not only from the labor of settling the land but also from escaping the threat of Hamas rocket attacks: "Each time there was a bombardment of mortars or loudspeaker warnings everyone would literally run to the shelters for safety"—as if the daily mobile reminders of their settler status were a form of physical training. In contrast, the stasis or immobility from the "the loss of land-based work" with the "prolonged and continuing state of temporary housing, with unclear prospects for resettlement," led to cases of depression and anxiety, which were in turn linked to increased "morbidity of chronic disease" (Kory, Carney, and Naimer 2013, 140).

For Perugini, the aftermath of disengagement saw "the clinical, moral and political suffering" fused, if not "welded together" (2014, 57) through moves to try to diagnose and treat the evacuees and reconstruct an understanding of the event through a wider locus of Jewish national identity and homeland. To this extent, the evacuations were persistently labeled as traumatic, an ideologically and theologically induced trauma (Sheleg 2007; Sheleg, Bloom, and Sapir 2004) produced by the evacuation's sharp disconnection from the settlers' reality (Elad-Strenger et al. 2013, 61). Representations of the evacuees as "extortionists" (Sheleg 2007), which followed the disengagement and the processes for acquiring compensation, also continued this break with legitimacy. The Commission of Inquiry report on the evacuation continued to articulate a variety of registers of what we have been calling *ungrounding*, *unsettling*, or *colonial vertigo*. It found that the settlers were "torn from their homes and from the center of their world," causing, it argued, a "profound emotional trauma" and "a deep religious, family, economic and ideological crisis" (Israel State Commission of Inquiry 2010, 2). In marrying the evacuation with psychological trauma and crisis, the report recommended treatment as if it were no different from a disaster situation—a common analogic response to emergency, as we have seen (Ghamari-Tabrizi 2013). Medical professionals were recommended who had experience in "sociopsychological treatment in crisis and disaster situations," the idea being that they could "formulate a professional plan for sociopsychological rehabilitation and, later on, coordinate this rehabilitational activity" (Israel State Commission of Inquiry 2010, 33).

Was the disengagement a crisis, or was it disaster, an uprooting or break, a suspension or stasis? For some it was a kind of repeating loop of trauma and displacement that the Jewish community had been subjected to for thousands of years. Another recursion. The vertiginous anxiety of a search for a place and some kind of temporal moment saw the disengagement become a return to a troubled past, in the manner of a cycle, an endless refrain. The evacuation was compared to the evacuation of the Sinai Peninsula in 1982, and even the Holocaust. As Shukrun-Nagar explains so eloquently, for the settlers, as well as others who opposed the disengagement, there was an "emotional linkage between the process at hand and the serious disasters suffered by the Jewish people, led by the Holocaust. The message is that the given process is nothing but a link in the chain of serious historical injustices against the Jewish people" (2008, 351).

Those who identified with this perspective used terms like *deportation*, which meant that the lexical processes involved in the disengagement only added to this feeling of a continuity of dispossession. Exemplified in a protest poster with the same terminology, clearing the area from Jews was another kind of *Judenrein*—a Nazi term translated as "clean of Jews," which carried with it the violence and stigma of the Nazi deportations from German and other European national territories discussed earlier. By painting numbers on their arms and wrists, by using the yellow badge and other Holocaust symbology, many settlers rearticulated this sense of historical déjà vu in what Perugini called an "endless repetition" (2014, 59–60). Many settlers described being unable to separate the events from the history of Jewish persecution. One stated, "Our world of associations is so tied up with the Holocaust. . . . I felt as if the Nazi soldiers were on their way" (Dekel 2010, 393).

Many welfare charities were hugely critical of the government's actions. On August 30, 2005, the chairperson of the nonprofit social services charity Lema'an Achai described the government authorities as "*incapable or uninterested in providing effective care*" (quoted in Perugini 2014, 66; emphasis original). Volunteer psychologists and counselors were dispatched to hotels and temporary accommodations for the evacuees, and a dedicated fund was set up for "mental health and crisis services" to help the settlers "come to terms with their trauma and begin to rebuild their lives . . . to regain their balance and to rebuild their lives again as self-sufficient, industrious and motivated Jews" (Morris 2005). In 2006 a group of volunteer women and playwrights worked with Lema'an Achai to produce a musical titled *With an Outstretched Hand* to raise funds for their Gush Katif fund but also to explore the experiences of the Gush evacuees for the "ongoing crisis situation of the refugees" (Fendel 2006). The play took its title from the biblical story of Exodus and compared and intertwined the two evacuations.

The trauma was not limited to the evacuees. Less than a week before the disengagement operation began, the IDF and police had begun a short period of exercise scenarios in Kibbutz Kerem Shalom in southern Israel (mentioned above), as a model settlement for the soldiers to practice on. The exercises were preceded by what Perugini describes as a month-long "orientation program" to address the political and legal aspects of the situation. This part of the program was about guiding and legitimating the IDF's actions and demonstrating the IDF's "sensitivity" to the process, "test[ing] our ability to demonstrate sensitivity towards the evacuated population, while still demonstrating the firmness needed to carry out this complicated task, using the most difficult opposition scenarios," according to the Southern District commander, Uri Bar-Lev (Somfalvi 2005). The scenarios were about demonstrating the "*sensitivity and assertiveness*" (Perugini 2014, 64) we have already discussed, to prepare the soldiers for the test they were to meet. Critics have suggested that the IDF's training transformed the soldiers into robotic automatons: an "army of expulsion." As Michael Feige writes, "They were to deaden their senses, or at least their reactions, and absorb passively whatever logical arguments or foul language the settlers would throw at them" (2009, 267). As the IDF plans were leaked around Gush Katif, a protester suggested, "They will look serene, will come with friendly faces, and they want to see us all out" (Weis 2005). What was clear was that the disengagement created remarkably conflicted feelings among the IDF, who were called to remove members of their own community and those that shared their religious, ideological beliefs.

The trauma was not limited to humans. Many settlers had to move so quickly that they did not think of or have time to make preparations to leave with their pets and companion animals. Some animals simply ran away during the chaos of the evacuation. Others required "rescue." Several animal rights charities organized themselves, obtained permissions from the IDF, and entered Gaza to perform evacuation rescues of the "disengagement dogs" and other animals who had been left behind. Thirty animal cages were brought to Gaza, including a mobile spay/neuter clinic, humane traps, and experienced vet and assistants as well as volunteers. The stranded animals were compared to children: "You wouldn't leave behind your kids, would you?" (Jewish Telegraphic Agency 2005). As a population of Gush Katif left behind by the evacuation, pets were found "inside homes about to be demolished, as though waiting for their owners to return," and were cast in terms of neglect and abandonment, some "caged in the sweltering heat" (Jewish Telegraphic Agency 2005). The operation was referred to as Operation Noah, and the mobile unit was called Noah's Ark. In the same way as the land was putatively left to ruin, the Tel Aviv–based animal rescue group Hakol Chai (lit. Everything/Everyone Is Alive) and Tsa'ar Ba'aley Chaim (lit. Animal Grief) described finding animals in a state of vulnerable stasis, waiting to be rescued by their absent owners. As the Hakol Chai's director argued, "Israel's expressed intention to be sensitive and behave responsibly during the disengagement should apply to all living beings" (Hakol Chai 2005). Hakol Chai would imply that the left-behind animals had no way of surviving the "extreme" conditions during and after the disengagement, unable to imagine life without their owners or carers, or life within the Gaza Strip with other owners or coinhabitants: "Cats and dogs left behind by departing settlers have no ability to survive. . . . When all that remains is dust and ruins, those who escape the massive bulldozers will die of hunger, thirst, and injuries" (Hakol Chai 2005). The charity encouraged the wider public to adopt the animals, described as "refugees," into "your home and heart to prevent their needless suffering" (Hakol Chai 2005).

Israel's own bulldozers and bombs may have presented a greater threat to the pets than the Palestinians, who are absent from many of the reports and stories. The disengagement evacuations were consistent with a longer recent history of animal evacuations from the Gaza Strip as well as Israel's violence against them. While Darryl Li (2008) would describe Israel's treatment of Gaza as a kind of "human zoo," during and after the disengagement, multiple animal charities campaigned to gain access to several of Gaza's zoos in order to evacuate animals and compel the zoos to close or adopt international welfare standards (Four Paws 2019). In May 2004 the IDF continued its strategy of bulldozing homes by bulldozing a path through Rafah Zoo as part of its assault on Gaza. By 2014 the Khan Younis Zoo was closed and featured in a *National Geographic* supplement (Tenorio 2016) about the "World's Worst Zoo." In 2019 the charity Four Paws evacuated forty-seven animals from the Rafah Zoo, including lions, monkeys, wolves, foxes, and emus, which passed through Israel into Jordan, where several of the animals were eventually settled—two lions went to South Africa.

For Perugini (2014), trauma is a kind of "affective glue" that produces the "settler subjects" as "traumatized victims" and calls on the nation to recognize the trauma of the "evacuated settlers." He suggests that this "corresponds to reaffirming the legitimacy of the entire colonization enterprise" (52). To produce trauma is to sustain the settler colonial project where the dispossessions of Jews during the disengagement evacuations reaffirm recursions of the dispossession and "evacuation" of Palestinians from other territories and spaces in an uneven and aesthetically one-sided politics of mobility—as if they hadn't happened before. As a researcher from the Israeli Policy Institute argued, "The pictures of the evacuation of Jews from their homes were heart wrenching, and there are those who are voicing the idea that if it was possible to do this to Jews, then Arabs can certainly be evacuated as well. From this point of view, evacuation is perceived as legitimate and easier" (Sheleg 2008).

The experience of the disengagement measures, as we have seen, was fraught and, for many, highly traumatic. The police and IDF were criticized for acting robotically, yet the IDF performed interesting forms of fraternal solidarity and empathy with the settlers they were tasked with evacuating. Of course there were many demonstrations, but there was less active resistance or violent opposition than they expected. On the eve of the first evacuations, thirty-four IDF reservists signed a petition "promising to refuse orders to take part in the evacuation" (Y. Levy 2007, 395), expressing how morally torn and hurt officers would feel in displacing their own communities and breaking from the Zionist ethics of settling the Holy Land. There were issues of both military and nationalistic discipline at stake. Rabbi Schlomo Aviner even warned of disobedience *as* disengagement: "To leave the army is to leave the nation. It is disengagement from the people!" (Y. Levy 2010, 199). From the IDF's perspective, their success could be attributed to their attempt at sensitivity. Footage demonstrates outpourings of grief and emotion *with* the settlers, perhaps not the sensitivity IDF commanders insisted on, but rather interesting realizations of masculine fraternity and sympathy—less the embarrassment or indifference seen in the training exercise with female protesters.

The military-spiritual imagination of Gush Katif mentioned above played an interesting role in the unsettling and ungrounding of the settler project in the Gaza Strip. We might turn to an open discussion of strategy and philosophy that took place for the IDF journal *Ma'arachot* with the IDF anthropologist Dr. Asaf Hazani in conversation with Major General Gershon Hacohen, and reproduced in Ori Goldberg's (2018) *Faith and Politics in Iran, Israel and the Islamic State*. The interview is not what one would expect, peppered with Hacohen's opinions on nonlinear dynamics and even the influence of post-structuralist philosophy. Hacohen appears to have been greatly influenced by the Brigadier General Shimon Naveh, former head of the IDF's Operational Theory Research Institute, first established in 1995, in which Hacohen was one of the institute's star students, being one among several other acolytes (including Brigadier General Aviv Kochavi, the last military commander to leave the Gaza Strip in September 2005) who "went nuts over" Gilles Deleuze and other continental post-structuralist philosophers they were encouraged to read (Baigorria 2009; Y. Feldman 2007). As Eyal Weizman (2006, 2012) has explored at length, the exposure of the group was influential in Israeli military planning and operational strategy.

In the evacuation Hacohen conceives of a kind of third space, not the ethereal Gush Katif mentioned above or the physical world, but an affective, experiential space of doing and performance that he and the IDF and police officers involved in the operation were seeking to design or engineer (McCormack 2008). In recalling how the police, IDF, and settlers sang the national anthem together in the Netzarim synagogue, where speakers had been set up on car roofs outside the synagogue to make the evacuation almost a moment of collective celebration and mourning, Hacohen explained, "You didn't need to understand! This is not an event for understanding. What activates the person in the event is not their cognitive consciousness. That is not how you design the event. This is an experience that unfolds in an emotional environment. For example, We played 'mood music.' We played a religious song asking God to have mercy on Israel, his people" (quoted in Goldberg 2018, 130).

The interviewer, Asaf Hazani, then responds, reemphasizing the sense of the IDF commander as a kind of architect of the aesthetics of the event: "It is all about feeling and sensing?" Hacohen seems to be articulating not an abstract space of geometries, or "physics" that would shape other parts of the disengagement, and other evacuations in this book, nor the spiritual Gush Katif he required his soldiers to separate from the physical world they were removing, but rather a lived phenomenal space of the sensual, affective, and atmospheric—of "music" and "demeanours," "in the middle of this thing" (quoted in Goldberg 2018, 130). The disengagement seems an event that merged the IDF's post-structuralist thinking, focused on disrupting and dominating not Palestinians but Israeli settlers through an operational and phenomenal logic of engineering encounters of sympathy and embodied solidarities.

Ori Goldberg uses the terms *force* and *fraternity* to describe Hacohen's attempts to find some kind of abstract equilibrium point that could be realized in a "forceful encounter between brothers" (2018, 117). Operation Brotherly Hand was even the title given to the disengagement operations on August 15 and 16 that gave settlers the military decree ordering them to evacuate. This followed the handing out of evacuation warrants to the Gush Katif settlers at the Kissufilm crossing checkpoint—a main entrance to the Gush Katif settlements—a week earlier, on August 9. Under Brotherly Hand, the notice, which was handed out by mixed soldier and police teams, was given to the settlers, who would be offered help to move if they needed it (Greenberg 2005).

Hacohen mobilizes a spatial imagination that refuses more general military spatial strategies of lines and points—ways of seeing evacuations in the abstracting diagrams of simple vectors that we have seen. Instead, the disengagement evacuations were conceived as more like a chaotic mixture of encounters, which Hacohen expresses as a kind of space of potent sharing, a "'mixture' or 'encounter created at each and every home'": "The evacuating forces stood before the people about to be removed from their homes outside each and every house—with all the anger and the pain, the eyes of Jews sharing the pain met (across what would have been a divide). Even if the intensity of the pain was not identical, the feeling was. We spread the forces throughout the physical space in a way that created points of friction over nearly all the space. We undid the linear order and transformed it into a nearly chaotic mix" (Hacohen, quoted in Goldberg 2018, 121).

This approach is reminiscent of swarming. As Weizman explains the strategy, the swarm was borrowed in part from post-structuralism and a longer genealogy of guerrilla warfare, to have "no form, no front, back, or flanks, but moves like a cloud" (Shimon Navah, sounding like T. E. Lawrence, in Weizman 2006, 61). Read in this context, Hacohen's approach to the disengagement evacuations was to swarm from multiple points in order to mobilize shared feelings of encounter (Hacohen 2018; see also, recently, Harel 2017). For Hacohen's interviewer Hazani (in Goldberg 2018), while the disengagement could signify a kind of rupture or separating action—what he even compares to an amputation—Hacohen conceived the promise of a new kind of shared space and experience. Evacuation was generative of an experience akin, he argues, to a kind of "fold."

Footage of the evacuation of the Neve Dekalim Synagogue shows singing and, as the Torah scrolls are removed by an IDF soldier, moving displays of grief, faith, and solidarity with the settlers they are trying to move. Much of the footage shows not only individual and collective anguish but the intertwining of IDF and settler bodies, especially of men, in a consoling embrace. The simulation drills discussed earlier had actually envisaged evacuees being carried by stretcher or in the arms of soldiers forcibly away from their homes after being pinned to the ground—indeed, some of the scenario's players even complained that the actors/evictors were much too violent and aggressive. A female soldier pretending to be a protester was knocked to the ground, causing the drill to be stopped (Harel and Shary 2005). Yet in the synagogue and outside at Neve Dekalim, the soldiers and settlers became almost indistinguishable, shoulders and arms entangled—not hugely unlike the sympathetic bearers carrying each other from the battlefield in chapter 2. On the front page of the English edition of *Haaretz* (2005), titled "IDF Aims to Complete Gaza Evacuation within 10 Days," a settler and IDF officer console one another in tears. Perhaps these more immediate expressions of gendered solidarity also belie other expressions of sororal agencies in the years following disengagement. The play mentioned above came about through a group of female playwrights, volunteers, and musicians. The Lema'an Achai musical about the disengagement focuses on the "outstretched arms" of women who had to "hold their families together" and who were stuck in stasis in hotels, apartments, tents, and guesthouses (Wisemon 2006).

The disengagement evacuations seem to promise other kinds of movements and belongings. It is difficult to really know how engineered these were, but evacuation is another a kind of rupture from appropriated territories of colonial possession even as the protests and memorialization of the events reinforced those ruptures through aestheticized ideals of settler life. Manifested as a kind of vertiginous trauma, the vertigo of the evacuation is very different from that of the high-rise evacuations we discussed in chapter 1, except that they rupture once more the aesthetic of evacuation as an individualized practice of emergency escape or removal and complicate the evacuated/evacuator bifurcation of the removed as disempowered and the remover as a military force—especially when this was in a context of already existing and past expulsions and displacements set against the historical memory of anti-Semitic violence and expulsion. The embodied subjects of evacuation and evacuator become much more complicated and enfolded into bodily forms and affects that solidified across faith, community, and gendered allegiances.

SEEING EVACUATION
LOGISTICALLY

Our starting point is a low-resolution map the BBC used to illustrate the rush to repatriate and evacuate foreign nationals at the beginning of the Libyan crisis, the imposition of the United Nations (UN) no-fly zone, and the eventual civil war in 2011 (figure 7.1). The image gives us a window into the other side of the circuits of migration, nonintervention, and pushback in the Mediterranean that would become the hallmark of the so-called migration crisis facing Europe's southern borders. We might even suggest that the mechanisms of "spatial organization and physical instruments, technical standards, procedures and systems of monitoring" that Eyal Weizman (2011) and Adi Ophir (2007) have in mind with regard to "the humanitarian present" (Weizman 2011, 4; see also Pallister-Wilkins 2015) would whisk Europeans away from Libya by opening up some pathways of mobility (some of them old) while closing down others.

The BBC image is not a classic infographic of lines and arrows designating evacuee pathways, as we have discussed of the diagrammatic aesthetics of evacuation (Barry 2017). These would be seen elsewhere during the Libyan crisis, and much later in the problematic representations of pathways and routeways Frontex and other European Union (EU) national governments and the media have been fond of showing to depict migration into Europe (Houtum and Bueno Lacy 2019). Instead of showing the abstraction of evacuation mobility (Cresswell 2006), the BBC's source is Drillinginfo

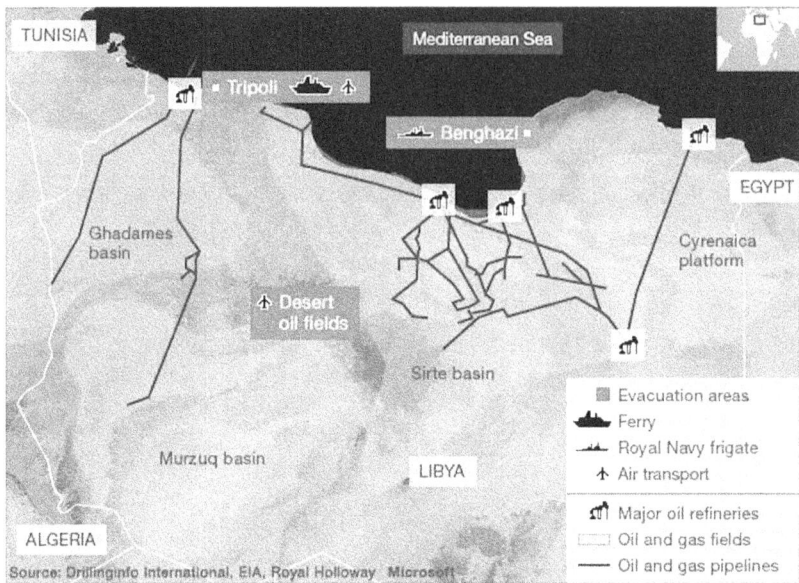

FIGURE 7.1. The BBC's map of Libyan oil fields and gas refineries, 2011. "Key Maps of Libya," https://www.bbc.co.uk/news/world-africa-12572593.

International (and, curiously, my own institution, Royal Holloway). The map outlines the different locations of Libyan oil refineries, oil and gas fields, and pipelines with a rather vague icon of "desert oil fields." This cartographic image signaled the locations of the oil industries with their resource workers, some of whom were British people who worked for companies like BP in exploratory operations and for other oil and gas outfits. A more familiar and popularized image, to many during the crisis, was one of male professionals, adorned with tool kits and heavy satchels, trailing their way across a baked desert surface to a Royal Air Force (RAF) Hercules plane manned by British special forces. Evacuation was a sharp dividing line. Against the vast humanitarian crisis of those seeking to or unable to leave the country, foreign nationals lucky enough to become evacuees were whisked away from ports of entry like Tripoli's airport and other distant airfields. Military aircraft, ships, cruise liners, ferries, charter flights, commercial airliners, private jets, and SUVs took part. Many were stranded in the besieged Misrata. Others were placed in temporary camps at the Tunisian border. And some took precarious journeys across the Mediterranean only to drift or sink (Heller and Pezzani 2014), even under the eyes of the new European surveillance regime (Tazzioli 2018).

The fast evacuation of people to safety tended to be reserved for an elite of skilled foreign nationals, VIPs, and embassy and government personnel. It revealed something of the urgency, "speed and dexterity" (Hyndman 2012, 244) with which evacuation is conducted for particular kinds of valued life—in which citizenship is a key defining category. It starkly demonstrates the ability of some individuals to be bestowed with a kind of mobility evacuation potential, their capacity to be snapped back at the sign of danger. It also reveals a crucial nexus of evacuation mobilities: of geo-economics, the circulation of oil, capital, other commodities, and expertise; the deployment of geopolitical, diplomatic, and military force; and the organization of violence as well as protection. The UN no-fly zone and the humanitarian military operation over Libya enforced by NATO (North Atlantic Treaty Organization) appeared as a new kind of international mission of protection. Yet for Paul Amar, the interventions were another type of recursion, "like a return to the past." "How could," Amar questions, "a colonial-era style bombardment of North African cities by French, Italian, and US air forces fit into an age of multipolarism and 'liberal peace'?" (2012, 2).

As Deb Cowen (2014) has argued, logistics systems (re)articulate colonial networks and alliances and conform to contemporary geo-economic imperatives that align with foreign investment in infrastructure and resource industries such as in Libya's oil and gas fields that were portrayed by the BBC. Logistics revolve around the performance of military power and law too. Humanitarianism cannot be very easily detached from the arts and science of military logistics, and as we have seen, neither can evacuation. We saw something of this in the emergence of mobility infrastructures and practices for moving the wounded out of a war zone and back into circulation in chapter 2, as well as the significance of particular vehicles and vehicular environments for mediating evacuation, its experiences, its politics, and crucially the way evacuation is seen. But whereas humanitarianism has been characterized as a watchword for imperialism and colonialism, what Antonio Donini describes as coexisting "in parallel" with and "sometimes functional to the logic of Empire" (2010, s223–24), as well as where "contemporary forms of logistical dispossession rear their head—in, for example, the construction of pipelines through Indigenous territories"—evacuation has tended to elude these associations (Chua et al. 2018, 620). What Rafeef Ziadah (2019, 1685) calls the "cross-border spatial cartographies of military operations, humanitarian aid delivery, and private logistics firms," Wesley Attewell has argued serve "as the condition of possibility for development and humanitarian activities" (2018, 720). Similarly, Attewell exposes the spaces of distribution of ports and airports and sees logistics as a "flexible tool of imperial control" (2018, 720). In this chapter we examine evacuation through these circuits of vehicles *and* pathways, technologies and practices of organization, which indelibly shape the way evacuations are seen, visualized, and represented.

The chapter creates an unlikely juxtaposition of the Libyan evacuations mentioned above with the events of the 2010 earthquake in Haiti. By understanding evacuation logistically *and* viapolitically, it explores how the evacuations pulled together medical-military evacuation and international humanitarian and logistics efforts, relying on the legacies of (neo)colonial policies of trade and global extractive, infrastructural, and logistics pathways. What resulted was the affirmation of citizenship within evacuation propensities between foreign militaries, diplomatic and medical officers, nongovernmental organizations (NGOs), and humanitarian agencies. Evacuation seemed to be so much more possible through postcolonial modalities of intervention, involving humanitarian aid and help, as well as logistics, coordinated by wider global circuits of international influence. Haiti's emergency unfolded around US military and UN assistance in response to the earthquake that struck on January 12, 2010. The UN's assistance and stability mission in Haiti (known by the acronym MINUSTAH, for the French elongation Mission des Nations Unies pour la stabilisation en Haïti) had been ongoing since 2004, succeeding the prior "stabilization" missions. The country's long status as a pariah lacking stable good government was a shaky platform undergirded by centuries of colonial occupation by Spain, France, and the United States, along with the presence of the UN and other international institutions. With the US Department of Defense's SOUTHCOM supporting the US Agency for International Development (USAID) and the Joint Task Force-Haiti under Operation Unified Response, the United States led a massive multifederal sector response to Haiti focusing heavily on infrastructure and logistics. Their first response was to get the airport up and running in order for aid and personnel to begin arriving in Port-au-Prince. It meant that American citizens could get airlifted out. Thought viapolitically and logistically, evacuation becomes a tool through which states exert power but is also where that power is undermined. It is a highly contestable dividing line between those who are able to be evacuated in permissible, legitimated formal ways—such as by aircraft, improvised ferries, and military ships—and those who are not, who have to try less regular and more precarious forms of mobility and migration.

Libya and Haiti are of course worlds apart, and I want to be clear that their juxtaposition is not intended as a way to categorize evacuation within the Global South or make any blunt characterizations of fragile states undergoing multiple emergencies. Instead, I pull them together to demonstrate the characteristics of their (non)evacuations—which occurred within just a year of each other—which offer some of the most visible and comparable examples of large-scale international diagrams of mobility, particular bodies and citizens, and infrastructure, as well as alternative ways in which publics represent and interpret evacuation diagrams. This chapter oscillates between evacuation practices in the very different settings of Libya and Haiti and deliberately follows the particularly distanced, aesthetic rationalities and logics of logistical operations (Grappi 2018) and humanitarian reasoning and practice. These converge in ways that blur civilian, military, medicalized, and corporate space (see also Sheller 2013), especially along the infrastructures and pathways of colonialism. First, the chapter examines how both crises can be explored viapolitically and logistically, primarily through state-led responses, before turning to the divestment of evacuation and the way evacuation can be conducted and seen beyond the state. Finally, there are some possibilities for disobedience, however, and the chapter examines other divested, private, and "civil" attempts at organizing and seeing evacuation otherwise.

SEEING EVACUATION LOGISTICALLY
AND VIAPOLITICALLY

As with medical-military evacuation, logistics (Cowen 2014) beckons our attention to flows and circulations of peoples, capital, and things in ways that follow a colonial, imperialist, and, as we saw earlier, viapolitical logic. In the last week of February 2011, in Libya, as the situation deteriorated, the first major groups of people to leave the country went by air; these mobilities largely involved citizens of foreign countries and foreign nationals working for the oil and gas sector. The flights were slow in coming. By February 22, 457 landing permits had been applied for. Canadian charter flights were refused landing at Benghazi before returning to Istanbul. Other planes left Tripoli empty. By February 27, Kenya's charter to Cairo and then back to Nairobi was delayed too. When charter flights finally got into Tripoli, British people and other foreign nationals effectively cut in line.

If we turn back to the BBC map once more, we realize a "profoundly imperial cartography" (Cowen 2014), for in showing us where some of the evacuated would come from, we see the patterning of foreign investments in North Africa from Total, Eni, Shell, (Norwegian) Statoil, Repsol, and

Gazprom. China also held large investments in the form of infrastructure projects and a railway from Sirte to Benghazi. If oil was key to what John Urry (2013) called a mobile century, it was also the slippery lubricant that greased the wheels of those workers' evacuations well before governments had decided to act. Having already organized two charter flights, three commercial aircraft, and two ferries to complete their evacuation of personnel, BP planned to continue its offshore activities unscathed.

These first evacuations from Libya really tell a story of historic partners, strategic oil relationships, diplomatic influence, and hardwired networks and infrastructures that were able to coordinate far beyond Libya's borders. State vessels like the HMS *Cumberland* of the British Royal Navy sat waiting outside the port of Benghazi, as would other ships that hopped back and forth to Malta, where the airport became an international hub for the practical coordination of the evacuation effort, a remnant of British colonial power given the island's site as a naval supply station and hospital when it was a crown colony and a crucial base for Allied supply lines during both world wars as well as for casualties who would eventually make their way to Netley hospital—that networked icon of Victorian imperialism. Libya's evacuations would be underpinned by Malta's embeddedness within a "complex set of transcalar interactions," what Nathalie Bernardie-Tahir and Camille Schmoll (2014, 44) have called a "counter-islandness." Malta's eventual coordination of the evacuation can be juxtaposed with its efforts to control the "irregular migration" of sub-Saharan African migrants, especially since its accession to the EU in 2004. Airports were crucial exit doors for evacuation as well as for the influx of aid, governmental, consular, and humanitarian support.

The 2010 Haiti earthquake saw evacuation mobility heavily bifurcated by vehicles, logistics, and infrastructures that additionally bestowed the legitimation of passage. Access to the international aerial mobility afforded by the airport was split. It was secured away from the Haitian population, who were more generally immobilized within the destruction of so much of Port-au-Prince's built infrastructure, yet accessible to the international mobilities of a global elite who had far freer access to the logistical capacity to leave or access resources via aircraft. For Mimi Sheller, "the ease with which foreign humanitarian responders were able to mobilize passports, vaccinations, air tickets, and other network capital to arrive in Port-au-Prince stood in stark contrast to the wide array of insurmountable hurdles that made it nearly impossible for most of the earthquake-affected population to leave" (2013, 194).

Alongside the circuits of international mobilities overbooking Port-au-Prince's landing slots as American citizens were snapped back to the United

States and congresspeople and celebrities carried out overflights and short visits to the country, some Haitians were able to leave. Many children and minors were evacuated and adopted by new parents in other countries, especially the United States and France. Officials from the United States' Department for Homeland Security, Immigration and Customs Enforcement (ICE), and Customs and Border Protection—usually the "petty sovereigns" of border rejection and deportation (Butler 2004)—landed in Haiti on January 23 to facilitate the process of evacuating American citizens and Haitian children who were being considered for adoption to the United States. Agents of ICE actually helped drive children being processed for adoption from the US embassy to the airport for evacuation. Media reports began to tell of an orphan crisis emerging (Larsen 2010; *NBC News* 2020). The Department of State was keen to emphasize that the embassy had fitted out an orphan waiting room with "food, water, mattresses and blankets, bathrooms and a diaper changing station."[1] An email sent from the US Department of State to an aide for Hillary Clinton reported that although the area was restricted to US government officers only, it offered affective respite, as they loved "to visit the orphans and get a hug from the affectionate children."[2]

In Port-au-Prince, the Toussaint Louverture International Airport marked one of the sites of the sharpest differentiation. The airport was a location of key US and international aid and military assistance from the coordination of the United States' Military Southern Command (SOUTHCOM) headquarters in Miami but only once Air Force Special Operations Command had sent a unit to the airport to take control of the site and facilitate the influx of not only aid and sanitation facilities but also weaponry and armor. Given that the control tower was out of order, the airborne wing established its own temporary air traffic control tower to manage aircraft movements and Haiti's airspace. This meant rapidly increasing the number of aircraft movements from an average of thirty per day to over a hundred; aircraft movements were apparently restarted only twenty-eight minutes after the arrival of the US personnel. Another group of US military logistics experts in the form of the 621st Contingency Response Wing deployed soon after and are credited with moving the airport from "rudimentary control and cargo processing to around-the-clock operations of over 60 flights per day," and in the following days to 200 helicopter and 140 airplane flights per day (D. Fraser and Hertzelle 2010).

Alain Joyandet, the minister in charge of France's humanitarian relief, was blunt in his criticism that the United States was performing a military occupation. "This is about helping Haiti, not about occupying Haiti" (quoted in Stephen 2010), while the French foreign minister described the airport as "an annexe of Washington" (Carrol and Nasaw 2010). Attempting a climbdown in the language, a UN spokesperson was reported by the *Wall Street Journal* (2010) praising the US logistical expertise: "In the sphere of logistics, we really have to thank them. Without them, the airport wouldn't work." France's intervention signaled a strange collision of the logistical logic that guided the evacuation but threatened Haiti's sovereignty and a relatively embarrassing admission, that some of the evacuations of Haitian nationals were part of the adoption of children out of Haiti, which had not followed proper processes. The adoptions threatened both the United States' and France's credibility as nonoccupying postcolonial powers. President Sarkozy visited Haiti to announce a package of aid and debt cancellation and suggested that France and Haiti were a postcolonial family of sorts. While they had their problems, they were ready to reconcile:

> Of course, we are at your disposal to take in more students, to help reunite bereaved families here or in France. To take in young orphans, provided of course, Prime Minister, that the legal conditions are fulfilled in Haiti. Provided they are genuinely orphans and not children taken from their families. And provided the conditions for bringing them to France and especially the transition phase have been complied with. . . . [Y]ou can be sure that France is open for Haiti's young orphans, but in agreement with you. You understand that this is very important! The vital objective for the children is for them once again to have a future, to be happy! It's the only one, no others are important.[3]

Beyond this futural logic of evacuation and adoption's promise, in the next sentence Sarkozy concluded, "I have asked the Minister for Overseas France to ensure that all the logistical resources in this Caribbean, which is dear to us and we share, can be made available to Haiti."[4]

Through combinations of aviation logistics and evacuations, for Sheller, an "imperial-aerial gaze" enabled "neo-colonial governance not only through the mapping of territories and the regulation of mobilities across borders, but also through the bio-political management of racialized bodies" (2013, 194). This was about the evacuation of US citizens and the provision of humanitarian welfare and, under that guise, was "a means to prevent the mass movement of refugees off of the island" in order to prevent an influx of migrants to the US mainland by aircraft, boats, or other means (194). In this sense, the logistics system involved a simultaneously mobilizing and immobilizing machine—what Sheller called an unequal "islanding effect." Some movements were mobilized "under the justification that rapid, efficient circulation is necessary to the welfare of the economy, the state, and its people," while others were limited according to a "calculative rationality and a practice of spatial ordering" (Chua et al. 2018, 624). For the Black Kreyol-speaking communities waiting at the airport's gates who found "their access to passports, visas, and ease of passing through a border checkpoint" (Sheller 2013, 193) utterly curtailed, this feeling of islanding was visceral, critical, and confusing, especially as they met military and security barriers.

The military-imperial-aerial perspective continued in the evolving access to Haitian airspace. The US Coast Guard sent several c130 aircraft from Florida on reconnaissance flights over Port-au-Prince to survey the damage. The Federal Aviation Administration and Joint Task Force-Haiti negotiated a memorandum of understanding with the Haitian government to assume temporary control over Haitian airspace. The US Air Force and the Federal Aviation Administration established a Haiti Flight Operations Coordination Center in the 601st Air Operations Center at Tyndall Air Force Base, Florida, to control and manage airspace and the flow of aircraft in and out of the airport, modeled, fittingly, on a process determined for the control of airspace during Hurricane Katrina.[5] The first use of Predator drones provided full-motion aerial viewpoints over Haiti to support the task force. The drones were operated in flight from Creech Air Force Base—the US Air Force's infamous center for drone operators remotely piloting drones in theater (Gregory 2012). The drones were landed and made airborne by local drone operators at a civilian airport in Puerto Rico, Aeropuerto Rafael Hernandez—marking a then-unprecedented convergence of military drone technologies and civilian infrastructure (Broshear 2010; see also Kulczuga 2010).

Evacuations carried out by foreign agents or organizations compete with sovereignty, rendering it "contingent" (Elden and Williams 2009). Under international law, the intervention to evacuate nationals using militaries is justifiable only on the grounds of national self-defense, or the right to protect (Article 51 of the UN Charter) should no effective state exist to permit extraction or if the state is unwilling or unable to. In the Libyan context, little critical analysis of the legality of evacuation seems to have been performed (Grimal and Melling 2011). During the first weeks of protests and armed uprising that would turn into the first Libyan civil war, Britain would play a coordinating role in an international response that came, purportedly, "too late." A joint Non Combatant Evacuation Operation Coordination Cell

FIGURE 7.2. Canadian photography and cartography pictured during the non-combatant evacuation cell in Valetta, Malta (C. Taylor 2011).

(NEOCC) for twelve countries was formed in an office within the British High Commission in Whitehall Mansions on the seafront of the Maltese town Ta' Xbiex, outside Valetta, its capital (see figure 7.2).

The Maltese government set up their own evacuee-processing hub, sorting over eight thousand people from over fifty countries in seven days, who arrived by aircraft or ferry, arranging for the evacuees to eventually be flown elsewhere. Malta's processing of the evacuees shadowed a longer-term system not of evacuation but of detention, deportation, and temporary exclusion. Hal Far was the old British RAF airfield whose hangars and runway infrastructure made for an effective industrial estate. Well before the Libyan crisis, Hal Far had begun to be converted into accommodation to receive asylum seekers and undocumented migrants in detention such as Lyster Barracks—dotted by the British legacy of Nissan huts and accommodation blocks named after British navy aircraft carriers—who were then released from detention into "open" sites. From March to June 2011, over 1,500 "irregular" migrants arrived in Malta. Postdetention, people were placed within a large hangar in the Hal Far Open Center and the Hal Far Tent Village, where prefabricated "container houses"—named after their resemblance to shipping containers—replaced fabric tents. The cramped, unsanitary facilities, so close to the airport, segmented the migrants from wider Maltese society despite the "open" nature of the site. In one sense, these (im)mobilities demonstrate the intersecting, nested scales of movement that perform Malta's role as a buffer island and redistribution and detention center on the periphery of Europe (Bernardie-Tahir and Schmoll 2014), where "care and control are intermingled" (Lemaire 2019, 724). Malta's role in the evacuation seems duplicitous given how it treats different categories of mobile subject, ranging from embrace and facilitation to detention and deportation.

Airports set the scene for dramatic escapes and emotional reunions; from airports, states wielded evacuation aesthetics as a mechanism to gain or regain political capital. On January 20 a flight carrying more than a hundred Haitian children arrived in Eindhoven Air Base. The children were carried by army personnel in camouflage, over which high-visibility tabards, balloons, and blankets adorned their uniforms. The Netherlands' experience was repeated over and over again in what some critics would call "the moment," manufactured in US media coverage of adoption flights to produce emotional moments staged on the apron tarmac of US airports and airfields to unite the so-called orphaned children with adoptee parents. While orphans seemed to be the unlikely and chaotic variable in the equation of American military logistics—"orphans and that kind of stuff that you just have to deal with," commented one US commander (Wallwork 2010, 7)—tensions over adoption were not new to Haiti at all (Bergquist 2009; Selman 2011). Moments of crisis frequently see calls for "heightened protections and lowered barriers," where the evacuation of children for adoption may result in the circumvention of ethical norms and legal processes, as well as confusions over the status of evacuated children as "refugees or orphans" (Bergquist 2009, 625). Intercountry adoptions and evacuations between Haiti and other countries served to perform particular versions of family that underpinned "neoliberal moral logics of reproduction" (Cheney 2014, 256). These reinforced the notion that an adoptee's family was unfit to bear children, while the adopting families are only completed by the addition of a child, thereby reproducing heteronormative conceptions of familial and surrogated reproductive futures. The relation seen in Haiti and elsewhere was based on "inequality," to the extent that the poorest "can and should service the reproductive needs of wealthier people" (256). For Laura Rose Wagner, a general incognizance of "Haitian patterns of life, family, and mobility" and the "fluidity of the non-biological kin networks in which the children

FIGURE 7.3. Protesters demand evacuation/adoption at the French Ministry of Foreign Affairs, 2010. Photo by Boris Horvat (*Libération* 2010).

were situated" (2014, 146) had meant a history of misunderstanding by foreign governments, NGOs, adoption agencies, and willing adoptees of the actual status of Haitian "orphans" (figure 7.3). The adoptees may not have actually been orphans—parent-less or guardian-less—but may have been put into an orphanage temporarily. Many Haitian parents believed that their children were simply being temporarily evacuated before they would be returned to Haiti.

In Libya, evacuating its peoples through commercial and stately partnerships in its new diplomatic framing of "overseas citizen protection," China involved the use of external military logistics, dispatching a Jiangkai-II class frigate, the *Xuzhou*, from the Gulf of Aden—repurposed from counterpiracy activities—to Libya to combine with cruise ships and provide logistical security, redeployed for evacuation and humanitarian protection. The operation evacuated fifteen thousand Chinese citizens, repatriating many of them without passports or IDs. Canada's noncombatant evacuation effort soon morphed into its combat activities within the NATO joint task force in March—both operations were titled Operation Mobile. Canada's military effort involved bombing ground targets and providing maritime surveillance. The events marked the first participation of the Chinese People's Liberation Army (PLA) navy in a noncombatant evacuation operation. The deployment of four PLA Air Force transport aircraft to the south of Libya to extract citizens was also unprecedented.

Chinese state media sold the evacuation as a success, utilizing the frequent and colliding tropes of evacuation as a "massive, orderly and extraordinarily efficient evacuation" along with those of logistical timing, an example of "running the government in the interest of the people" (Zerba 2014, 1094). A US congressional report saw the deployment as "largely symbolic," suggesting the evacuation would improve the "PLA's capability to operate at greater distances from the mainland" and portend a "global military presence" (Department of Defense 2011, 59). In light of other Chinese evacuations, from Nepal and Yemen, other analysts predicted the evacuation as a sign of future power projection. Military long-haul transport aircraft flew continuous flights, over Pakistan, Oman, Saudi Arabia, Sudan, Libya, the Arabian Sea, and the Red Sea. Stopovers for refueling were made in in Khartoum and Karachi at commercial airports, reflecting Sudan's greater importance as a Chinese ally in the region. Others point to China's Communications Construction Company's intention to begin a billion-dollar contract to build a new airport terminal in Khartoum.

The UK government's immediate response had been to send a rapid deployment team (RDT), which reached Tripoli airport on the morning of February 23. A decision had been made not to request a visa to enter Libya but to work airside with the consulate team to process potential passengers. Ironically, they set up outside the VIP terminal formerly used for private jets (in 2017 the new Tripoli airport opened in the mainly undamaged VIP terminal). The embassy provided several embassy vehicles and three buses as a form of refuge amid the "chaos and some danger" (Foreign and Commonwealth Office 2012, 35) for personnel and sheltering British citizens. Officials from the RDT waved the Union Jack flag to attract British citizens and allowed them to push through the crowds despite the deteriorating conditions, which were described as "highly charged and violent" and saw the airport windows smashed, a "stampede through the terminal," and excessive lines even for basic foodstuffs (2012, 35). As Ziadah (2019) emphasizes, the continual valorization of logistical expertise is a particular form of privilege. As with Haiti's airport-like camp of humanitarian, government, and logistical equipment, personnel, and expertise, Tripoli's airport became an "'island of stability' and 'efficiency' in otherwise chaotic neighborhoods" (1689), fortifying international state, humanitarian, and military agencies acting extraterritorially.

It's a big and unpredictable world out there. . . . What's a big corporation to do when staff in its far-flung offices need to get out of Dodge—often while curfews, martial law, road closures, armed rebels and sundry other unforeseen challenges conspire against them?

—ERIKA FRY, "THE ESCAPE BUSINESS"

The extension of evacuation networks along logistics and resource flows and infrastructures, through postcolonial partnerships and alliances, through nodes of ports and islands, and islanded extraterritorial air bases and places of investment, meant for Libya and Haiti that some people could evacuate while others could not. Evacuation mobilities followed the nodes and links of strategic economic and military interests, but they elicited other, more disobedient and civil visualities too (Pezzani and Heller 2013).

For Western audiences, the evacuation flights from Libya were made visible through maps and other visual tropes by the media. The route of China's frigate mentioned above was reconstructed by a blogger. Understood to have been operating at the mouth of the Red Sea deploying helicopters to deter pirates threatening a South Korean ship, the frigate had replenished at Oman and then traveled up the Red Sea and through the Suez Canal before entering the Mediterranean, an apparent sign of port basing agreements. Thought viapolitically, the routes and vehicles of evacuation become an important lens with which to understand geopolitical shifts. Images and videos quickly emerged in the media showing the resource workers making their way to RAF aircraft and arriving in Malta to waiting press photographers in the theater of an airport apron bus. These were similar moments of fairly banal airport processes and performances structured in a more grounded form of airports' "imperial-aerial-gaze" (Sheller 2013; see also Adey 2010). The media reveled in the heroic stories. A smiling RAF pilot was pictured in the press and told tales of a speedy landing in a desert airfield. The story was led by a black-and-white image from Google Maps of the location. Narratives ranged from daring improvisation to Human Rights Watch warning the more active governments that while their own citizens may have benefited from these actions, other countries had been unable or unwilling to act (Human Rights Watch 2011). The oil and gas industry's reliance on cheaper labor from Asian and sub-Saharan countries meant hundreds of thousands of workers, stuck in the besieged Benghazi or at the border with Tunisia, were at risk from persecution, especially given reports of sub-Saharan mercenaries paid by Muammar Gaddafi's regime to quash public protests.

If evacuation and logistics became a way through which power was exerted, they were also particularly vulnerable to visual scrutiny and critique, reminiscent of the public concern for the treatment of the wounded on the ambulance trains of World War I. In Britain the government's response was reliant on existing scheduled flights from Tripoli to take British citizens to safety. These were eventually canceled as landing permits were revoked or not granted. Several planes chartered by the Foreign and Commonwealth Office (FCO) were very slow to materialize. One even broke down. Public opinion determined that the government was not acting; decision-making seemed inept or paralyzed. Stories even emerged that British consulate officials were bribing workers at Tripoli airport to process British passport holders more quickly. Compared to the corporate response, the government seemed slow. The expeditionary nature of Britain's diplomatic teams is perhaps telling of the islanding and fortified nature of consular and humanitarian care. The crisis led to a major restructuring of decision-making in a review of arrangements within the FCO to implement crisis management procedures in the aftermath. This review sought to determine preplanned protocols or guidelines such as charter agreements for aircraft and ships (which the government had decided to put to a tender process in the midst of the crisis), relationships with aviation brokers, funding, and a policy on Commonwealth nationals. The FCO had felt embattled in the midst of other evacuation procedures of what it would call a "wider and unfolding crisis" in the Middle East and North Africa region, requiring evacuations of British citizens from Tunisia and Egypt. In addition, FCO and consular officials were stretched by the Christchurch earthquake.

Canada had directed its own operations in late February to rescue its "Canadian entitled persons" (CEPS), flying Globemaster and Hercules aircraft from Malta into Tripoli. Bloggers made comparison to Britain's more covert entry, suggesting the Canadian public did not have the same political will to aggravate Libya's sovereignty. They also investigated the Canadian military's photography and its diagrams of evacuation in quite interesting ways. One blogger scrutinized an image from the cockpit to decipher that the aircraft was landing at runway 27 of the Tripoli airport, based on their old pilot's navigation map of the airport (C. Taylor 2011). They reoriented and reconstructed a map depicting the location of entitled personnel from a photo of Canada's NEOCC cell in Malta, and based on their knowledge of the flight times, they produced their own diagrammatic evacuation map of the

Canadian operation. This was perhaps a civil and "disobedient" (Pezzani and Heller 2013) set of networks of nonstate expertise, enthusiasm, and aesthetics necessary not only to make evacuation mobilities visible but to question the government's prioritization of certain "entitled persons." The blogger writes ironically, "CEP is the acronym for Canadian Entitled Persons, the 'entitled persons' being individuals who are entitled to ask their government to help them leave a foreign country on the taxpayer dime when a crisis arises. Wait, isn't everybody entitled to have the government spring to their rescue when a crisis arises? Theoretically, yes, but CEPs are those that the government considers it more or less mandatory to evacuate; the evacuation of everyone else happens on a 'best effort' basis. If non-CEPs make it out too, that's great; but if not, oh well" (C. Taylor 2011).

These alternative narrations of the evacuation were important. They formed a kind of rupture through which the principles of the evacuation could be contested, especially the determination of who was deserving. These alternative narratives to the line of nation-states even shifted policy. China was moved to act following a story shared on Weibo, that of Happy Xufeng, a worker describing his experience for the state-owned railway company, questioning why China had not acted sooner. Xufeng and other social media bloggers put a public and global spotlight on the plight of Chinese overseas workers, and especially the Chinese government's (in)action. There was significant evidence of cooperation with other countries in terms of both the requisitioning of Greek ferry ships and the fact that many Chinese were evacuated without passports or IDs. Most were taken to Greece, Turkey, Egypt, Malta, Jordan, and elsewhere. The Chinese Ministry of Foreign Affairs also simplified their entry procedures and organized the internal logistics for moving the workers back to their homes. By March 3, however, reports suggested that of the over fifty thousand Bangladeshi citizens working in Libya, only five hundred had been evacuated by the International Organization for Migration (IOM).

The postdisaster events in Haiti followed a similar path of national favoritism and protection and other forms of civil resistance and counter-mappings. The earthquake revealed an uneven power geometry of multiple circuits of mobility of international aid workers, militaries and diplomatic staff (Sheller 2013), contractors and military personnel, celebrities, and politicians. For the most part, few Haitians were actually able to evacuate from Haiti or its capital, Port-au-Prince. Some would travel across the border to the Dominican Republic or make their way by boat to as far away as Florida.

Sheller notes the obscene inequalities of movement and visuality. This was well demonstrated in SOUTHCOM's operation of what they would call "overflight tours" of the country for visiting officials (perhaps akin to those of New Orleans, mentioned in the previous chapter) and the continued and curious convergence of evacuation and emergency mobilities with tourist visualities (Lisle 2016).

Department of State emails and cables released by WikiLeaks focused on the logistics of aid given by the "whole of government" approach to the disaster but also VIP travel, presumably because of the political celebrity of figures such as the actor Sean Penn, the filmmaker Paul Haggis, the musician Wyclef Jean, and other US political elites (Guha-Sapir et al. 2011). By January 16, when it was increasingly difficult for US citizens and other nationalities to reach the airport—a congested, "chaotic and unsecure environment"—US military teams expanded the privileged zone in order to give more space to the processing of American citizens for evacuation.[6] It was eventually agreed that USAID, SOUTHCOM, and the UN could begin vetting US and other international flights in order to work through the backlog of flight requests, which by January 22 stood at around 1,400 requests for landing slots, ten times the usual number. The previous day, 16,811 American citizens, according to the Department of State cables, had been evacuated through the airport while an estimated 500,000 Haitians lingered on its doorstep. The Haitian Civil Aviation Authority began plans to allow American Airlines flights into the airport on the condition that the US government would support the rebuilding bill to repair the west end of the airport terminal.

Evacuation had been already divested from the state to private industries and business. The International Association of Oil and Gas Producers (IOGP) has its own guidance around country evacuations, as produced by its security committee in September 2012, largely as a response to the experiences of Libya in 2011. The logic of evacuation for these kinds of sectors, which include wealthy businesses and self-employed contractors, is insurantial and dependent on risks secured by insurers. The one flight the British government had chartered actually broke down, and the brokerage firms the government had hired the planes from raised concerns that the security situation would invalidate their insurance. Others in private and professionalized firms relied on expatriate and private evacuation insurance, a form of what Conor O'Reilly calls "elite rescue." As O'Reilly explains of the "guardians of global mobility," "they secure, enhance, and where necessary restore, the (primarily) profit-oriented circulation of their clientele," thereby consolidating the "elevated social—and mobility—status of those to which they are promoted" (2011, 179).

Clements and International SOS—dominant players in the market—provided evacuation capabilities for high-value clients during the Libyan crisis, their talents solicited by the FCO during the review of their procedures and advertised by thankful clients in the aftermath. International SOS claimed it had evacuated 760 clients from Tripoli, Benghazi, and the oil fields between February 23 and 28 via charter flights. These businesses build on logistical networks drawing on former military and security expertise, which they market toward expats, business travelers, and middle-class elites taking international vacations (such as safaris)—including colleges, universities, and schools that encounter "the unpleasant consequences of wandering in the global 'badlands'" (O'Reilly 2011, 183).[7]

The confluence of the Libyan crisis with the Bahrain antigovernment protests (and brutal military and police interventions) and the general instability of the Arab Spring presented opportunities for firms like the Texas-based geostrategic intelligence agency Stratfor. Global Rescue and International SOS were several of the many global firms Stratfor would recommend. Listening to Stratfor's intelligence and analysis could help firms reposition assets and individuals—investments and capital might be evacuated too. The industry emphasis is on the individual and on primary and secondary evacuation routes and plans; they advise you to not "expect your government, or your company to help you for 48 to 72 hrs" but to "make a plan for yourself" (Stratfor 2011a). In a video about "how to travel safely" in July 2011, a Stratfor executive discusses the travails of the global business elite. The video recommends evacuation policies: "They will help you get out of the country, they will medevac you, they'll physically send people to help you," but first they present a series of objects that might be easily stowed within the travel bag of the global executive. This includes a Maglite useful for blinding an attacker at night, or finding one's way down an emergency stairs in an emergency, and even a paracord, "in the unfortunate event, in case you need to attempt to rappel off your balcony" (Stratfor 2011b). For O'Reilly, this is all part of the "divestment" of "emergency disaster assistance" (2011, 191). Evacuation insurance could be compared with kidnapping insurance, where O'Reilly argues much of the industry had to cut their teeth on dealing with elite rescue. Both operate in both a preventative and a reparative manner following the circulatory notions of evacuation we began the book with. The intention seems to be to put the valued subject, who in that moment is threatened not only in terms of their life but also as a productive agent of capital circulation and economic productivity (Lobo-Guerrero 2007), back into circulation through evacuation.

DISOBEDIENT EVACUATIONS

The adoption/evacuation confusions seen in Haiti continued the neoliberal divestment of not only the responsibility of states for the protection and well-being of a population but the reproductive labor and childcaring capacities of parents and families. The convergence of faith-based emergency inter-ventionist leanings and neoliberal imperatives of US social welfare and its foreign aid and development activities (M. Cooper 2015) may explain some of the ease with which religious and faith-based organizations and institu-tions were able to extend their reach into Haiti. In tune with the evacuation's many agonisms, divesting evacuations to the private and civil sphere does not necessarily mean they are done better. Metaphors of childhood adoption were caught between "a savior catching a child falling in midair and bringing him to safety or the darker image of someone's offspring being snatched away from her family and home" (Joyce 2013, xvi–xvii; see the overtones with the myths of the caught babies discussed in chapter 1; see also G. Adams 2010). France's involvement was particularly cautious given the involvement of French charity workers in the Zoe's Ark scandal of adoption-evacuations from Chad in 2007.

The most infamous case of irregular adoption involved US missionaries who were stopped at the Malpasse border with the Dominican Republic with thirty-three children. Members of a Baptist congregation from Idaho, who had formed an American-led charity, were arrested and later charged by the Haitian government with child abduction. In the aftermath of the event, the Hague Conference produced advisory notes to governments warning that the principles of the Human Rights Convention on the Rights of the Child were being flouted. Haiti's children's "best interests" should be given priority. They complained that "primary concern should be the safety of these children and some consideration is being given to the possible need for evacuation," but, they went on, "evacuation should not be confused with intercountry adoption which is a more radical measure changing the parenthood of the child" (Hague Conference on International Private Law 2010, 1–2). From evacuation, to adoption, to abduction, once more evacuation is confused as a term and practice. The legal and political vacuum meant processes of adoption could be easily circumvented through the expediency and urgency of evacuation. Evacuation, by being confused with adoption, whether deliberately or not, could lead to the other, both a means and a motive for adoption following the teleological assumptions we have seen elsewhere. Evacuation becomes a sort of slippery slope, or as we saw in chapter 5, it has a kind of momentum. The desire to evacuate a child urgently could short-circuit the lengthier decision to enable adoption and result in more "unsafe" adoptions (Hague Conference on International Private Law 2010, 2).

In Libya many third-country nationals who had escaped to the countries bordering Libya, such as Egypt and Tunisia, would not be repatriated for some time, residing in camps on the Libyan-Tunisian border. Other nationals were evacuated much more slowly, some only with the support of the IOM. Those who came aboard the Greek passenger ferry *Ionian Spirit* in one journey carrying nine hundred were evacuated by NGOs in the form of evacuation ferries. The *Spirit* was chartered by the IOM to evacuate African workers, Filipino educators, and wounded Libyans caught up in the fighting from the besieged city of Misrata to the rebel stronghold of Benghazi in a twenty-hour journey by sea (Chivers 2011). We can see an indistinct line between the logic and logistics of evacuation—the conduits and legal mechanisms for moving people across borders—with paralleling logics and mechanisms that produce lines, that reinforce borders. Sometimes the same actors with the same machinery are involved. While some states appeared to preserve almost a monopoly over evacuation, ready to flout territorial integrity, international law, and even their own border and immigration practices, there are other less state-centric, civil evacuations. Of course, the long aftermath of the unrest has seen even more people seeking to travel to Europe. Civilian evacuations alongside NGO-performed emergency rescues can short-circuit state-centric deterrence and the pushback and closure of borders that have led to hardship and death at sea.

The ferries and other humanitarian vessels that evacuated the vulnerable and wounded within Libya demonstrate not only that evacuations are possible but that the agencies of civil and public collectives and solidarity movements may provide one model in advance of more permanent and humane ways to support the displaced. In 2015 the migration network WatchTheMed called for evacuation ferries that would "consider the freedom of movement not as a distant utopia but as a practice," a ferry that "should travel to Libya and evacuate as many people as possible." Evacuation would be the first step, and a pathway to be "granted unconditional protection in Europe" (AlarmPhone 2015). As Charles Heller has commented, "Merrily envisioning this possibility denaturalises the daily images of migrants cramped in

FIGURE 7.4. The empty plane preparing to evacuate two hundred migrants from the Moria camp in Lesbos, 2020. Photo by Georg Gassauer (Twitter, "On an empty plane heading to #lesbos with @WeGaanZeHalen will arrive in 3 hours . . .," October 5, 2020, https://twitter.com/GGassauer/status /1313013995515924481).

small rubber boats, and presents us with what might be one of the concrete manifestations of the institution of freedom of movement: the banal movement of people on board a ferry, such as those that have long connected the shores of the Mediterranean" (2020, 15).

In October 2020 the Dutch organization We Gaan Ze Halen, which began in 2016 to call for the Dutch government to relocate refugees from detention camps in Greece and Italy, tried to land a plane in Greece on the island of Lesbos with the hope of evacuating what they called a "decent proportion" of the Moria camp, where refugees had been placed and confined in intolerable conditions (figure 7.4).

Once they had hired a Boeing 737, the NGO wrote to the Dutch state secretary with their plans and asked for permission "for the evacuation of the first 189 refugees from the island of Lesbos to the Netherlands" (We Gaan Ze Halen 2020). The vehicle of evacuation and solidarity was an aircraft in this instance, perhaps not such a banal form of mobility, even if the 737 has been the workhorse of the low-cost airline industry. The plane was redirected by the Greek authorities thirty minutes before it was to land in Mytilini Airport and sent to Athens for political or "security" reasons (van der Kolk 2020). For the campaigners the aircraft's image was important; "the picture of the plane was too powerful" for the Greek authorities "to ignore," even if that meant the denial of their project through the sovereign yet technocratic withdrawal of their permission to land on the island (Forschungsgesellschaft Flucht und

Migration 2020). In another way, the possibility of an evacuation by air was the only proper response to the urgency of the situation for those enduring the camp conditions and a recent fire that had displaced over twelve thousand migrants to tents, parked ferries, and other temporary homes.

CONCLUSION

The logistical and vehicular processes we have explored play a crucial if divisive role in the evacuations of humanitarian disasters. This is not the way most people evacuate or flee disasters of this scale. In the viapolitics of evacuation, certain logistics and modalities of mobility seem to be essential to what we think of and designate as an evacuation, while those that fall outside of it appear unruly and unlicensed. Access to these privileged evacuations depended greatly on the actions of states, the right documents and paperwork, and of course the markers of citizenship, such as passports. Vehicles and logistics are important ways that evacuation is seen or not seen—evacuation is logistical—while logistics and vehicles shape how evacuations are given meaning and interpreted, often across blunt categorizations and divisions of permissible or legal, formal or informal, regular or irregular, legal or illegal.

What is seen of evacuation and how it is seen is a crucial geopolitical project. Sovereign legitimacy is given and questioned in the act of evacuation. How well states and governments do at intervening to evacuate their own populations, and others, becomes a crucial test of a government's capability to protect, a yardstick of its performance and prowess at statecraft, logistics, and care, just as we saw in chapter 2. In this sense, evacuations become a viapolitical lens through which (im)mobilities, and the way they are controlled, governed, and lived, are contested. And states and militaries do not so simply control the way evacuations are seen or made (in)visible. We saw instances where the logistics and military infrastructures of evacuation were articulated by NGOs, bloggers, and nonstate actors drawing on similar aesthetic practices to visualize and investigate evacuation (im)mobilities and contrast state messages more *disobediently*.

We saw alternative evacuations in collective acts, in which organized civil evacuations struggle against states. To draw on Bonnie Honig (2009; see also Anderson 2017), these seem crucial for a more democratic politics of emergency, where emergency and its claims and conditions are divested from the hands of the state and sovereign power. For We Gaan Ze Halen, WatchTheMed, and other advocates of open borders, evacuation is a far more effective and perhaps first and urgent step to humane treatment, relocation, or resettlement in emergency. Of course, this is highly agonistic. The urgencies that evacuation and emergency politics bring can cut a route through the slow, deliberative, and deliberately obstructive policies that deter and prevent the relocation of migrants to EU countries. This was precisely the critique leveled at the adoptions from Haiti. Social media commentary on We Gaan Ze Halen's expedition compared the event to a form of human trafficking, suggesting that organizers of the wasted flight were putatively unaware of the need to limit aircraft movements in the context of climate change.

BURN

In Melbourne Museum stands a tall, quite ornate, corbeled redbrick fireplace—replete with age, cracks, and worn edges on the brick faces (figure 8.1). Surrounding it are many tall mountain ash and other eucalypts tarred with black or burned encrusted skin from a forest fire—a bushfire. As Tom Bristow and Andrea Witcomb have suggested, the space offers a kind of "affective lens of loss, destruction and displacement" (2016, 73)—a melancholia. The exhibit is a memorial of the 2009 Black Saturday bushfires, which were widely touted as the worst natural disaster in Australia's modern history and now a harbinger of the terrible bushfires and wildfires that have devastated the Amazon, Greece, California, and Australia's south and east coasts since. The relic has been evacuated to safety—in a manner—preserved in the museum's vast open space. Bristow and Witcomb see in the exhibit an "an emergent museological amalgam," where within the densely woven layers of bark and brickwork are "compressed relations between indigenous knowledge, settler practice and different forms and frequencies of clearing, regrowth and regeneration" (2016, 72–73).

The exhibit is also an interesting symbol of the movement of materials and energies. The chimney has exhausted the fumes and carbon of the eucalypts, coal, and perhaps other waste up into the air from the inside of its volume. The eucalypts nearby respire carbon back into their taller bodies, root systems, and mycelial networks into the soil. The gallery reminds us that the

FIGURE 8.1. The "Uplands" Kinglake chimney in Melbourne Museum's Forest Gallery. Photo by the author, 2017.

colonial and postcolonial appropriation of Victoria's forests upset a delicate ecological balance. The chimney was taken from a property destroyed by fire known as "Uplands," on the Whittlesea-Kinglake Road in Kinglake—where one of the Black Saturday fires centered. A nineteenth-century homestead and social gathering place for the community, Uplands sat within a bucolic atmosphere, positioned within the bush with beautiful views over the Yarra Valley region in Victoria. The chimney had been faithfully taken apart—each brick numbered—and relaid. The intent was to make it appear as if it had been lifted directly from its original site. Chimneys such as these are commonly seen in the Victorian landscape as the few remnants of homes that tend to survive bushfires. Peter Read's evocative book on the meaning of "lost places" recalls a woman screaming at a fire crew about to demolish a chimney—the last standing relic of her home, which had been destroyed in a bushfire in 1983 in Gisborne in the Macedon Ranges—"a site of collective family memories" and a reminder of trauma (1996, 107). In another strange but performative remnant, in the aftermath of Black Saturday, the attorney general noted in the Victorian Parliament that while houses were lost, the children were back at school playing, "evacuating in safety" in their "lunchtime games."[1]

On the evening of Black Saturday, a wedding party celebrated in De Bortoli, an exclusive vineyard in the Yarra Valley (Langmaid 2009). An image captures an odd moment as the wedding couple pose, floodlit in the parking lot of the vineyard, seemingly indifferent to the fires blazing behind them.[2] Framed not by nature but by automobility and its infrastructures, parking lots, signs, the glare of a floodlit mobility space—society's apparent escape routes—they smile while the Kinglake ranges burn in the background. We presume they were unaware of the death going on behind them, the photo probably a fleeting moment of irony. De Bortoli was saved from the fires, although some of its vines and other parts of the Yarra were not. The snapshot seems to cement an apartness from the ravages of fire. (Auto)Mobility will protect us.

In several phases between 2017, 2019, and 2022–23, I lived with my family in Kinglake. It was a strange, humbling, dissonant experience. The town is full of reminders of fire. Blackened bark darkens the trunks of many of the trees, including those in our garden. The owners, following the destruction of their own home in the fires, built the house according to a vernacular of low-profile construction on stilts. It has a large timber deck, a galvanized corrugated roof, and vast concrete water tanks. Some charred stumps remain in the garden. In the town center, plaques denote recovery monies and new investment. Many new houses followed, some benefiting from two class-action suits that found the utility company SP AusNet liable for the fires because of a sparking power cable. Streets like Pine Ridge Road were widely reported as being wiped out entirely. Some of the emergency responders who acted bravely during the fires were one and the same individuals recognized for their bravery at the ceremony that Julia Gillard attended at the Lobby restaurant—and was clumsily evacuated from—a few years later in the events we opened the book with.

Some of the people I spoke to, even just in passing, did not really know what memories and feelings they held, or what would surface when caused to remember them. I will never forget talking to a resident from a nearby community with my wife. While looking at the burned trees bordering his property, he began telling us about how, after saving his own home, he drove up to Kinglake to provide the emergency services with some lunch and saw many cars lining the roads. He stopped his narration and couldn't continue. He was choked with emotion. After a few long seconds, he explained that the cars he had driven past, he came to realize, were full of people asphyxiated by the fire's fumes. His initial narration of the event couldn't fathom the depth of feeling his retelling would cause to surface. He was surprised that automobility, evacuation, and fire could be so affecting. In one way, this was a strong cautionary warning of the dangers of clumsily talking to people. In another, it reminded me that an evacuation and the emergency around it, while sounding like an abstract, inhuman process, is never just that. The car, so important as an evacuation machine in this book, was seen in this instance as a deadly modality.

Kinglake is a horrifying example of the tensions around evacuation, but the fire was not the first to befall the Victorian ranges surrounding Melbourne, nor was it the only time when evacuation (im)mobilities have been controversial. They are both rooted in the history and ecology of the region's experience of fire, colonialism, and mobility, as we saw in William Strutt's vast canvas hanging in the State Library Victoria in chapter 5. Framed most directly within the complex (post)colonial and Indigenous politics of fire ecologies, the legacies of colonial transformations of space and nature, those considered as part of nature, and gendered policies and practices of evacuation governance, the chapter draws some of the book's wider concerns into the bushfires, walking or perhaps driving back in time from the wedding scene and forward. The chapter centers on the scene, wider historical context, and longer-term consequences of the 2009 Black Saturday fires, and especially the town where the worst of the disaster was located: Kinglake. It does this to explore the conditions and consequences of the Black Saturday fires and the wider context of bushfire evacuations, which cannot be separated from the recursions of colonial forms of mobility, power, and the governance of life, which we have been tracing so far. Bushfire evacuations in Australia have emerged at the interstices of not only climate change but expulsive colonial land use practices of appropriation, forestry management, and fire policies, which have increased the danger and frequency of bushfires. Meanwhile, ideological and gendered values of masculinity, family, and individualism have shaped Australian notions of home and (auto)mobility in ways common to the liberal imaginations of evacuation mobilities seen in the United States.

First, the chapter explores the nexus of bushfires with Australia's settler colonial practices of forestry, landscape transformation, and automobility, along with the history of Indigenous displacement and "preservation" in Victoria, and the way the displacements of Victoria's First Nations peoples into reserves mirrored strange recursions in the emerging evacuation practices of emergency escape. It moves on to examine the evolution of evacuation policies and practices in Victoria following the Black Saturday fires and the dubious assumptions about the role of the car and automobilities in evacuation common sense. Evacuation from bushfires in Australia is shown to be particularly contentious, especially with regard to gendered notions of the home and its defense. I turn to my own experiences of living with the threat of bushfires and being prepared to evacuate our home in Kinglake in relation to an array of emergency guidance, diagrams, and apps, before the chapter concludes with the more successful evacuation of nonhuman animals during the same fires, which threatened to overwhelm an animal sanctuary. That sanctuary, in Healesville, had grown close to, and on the former lands of, the Coranderrk Aboriginal Reserve.

BUSHFIRE ECOLOGIES

The writer Sophie Cunningham's (2012) paean to Melbourne begins somewhere near the 2009 fires and another wedding. She describes the bizarre scene of attending a wedding in Melbourne's Zoo, at Royal Park. The lions are "lying on their backs, paws in the air, sprinklers cooling them." Thirty miles away, at Healesville, and not far from De Bortoli, the situation was very different. You can drive to the Healesville Animal Sanctuary from Kinglake in about thirty minutes, down through a twisting road of ferns and tall mountain ash on the Myers Creek road. You can see the powerlines now suspended by flexible cables in the air along the road and across the valley. As a site of refuge, study, and tourism for the State of Victoria's rare animal species, the sanctuary began its development in 1924, founded on the edge of the old rush mining town of Healesville. The sanctuary holds rare species and injured animals brought in from the bush. During the Black Saturday fires, the sanctuary decided to evacuate many of its animals and staff from the sanctuary site. Some of the animal keepers and administrators drove in convoy down the Melba Highway to Melbourne Zoo, which had agreed to temporarily look after the animals. Animal sanctuaries like Healesville's and public zoos are subject to the logics and technologies of emergency and contingency planning.

There are overlapping recursions of the histories and contexts of evacuation going on in the Black Saturday fires. I visited the sanctuary on many occasions, often with my family, who during one visit listened to an Aboriginal Kulin elder who was speaking at the sanctuary. He explained that some of the property actually sat on Aboriginal lands, which formed the original Coranderrk Aboriginal Station and Reserve after the purchase of seventy-eight acres by Colin Mackenzie and later tussles between the state and commonwealth governments. It is easy to forget a longer history of displacement amid other governmental logics and practices of protection and preservation.

The Coranderrk Aboriginal Station was a protected space for the Kulin Aboriginal clans, who had been driven and coerced from their ancestral territories in other parts of Victoria by European settlers into a reserve system under the Victorian Aboriginal Protection Board. The Victorian system was seen as a model for solving the "Aboriginal problem," becoming a "laboratory of colonial governance" (Boucher and Russell 2015, 4). Later came the Board for the Protection of the Aborigines in 1869, a paternalistic, colonial institution, managed according to biopolitical principles of regulating many aspects of Indigenous life, whose policies shifted from racial segregation to absorption. The Half-Castes Act of 1886, which required Aboriginal people with European ancestry to leave reserves, was pronounced at the same time as the Victorian public began questioning the validity of the reserve system, with "a growing sense that the reserves were no longer fulfilling their original function: to provide a sanctuary where the 'Aboriginal race' might live out its final days in peace" (Nanni and James 2013, 183). The Victorian elite was determined to disperse the reserve—apparently so that they might pursue some kind of wilding of the bush with European hunting species—which was initially met with powerful Aboriginal advocacy and a royal commission inquiry. Others saw the dividing up of Coranderrk as a more proper use of its potential land value for settlement. A deputation from Healesville, including a representative from the sanctuary, lobbied the minister of land to subdivide the area into blocks of land for people to settle in, especially for returning soldiers from World War I (*Healesville and Yarra Glen Guardian* 1927, 2). Amid these pressures, as the reserve lands began to be parceled off, the Colin Mackenzie Zoological Sanctuary opened as a nationally supported anatomical research center for animals, which came with continual support from biological and scientific societies lobbying the state government. The sanctuary could be seen as another kind of refuge from the settler-development forces that were threatening Australia's "continent-museum," a "natural history museum at large" (*Daily Mail Brisbane* 1921, 6).

It is obviously hugely problematic to see the animal sanctuary as some sort of repeat of an intervention performed on peoples considered less than human. Yet we can notice what Tom Griffiths has called "sinister continuities," or tendencies and consistencies across "these two reserves for indigenous life" (2001, 123). Both reserves were important and essential sites for the economies of tourism through mobilities and practices of looking and seeing that fed the post–gold rush fabric of Healesville and the Yarra Valley. Train journeys and motorcar tours for metropolitan Victorians and colonial travelers seeking the Australian bush were very popular, peaking in 1924 just as the reserve was being dismantled and the animal sanctuary was taking form. As Griffiths suggests, "Both enclosed 'remarkable survivals'" (2001, 123). Racialized and eugenicist notions of species-level extinction had been common lenses through which to make sense of European settlement but also the "preservation" of "full-blooded" Aboriginal people (Stephens 2003, 161), who were cast as an endangered species, what Marguerita Stephens calls "atavistic primitives necessarily bound for imminent extinction" (2003, 129). The boards saw the reserves as spaces for Aboriginal communities to become civilized, "rehabilitated" as respectable economic citizens in order to achieve the nonreproduction of Indigenous life and lifestyles in Australia (Stephens 2003). Some recall an "outing" to Coranderrk as a required leisure pursuit, as if Aboriginals were wildlife to be appreciated, even feared. People came to "look at the Aborigines" (I. Clark 2015), to look around the "black station." Artists recorded visiting Coranderrk as a diminishing resource. In the local press it was a "remnant of the original inhabitants of Victoria" to be "record[ed] with brush and pencil, pictures of life and scenery of this home (since closed), of some of Victoria's last-disappearing native population" (I. Clark 2015, 317), Coranderrk was perhaps the most photographed Aboriginal space in Australia during the nineteenth century (Braun 2008).

Fire-making was another curiosity (I. Clark 2015). Tourists marveled at the speed and dexterity with which the Coranderrk people were able to ignite a fire from only two pieces of wood. Tensions surfaced from the new Colin Mackenzie Animal Sanctuary, "jealous of the adjoining Aboriginal land," who had "complained that the remnant bush of Coranderrk represented a fire threat" (Griffiths 2001, 60). Petitions from the Sanctuary Committee argued that the whole Coranderrk Reserve should be turned into a zoological park (*Healesville and Yarra Glen Guardian* 1936). The closure of the reserve provided an event for the tourist board, who sought to take photos of the last "full-blooded" Aboriginals. Where the preservation system was perceived as a way to almost evacuate Victoria's Aborigines before they died out or were assimilated by the camp itself, the animal sanctuary afforded a similar kind of protection to sick or endangered animals.

In 1926 a set of bushfires in Victoria culminated in the Black Sunday bushfire of February 14. The press lionized the ability of the heroic settler to face down the fires and pick up the pieces: "In all the epics of colonisation there is, surely, no braver fighter than the settler, who, returning to the spot where his shanty, sheds, fences and wagons is in a heap of ashes amidst smoking-forest fires: squares his shoulders, and sets out to win back everything that fire has taken from him" (*Northern Herald* 1926, 15).

Such appreciation of settlers, the pioneers of the mining and log-felling industries of the Victorian uplands, digs into deep and felt narratives of national identity in Australia. These feelings draw on self-reliance, where evacuation has a tendency to be seen as a weakness when compared with narratives of ruggedness, masculinity, wildness, and danger, which still persist today. Local to Warburton—a mining and railway junction town hugely important to Victoria's logging industry—a song has been written about a team of loggers who buried themselves underground to escape a bushfire rather than flee it. In the morning they were seen walking back down the mountain. Stories like this lie at the heart of the accepted wisdom around fire evacuations in Victoria and wider Australia. As Lauren Rickards, Tim Neale, and Matt Kearns suggest, terms like *bounceback*, *dismissal*, and *endurance* are common-enough articulations of a settler-nationalist imagination to withstand natural emergencies and "'tame' the environment" (2017, 4).

Of the earlier fires to strike Victoria's forests, images adorning Melbourne's newspapers the *Herald* and the *Argus* show men battling the fires, using water guns, spades, and brushes to dampen the flames (see figure 8.2). Their heads are down or facing the fire, the reports read, to "renew their attack" on the blaze. Their bodies strain in exertion. Frequently, the proclivities of the newspaper's photographers were to find cars, usually in the foreground, with burning and smoking or fallen trees behind in the background. Different masculinities prevail. The enduring, surviving, muscular man is immersed in the conflagration. In another, the technocrat, from the position of the road, or the reader of the newspaper, examines the automobile—the windows having been blown out, its body twisted by fire or halted by an enormous bough or trunk crossing the road (see figure 8.3). This wrecked but still-standing car is a common symbol of the technological mastery and common tropes in Australia's settler colonialism, even if it can occlude Indigenous automobilities (Clarsen 2017). Women elude these scenes too, pictured elsewhere with children and the elderly, certainly outside of the car

FIGURE 8.2. Leaning into the fire. "Bush Fires Sweep the State," *Argus*, January 9, 1939.

FIGURE 8.3. A burned car in Toolangi. Press cuttings, January 10, 1939, Public Records Office, Victoria.

and often with teacups. Through these scenes, evacuation takes on other associations with war and the gendered divisions of labor, domesticity, and the home front (Grayzel 2012; S. Rose 2003). Evacuation is galvanized by women and family, who seem shorn of the heroism of men, who are determined to fight and for whom evacuation, or worse inaction, is a form of defeat.

Even the help given by the state to those burned out of their homes and mills, which could do damage to the self-reliance constructed here, draws on these tropes. In one cartoon the State of Victoria takes the form of a tall bureaucrat or politician placing a reassuring and fatherly hand on the shoulders of a young pioneer, probably a timber worker, whose house is disappearing into the conflagration (figure 8.4). A sole chimney stack and hearth remain, buffeted by the blaze. Intriguingly, a new policy of "Leave and Live," popularized in Victoria in 2014, five years after the Black Saturday inquiry, and discussed more below, featured very similar imagery (figure 8.5). In this instance the emergency management commissioner for the Country Fire Association stands in front of a near-identical chimney, sticking out from a twisted tin roof, and explains, "You are gambling with your family's lives if you don't leave early" (Victoria State Government 2014).

Judge Leonard E. B. Stretton, who oversaw a royal commission set up to investigate the 1939 fires, famously concluded in his detailed and evocative report that the fires were a product of no one single source. "It will appear that no one cause may properly be said to have been the sole cause"; however, he went on to claim, the fires were "lit by the hand of man." In the numeration of eleven main causes, and several subcauses in each, the judge argued that the main cause "which comprises all others, is the indifference with which forest fires, as a menace to the interests of us all, have been regarded. They have been considered to be matters of individual interest, for treatment by individuals" (Stretton 1939, 10). A report in a local newspaper described the fires in a slightly different way, seeing that the death brought to Victoria's green places came from the desolation of drought and dry hot winds from the inland which was "only as dead as the white man is making it" (Oriolus 1939, 2). Fire, in these interesting criticisms of the individualism usually associated within settler colonialism, is a commonly self-inflicted narrative.

Kinglake's settlement began during the gold rush of the 1860s. Extractive industries dominated Victoria as the timber-felling and wood-sawing industry grew to supply the gold rush with the required lathes for the tunnel excavations, timber for the homes, and bark for the roofs, which all gradually gave way to more industrial scales of tree felling and rough-sawn timber production (Vines 1985). At one point the Victorian forestry industry was

FIGURES 8.4 AND 8.5. "Help for Bush Fire Sufferers" (*The Age*, January 6, 1939) while the commissioner for the Country Fire Association explains how to "Leave and Live" (Victoria State Government 2014).

producing and shipping more timber than anywhere in the world. The increasing regularity of bushfires was a special product of not only those industries but the depletion of Victoria's indigenous forest made up of mountain ash, giving way to the common eucalyptus along with the displacement of Indigenous people (and knowledges), who had managed to live with both forests and fire for millennia. Judge Stretton's report echoed a longer history of criticism of Victoria's forestry practices. Timber was often wasted or left to rot. Sawmills left flammable material out, ready for ignition. And the convoluted management and administration of the forests had passed from department to department. Protection and conservation were extremely difficult. As the historical geographer J. M. Powell puts it so delightfully, "In colonial Australia recognizable *forestry*—I mean as a scientific and technical field—was . . . where lonely and frustrated inhabitants basked from time to time in the imperial vision" (1991, 37). Disparaging reports had been written by both Victoria's and the British Commonwealth's forest managers. They conflated "protection" with the "rational scientific management" that James C. Scott (1999) has elucidated within modernist state techniques to turn forests into productive spaces. In 1926 Owen Jones, the former chairman of Forestry Commission Melbourne, wrote from his new position in New Zealand, comparing the unruliness of the timber industry with the nonhuman animal inhabitants who populated the same space. "It is difficult to see how the timber workers and their horses, who can be kept under control, are a greater source of potential pollution than the wallabies, wombats and possums, which roam at their own sweet will. It is still more difficult to understand how a purer water supply is to be obtained from an area infested with partially burnt and decaying timber" (quoted in O. Jones 1926, 93).

Jones quotes heavily from an 1890 report by a forestry guru, Berthold Ribbentrop, from the famous Prussian school, innovators of the rational scientific management of timber cultivation and production (Scott 1999). Ribbentrop was then the inspector general of Forests India. His report followed an equally scathing report by Friederich Vincent, another German making his way in shaping India's forestry who was lured to Melbourne. The report would ripple through the empire, accusing Victoria's forestry of useless "waste and destruction" of such a scale and manner that "def[ies] all description" (Vincent, quoted in O. Jones 1926, 91; see also Beattie 2011; Barton 2000). The experts aestheticized the forests as an unmanaged chaos, a site of wastage, pollution, and decay, open to unruly mobilities and flows of effluent. Jones is equally critical of the ornamental planting of more temperate trees along irregular lines, in a "jumble," and therefore of less commercial or productive value. He seeks the revaluation of the forests and their utility in storing rainfall; regulating natural forces, such as the speed of streams; buffering "hot desiccating winds"; and binding the soil together to prevent landslips (O. Jones 1926, 95). Following the 1939 fires, a forest school was set up for the training of nonprofessional forestry personnel by the Forestry Commission (Ferguson 1947), located in Kinglake West in a camp built during the war to house prisoners of war. The curriculum was large, from plant botany to forest legislation, from plantation management to economics and fire protection—which would help in "cultivating self-confidence" (Ferguson 1947, 55).

By the mid-1990s, ecologists, historians, and geographers were still highlighting fire management problems, but now they dwelled on the effect of colonial forest practices. Fire vulnerability was increasing from the reduction of biodiversity and an increase in the fuel load. Controlled burns, a millennia-old Aboriginal practice, were difficult to achieve because of the increased private ownership of Victoria's ranges. Some have advocated for a less scientific and validatory approach to understanding Indigenous burning practices. According to Tim Neale in partnership with Aboriginal Dja Dja Wurrung practitioners of bushfire management, bushfire managers "have rarely sought to engage with actual living Aboriginal peoples about this matter" (Neale et al. 2019, 345). And yet they suggest that there is now evidence of Aboriginal fire knowledges and experiences beginning to be interwoven with contemporary understandings of fire and forest management practices (Eriksen and Hankins 2014; Neale et al. 2019; Weir, Sutton, and Catt 2020), even if in the past these have "borrowed, adapted or stolen fire practices" or been applied "just to reduce 'fuel' (or, dry organic matter) and mitigate the probabilities and consequences of bushfire impacts on property" (Neale et al. 2019, 345).

One of the most obvious ways that Indigenous and modern bushfire and forest management practices have differed has been over an ethics of control. The tendency and desire of settler bushfire practices have seen fire as a threat that needs mastering through prediction and management, echoing the scientific management approach. As Neale explores, the reality contrasts "dramatically" with ideals of "ever-increasing control and capacity" (2018a, 483). The difficulty is additionally one of narrative. As Peg Fraser explains of her own retelling of the Black Saturday bushfires, her approach avoids a straightforward narrative that might continue an "illusion of order and control." To somehow make "an event comprehensible," she concludes, "we add chronology, linearity and distance," an illusion compared to what she sees instead as a far more "messy history" (2019, 264). This seems common to writing about fire's elemental form, akin to the monstrousness described only a few chapters ago. But if a wildfire is "feral, 'untamable, predatory, beastly,'" there are problems with trying to tame it, and with the ways by which we understand and narrate those practices (Chloe Hooper 2018, 237). Similarly, as Neale puts it so thoughtfully, there is a tendency to try, even in approaches that appear sympathetic to Aboriginal practices, knowledges, and sovereignties, to stabilize an ideal past, to find continuities with a possible future, and to turn "an aleatory future prey to the obscure tempos of an inhuman climate and lively materiality of emergent ecologies" into a "technical problem of governmentality" (Neale 2018b, 86).

"FAMILIES WERE BLOWN UP": STAYING OR GOING AND EVACUATION POLICY

The 2009 fires were preceded by the Victorian premier John Brumby's advice for families to stay at home, delivered on the radio on February 6 at 6:10 p.m.: "If you don't need to travel, don't travel. Don't go on the roads. . . . If you can stay at home, stay at home" (quoted in Lauder 2009). Brumby's statement echoed the mantra of emergency advice, that the home was the safest place to withstand an emergency.[3] In the fallout of the Black Saturday fires, there has been a long but fraught debate over language. The royal commission set up to pore over evidence and testimony in their investigation of the fires could not help but draw on the reservoir of lexicon, cultural feeling, and practice that has been integral to fire advice and policy in Australia. One piece of testimony to the inquiry came from a citizen of Steels Creek, a small community southeast of Kinglake. The resident had survived the Black Saturday bushfire with their family. Five other families decided to stay and defend nearby properties, and only three of those families survived. The resident was asked to discuss his statement. He expressed how notions of home, responsibility, and resilience, which some authors have characterized as inherently masculine ideals of Australian manhood, have imbued understandings of bushfire preparedness: "I believe strongly I have a right, as long as I'm not being stupid, to stay there and defend my property. As long as I'm able to do it, I'll do it. I'm not quoting you out of movies, but your home is your castle. There is too much in it. I can't go there to be forced out. A cop came down to me the next day and he said something about, 'Maybe you should have been evacuated.' I said, 'You'd want a bigger gun than that, mate.' Simple as that."[4]

As author Chloe Hooper (2018) explores in the context of the 2009 Black Saturday fires, it is even common for bushfires to be the product of arson.

One man was convicted of starting a fire near Churchill, in the Latrobe Valley, not far from the now-demolished eight giant chimney stacks of the Hazelwood coal power station. Hooper's book makes a nuanced point about the ways fire and communities are nested within webs of social, economic, political, and global environmental entanglements that produced the conditions for the blaze and the blame and ostracism of a man and his family. A remnant of Victoria's coal industry, the Hazelwood chimneys stand as another "brutalist monument" to fire, forests, and society (237).

Testimony such as the Steels Creek resident's continues the militaristic discourse around fighting fire as a form of conflict. Defending one's home, from the state or the police and from a bushfire, relies on whether the home or property itself was in a "defendable position." Evacuation, one forced on them, would be the epitome of undermining the community's resilience to bushfires and might even make them complacent: "They would work on the basis that, 'Well, let's not worry about it too much. We're not going to be here anyway. We'll be evacuated. We'll be gone.' . . . So, you know, for want of a better word, creating a nanny state where you do everything for them and you make it all easy saying, 'Look, don't worry; you will get your house back.'"[5]

Living in the bush took responsibility, where partial evacuation was perhaps a more permanent state of affairs. After the main fire had come through and they had survived the ordeal, the resident sent his children to Melbourne to stay with their extended family, while he stayed at home to put out the smoldering material on his property. He admitted that they would routinely pass the family's photo albums, jewelry, and other valuables to family in Melbourne each year, just in case their house should be taken by bushfire. Their precious objects were evacuated before the family were, without all the fuss of the frequent and routine evacuations the guidance might suggest.

The language around fire, as Tom Griffiths (2012, 2) has discussed so cogently, is a common celebration of holding "'steadfast' and facing down the fire, in other words, not moving, not retreating." For Griffiths, the watchwords are "staying and fighting," certainly not "leaving early" or evacuating in cowardice. It is a rehearsal of ANZAC (Australia and New Zealand Army Corps) courage, the fire a front, the Country Fire Association (CFA) an "army of volunteers" (Cox 1998, 132), the discourse highly militaristic. "Courage," Prime Minister Kevin Rudd declared at a memorial service after Black Saturday, "is a firefighter standing before the gates of hell unflinching, unyielding with eyes of steel saying, 'Here I stand, I can do no other'" (quoted in Griffiths 2012, 2). Peg Fraser, a curator at the Victoria Collection in the Melbourne Museum, examined the fires within an oral history project in

Strathewen, in which twenty-seven people were also killed in Black Saturday. Picking up on these themes, and noting the masculine characterization of the settler, she wrote of the tendency to "celebrate drama and heroics" but, she complained, "not the quiet evacuations that also save lives" (2018, 231).

The testimony we are discussing bears on evacuation very negatively, for lots of the reasons discussed in this book already. The royal commission's interim report discussed some of these issues. It found that the policy of "stay and defend" was the common mantra. If one was to leave, then they should "leave early," and this was accepted knowledge. Deaths had happened because of "late" relocation. Such a basis would support the "extreme reluctance on the part of the CFA to recommend evacuation of an area threatened by fire" (Royal Commission 2009, 174). Automobility was a related issue. Existing guidance had tended to characterize the car as a kind of vulnerability, as did the resident at the start of the chapter. The facts, however, appeared to be contradicted by the Victorian fires, where most of those who died actually did so in their homes. The Black Saturday blaze ran against common knowledge and experience because comparatively few people died in their vehicles. According to the fire expert John Handmer, this could be down to the reliability of modern cars and the fact that some used their cars not to flee but to shelter. Akin to the practices encouraged in the Cold War rhetoric explored in chapter 4, car owners could shelter in the cars by moderating their climates below a "lethal temperature and sit[ting] out the fires" (Handmer, O'Neil, and Killalea 2010, 24).

The "go early" policy tended to be conflated with "going late," which meant that the tethering of evacuation to the image of families dying in their cars was tied up in a temporal commonsense relation with punctuality: "Don't be late." It signaled abandonment of home and community and aligned evacuation to a broader characterization of "late evacuation" as somehow tardy or neglectful, where evacuation through automobility became a symbol of vulnerability and maladaptive behavior. As a commentator on the 1983 Ash Wednesday fire inquiry implied, those defending their homes had tended to live. Others were found dying on roads and in cars. Griffiths even claims that the old demon of evacuation, panic, was inbuilt into the lack of warnings: "It was because of a conviction that late warnings would precipitate late departures, and that people are most vulnerable when in panicked flight" (Griffiths 2012). Authorities thought it was better (Kissane 2010) to "infantilize" the public by withholding deleterious information. The commission's interim report refers to a letter from the assistant commissioner of the Victoria Police, detailing emergency arrangements for the 2008–9 period, suggesting that most dangerous thing to do would be to hesitate; hesitation figures as a kind of volatile potential, where a "last minute change of plans and evacuation attempted immediately in the face of a wildfire" would be lethal.[6] Similarly, the CFA policy on wildfire evacuation recommended that "late evacuation" was "fraught with danger," subject to radiant heat, confusion, and poor visibility, all contributing dangers to the "late evacuee."[7]

Those "who should" consider evacuating "early," according to the royal commission's interim report guide, include those who are "unsure": about their own preparations, about their psychological state, about their ability to fight the fire (Royal Commission 2009, 177).[8] Doubt and insecurity again are loaded onto evacuation. The ghost of panic would even shape some of the commission's interim report, which quoted from the Bushfire Cooperative Research Center's own interim report on the 2009 bushfires: "People do not always react in a rational manner in high pressure situations: the flight instinct can be very powerful" (Royal Commission 2009, 200). People, in other words, should be careful not to succumb to their animal passions, to flee, to evacuate. Christine Hansen and Tom Griffiths (2012, 172) cite commentators on the "irrational evacuation mentality," worried that it might replace the desire for "patriotic pride in the development of a libertarian 'Australian approach' of community self-reliance," where, as Chloe Hooper (2018, 17) characterizes, "staying to defend your house is the Australian test of grit."

This enshrinement affected emergency decision-making during Black Saturday. Police Superintendent Rod Collins—the then–acting state emergency response officer—during his engagement with the royal commission reflected on the contradictions within the policy and guidance, especially around the phrasing of the "stay-and-go" policy, which seemed to want to hold together opposite things. Collins was asked about why his understanding of the "stay-and-go policy" conflicted with the advice around evacuation and even the Victoria Emergency Management manual. Bushfire, in Collins's understanding, was outside standard practice; it had "evolved outside the normal arrangements in relation to this 'stay and go' policy" and had absolved the emergency services of advising people to evacuate.[9] Collins explained that in comparison to other emergencies, where they would direct evacuations, in the context of bushfires the decision was in the hands of the residents. Reports from the evening of Black Saturday within Victoria and Australia's emergency management agencies show encouragements for residents to "activate their emergency plans." Collins declared, "My answer doesn't change. I'm accepting that it is in the manuals, but the policy that has evolved since 2003 has—those that are responding are of the belief that we don't evacuate.

People have got that opportunity to do that themselves, we give them all the information, we give them all the learnings that we have and they make that decision themselves. I acknowledge that and I acknowledge that it is not in the manual."[10] Collins's appearance at the commission was fraught. The text still bristles, particularly when his statements illustrate the complexity of emergency evacuation planning and the difficulty of communicating it.

The 2009 fires seemed to mark a rethink in what was the common policy, even though evacuation was still not commonly or publicly recognized as a legitimate practice, and this affected how even the word was (not) used. People I spoke to denied that the state had any kind of policy on evacuation. In many ways evacuation was, and still is, unthinkable. But in the Kinglake community action guide on "leaving early," produced after the fires, a graphic shows a car leaving the bush in good time, headed to the bright and safe lights of the city. The commission's findings changed the way the car could be seen, once more as a cocoon to ride out evacuation and emergency—in certain conditions—in safety, and it reinvented some of the language around evacuation as opposed to fire or disaster (Gough-Brady 2012; Griffiths 2012). This was not, however, to be understood as evacuating. It is "relocating." Earlier definitions of the word *evacuation* within Victoria's own emergency management statutes and manuals used the very same definitional text: "the planned relocation of persons from dangerous or potentially dangerous areas to safer areas and eventual return" (Royal Commission 2009, 174). The interim recommendations, however, use distance to separate people from the danger created by the emergency. They deliberately do not use the word *evacuation*. The word *evacuation* had the smell of "compulsion, of people being forced to leave their homes against their will" (174). The report urged caution because it could "involve a perception of a more organised exercise than simply leaving early." As in other contexts, evacuation seemed to imply a set of processes that are done to someone, while it was recognized as not merely technical but "an emotive, urgent word" (174). It was better to simply refrain from using the word altogether. *Relocate* was chosen instead. The report additionally built on a longer and wider set of legal discussions around evacuation and the home. Police, fire, and local authorities held ineffective powers to force anyone to evacuate if someone could provide a "pecuniary interest exception" in property, goods, or valuables, within land, buildings, or premises.

A common refrain through all of this lingered around the relations among evacuation from the home, of family, and once again of animals. When I spoke about evacuation with residents, public officials, and consultants, many wondered why anyone would want to evacuate. Evacuation feels like an imposition. Facing down the fire, or chasing into the fire to retrieve one's family, was an (im)mobility at the heart of what befell Kinglake. Common divisions of labor, and the organizational and physical geography of Kinglake, meant that many fathers and mothers were "off mountain." Children may have been at different schools. Families were split by space and between the decision to stay and defend or to leave, which fractured at roadblocks. Families were "blown up." Some have suggested that staying and defending was overwhelmingly a male attitude, while leaving with children to safety tended to be performed by women. The trends of bushfire fatalities support this, as Christine Eriksen bluntly explains: "Most women intend to evacuate, women predominantly die while attempting to evacuate or sheltering passively; most men attempt to stay and defend, men mostly die outdoors attempting to defend assets" (2014a, 39). Women's attitudes to evacuation are often maligned within official policy, practice, and some academic opinion as being ill informed. As Peg Fraser argues, the "woman's voice was silenced" and the "deferral to the masculine point of view" led many women to stay behind in the bushfire zone because they wanted to support their husbands (2018, 183). People would be "alive today," Fraser contemplates, "if the woman's desire to evacuate had not been overruled by the man's desire to stay and defend" (184).

Outside Australia "there is an understanding that a preference for evacuation is less likely to stem from ignorance, and more likely to stem from gendered norms of responsibility (e.g. care-giving)" (Tyler and Fairbrother 2013, 23). Kinglake's experience may well have cemented a feeling that questioned the masculine version of bushfire response. It is women who would support the communities and their families in the aftermath of the disaster, death, and evacuation, such as in the Firefoxes support networks.[11] In 2014 the CFA in Victoria launched a campaign "Leave and Live," where Fraser's focus on the "'masculine' preoccupations with protecting property and defeating a fire" (2018, 188) are somewhat diminished. Observing a fire on the horizon, a couple turn to each other, the woman asking, presumably a partner or husband, as children play with a ball beside them, "What do you think?" He responds, pensively, "I'm not sure." The message is to leave early, to "leave and live"—not to "stay and go" (Victoria State Government 2014). Under these conditions a resident should not be undecided about what to do. While the woman defers to her partner, the resilient and assertive masculinity discussed above is replaced with a far less decisive response. The authorities seek to suppress this equivocal autonomy and replace it with no decision at all.

The royal commission inquiry sought to comprehend how people had tried to leave Kinglake. They criticized the inaccurate and conflicting information that had been communicated to residents. The recommendations paved the way for widespread fire-preparedness plans that could equip households for fire and evacuation, and a statewide emergency app accessible from people's smartphones. Such measures appear to inculcate particular sequences of acting and decision-making common to emergency planning practices that we saw codified during the Cold War (Adey and Anderson 2011; Davis 2007) as a way to inoculate people from, once more, more maladaptive behaviors. The CFA and Victoria Emergency Management emphasize "trigger points" that use fire danger ratings to stimulate particular behaviors among households. They encourage residents to think about how those triggers might be dispersed within the household members in a checklist document "What Is Your Trigger to Leave" as if to cultivate "forms of sensory perception attuned to threat . . . the creation of an endangered community of sense" (Zeiderman 2020, 73; see also Victoria State Government 2018). The techniques rely on the neoliberal consumer subject as the ideal and flexible unit to respond to the fires. "Stay and defend" is still very evident within this guidance. It recommends "active" sheltering rather than "passive," which means a whole range of observing, maintaining, and defending practices to suppress sparks that might ignite. It implies an active sensing, thinking, feeling, preplanned subject. Guidance recommends the organizing and preparation of materials like an "emergency kit" or "go bag," with origins in conflict and a long and wide cultural legacy within the home (Garrett 2020). The emergency kit is annotated with helpful labels to identify its content, and wider advice is given on where it should be stored in the home. Notably, pets are included as part of a checklist, which requires a plan for where they should go during "relocation," and how they should be cooled and kept hydrated.

Real-time emergency information is provided by the VicEmergency app, underpinned by Phoenix Rapidfire, a fire-modeling program, to show emergency information using alerts, textual information, and maps that depict emergencies in a location with a particular spatial shape, adorned with other annotations of information (figure 8.6). This was developed partly for the public and partly for the provision of networked information to multiple agencies involved in bushfire preparedness and response, for whom maps of the Black Saturday fires became such a problem of communication. Research has shown that there is a lack of understanding of the "effectiveness

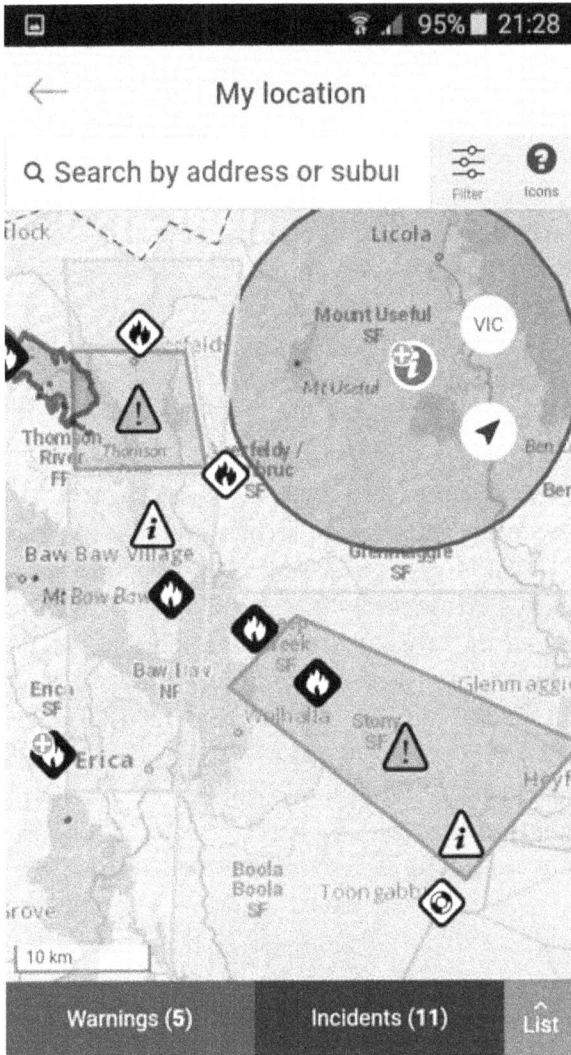

FIGURE 8.6.
Bushfire emergency polygons on the VicEmergency app. Screenshot by the author, 2019.

of existing wildfire warning maps for eliciting appropriate public responses," particularly around decision-making (Cao, Boruff, and McNeill 2017, 1478).

Living for a while with some of the habits and routines of life in a bushfire-prone town, all of this guidance, preparation, and warning enmeshed me and my family in an expressive and cartographic frequency, a network of lines of potential, triggered each time the VicEmergency app pinged, or the breeze seemed strong on a hot day, or we glanced the black, charred bark on a eucalypt nearby—a lone chimney of a long-gone house peeking above a fence line. These are not representations but more like potential gestures of what to do. The kinds of spatial referent dominant within the app are polygons drawn to express fire-alert areas. These are based on a model of how far the fire is likely to spread and its intensity, which the CFA uses to direct vehicles and personnel to fight the blaze. Overlays of weather information are possible to toggle on, and even representations of how many vehicles are attending. When the fire danger rating reached the orange (severe) and red (extreme) levels, the app became a constant companion. It pinged when bushfires were announced, especially if they were in my "location" or "watch zone"—radiuses of danger the app would warn me about depending on where I was, or where I'd set my "watch zone" to be. I got used to thumb-and-finger-pinching to see exactly where the blaze was located. The color-coded nature of this communication of risk and danger is of course effective and affective, and comparable to other warning systems, even when it seems you are constantly faced with different degrees of orange and red hues. We might think of the colors and cartographies in what Erin Manning has called a "a moving-with of perception in the making," a "feltness of seeing" (quoted in Beyes 2017, 64). Changes on your phone arrive with a ping and a pang of concern. The signs become a background radiation, impeding my feelings and thoughts carto-chromatically.

In mid-February, often the hottest and driest month in the state, the owners of the house we were renting messaged me just as I received a new update that a bushfire was only a few miles away and that I should be prepared to leave if necessary but that no "action" was needed. In other parts of the state, whole areas had been designated for early evacuation. In my research diary I recorded how the emergency diagrams of bushfire preparedness seemed to crisscross my body through my partner and our children, and through the different spaces of our house: from a suitcase not properly packed (there might be one thing in it), to passports and keys and underwear I'd uselessly placed in one section of a wardrobe, to a car and to the road we will take. The potential evacuation diagrammed us. Should the roads be blocked, we imagined driving into a vast field nearby where sheep had kept the grass low, although we'd read stories from Black Saturday about people perishing even in the middle of sports ovals, seeking shelter in the apparent open. We tried to figure out the likely distance needed to avoid the radiated heat from the dense bush that bordered our house.

On the severe days, we took the advice and ensured we had all left the town for the city by early that morning. We arrived back late in the day when the peak temperatures had cooled for a restless night. Our rhythm of leaving for the day and returning corresponded with many others', and the possibility of what is identified as "indecision evacuation fatigue," where leaving and coming back are a "drag," where you "have to sort out animals and valuables, drive in the heat, hang around in the suburbs" (Hyland 2015). On one drive back from Melbourne in the evening, after a week of forty-degree-Celsius temperatures and strong, gusty, hellish winds, we got an alert that a small bushfire had begun in our watch zone and that we should be prepared to leave. We paused in Hurstbridge, the last town before the drive up the mountain to Kinglake, hoping that things might have changed by the time we got there. The fire was being attended to but was not yet under control or "safe." Our options were to sit tight or to drive around while we looked for hotels in Melbourne's suburbs where we could stay for the night. As we were about to make a decision about booking somewhere, our app updated us that the fire was now "safe," and we began our journey back home. These methods of warning seem both real and somehow so abstracted. The polygons create little bubbles of danger. You feel sympathetic to those who fall within them. Relief that you're outside of them. Concern that they might be redrawn around you. This spatial orientation to bushfire as shapely contrasts with evacuation, which tends to be imagined as rather literal lines of flight— perhaps as ways to escape these shapes before it is too late.

CONCLUSION: GIVING SANCTUARY

In 2007 *Animal Keepers Forum*, one of the voice pieces for zoo-like organizations in the United States, published a special issue in which they acknowledged the pressing context of crisis and emergency, from bushfire to terrorism to emerging diseases (Baker and Chan 2007). As we have seen, stories around animal evacuations have emerged as a way of decentering attention from human catastrophe to an animal one. Animal evacuations can be contrasted with the spectacle and curiosa of animal escapees. Tokyo Zoo's emergency exercise for escaped animals became a social media meme as the zoo evacuated the public and then set about capturing the animals (Alan Taylor 2015). In Sydney's Taronga Zoo, several escaped tigers recently triggered an evacuation. The auditory announcement to "evacuate now" was repeated in a viral video of a lyrebird mimicking the announcement in their song (9News Sydney 2022). As Chris Philo and Chris Wilbert have argued, the pairing of animal escape and evacuation demonstrates a "deep unease often spurred by animal transgressions of human spatial orderings" and a "measure of resistant agency" from the animals (2000, 21).

Much of this has to do with a different calculus and set of practices for how life is cared for and valued in emergency, especially given that the evacuation of animals, in comparison to the visiting public, is something many zoos and sanctuaries consider doing only as a last resort. An interesting contribution to the *Animal Keepers Forum* special issue was authored by Dan Maloney, who in 2005 was head of the Audubon Zoo in New Orleans. During Katrina, Audubon had not planned to, nor did they, evacuate the majority of their animals. In fact, just the sea lions were forced to be evacuated to facilities in nearby zoos in Alexandria and Baton Rouge. Evacuations of endangered and rare animal species are considered more dangerous than the threat they might face by staying, often because of their intolerance to changes in the environment, especially their sensitivities to temperature. Maloney would later move to Zoos Victoria as general curator and was the lead manager when the Healesville Sanctuary eventually evacuated during the Black Saturday fires of 2009.[12]

The story of the sanctuary's evacuation has been told now in multiple forms, offering some esoteric fascination for an ostensibly human measure of protection afforded to zoo animals. In 2018 Zoos Victoria even produced a thirty-minute podcast documentary titled "How Do You Evacuate a Zoo" (Zoos Victoria 2019), which featured interviews and testimony from the sanctuary staff who experienced the 2009 Black Saturday events. The story is relatively heroic, the evacuation a curious example of human ingenuity and human-animal cooperation rather than a negative retreat, as human bushfire evacuations have been historically imagined. The animals were subjects of a protective apparatus but in very different ways to some of the competitive relations unpicked during and in the aftermath of Katrina. The Zoos Victoria documentary lingers on the human stories of trauma that paralleled the animal one as some people lost family, friends, and their homes. There is a convergence of the kinds of procedural sequences of preparation already discussed throughout this book, which are not dissimilar to the kinds of practices the sanctuary keepers performed, and now routinely exercise, in planning for the evacuation of the sanctuary today. When I discussed this with a sanctuary keeper, they explained how they sought to consider the specific habitats or lifeworlds the animals are so sensitive to. The weather conditions of Black Saturday had meant a high of forty-seven degrees Celsius, over twenty-five degrees higher than what many species could normally survive in. Once the animals arrived at Melbourne Zoo, they were placed in temporary enclosures, with their keepers staying with them.

As with so many other evacuations for which the car is the vehicle of neoliberal autonomy in emergency, but in contrast to some of the earlier common sense around bushfires, where the car was conversely an agent of danger, the automobile and the cooled van were crucial as a mobile environment of safety. Yet, while immobility might appear to characterize an animal's life in a sanctuary or zoo, in fact, they are moved and put into circulation relatively regularly. An animal might move because of breeding requirements, or go to breeding institutions, and this might happen yearly. Those who are postreproductive may move from one place to another. Some newborns may need to leave an institution so that they do not stay within the same gene pool of potential mates, avoiding inbreeding among a population sharing the same genetic base. It means that the assemblage of vehicles, transport boxes, and procedures for gathering or packaging up the animals was not completely alien to the sanctuary, or to the animals. Some animals in zoos have crate training, which means they can be called or lured into a transport container, using the skills familiar to the zookeepers attempting to guide the errant agencies and mobilities of a species like the Tasmanian devil (Philo and Wilbert 2000). In many ways some of the animals were already evacuation drilled.

Evacuating zoos and sanctuaries, should it be necessary, becomes about attending to microlevel and banal details in terms of ensuring particular foodstuffs are ready and accessible, travel boxes are easy to hand and locatable, vehicles are in the right location, bedding materials have not been forgotten, and the right screwdrivers are ready to hand. In this sense, particular materials and particular actions, and the ideal sequences in which those actions should happen, are codified into animal evacuations—as we saw in chapter 5—in an even more formulaic way than in human evacuations.

Somewhat incongruous with this kind of care, which seems both practical and mechanistic yet brimming with caring responsibility, the sanctuary renders a hierarchy of evacuation around biological and genetic value, which is quite unlike the systemic and informal proclivities that structure human evacuations. Zoos and sanctuaries have to predetermine whether certain animals should be prioritized, especially if they are faced with a situation of being able to evacuate only a limited number in a given period.

The bushfire planning for human residents of Victoria has been agonizing. The tensions over evacuating versus fighting fires and defending homes is a moving legacy of settler colonial practices and narratives of national identity and gender and control, where evacuation has felt like an imposition, the word loaded with meanings and feelings. The way guidance has moved toward evacuation is a quieter but important possibility of surviving bush-fire, yet evacuation is still resisted, and other technological narratives and practices of control are dominant. Evacuation of the sanctuary, however, is different. Compared to the necropolitical sort of momentum we saw in chapter 5, there is something more generous about the zoo's modes of evacuation and care, even if highly proceduralized. Evacuation of animal life seems to depend completely on relations to the humans, who could be going through their own personal battles, whose houses burned down, whose families were threatened and displaced. There is not only a different ecology of bushfire but also the ecologies that shape the conditions of emergency evacuation. This is an ecology of possibilities attuned to the narrow environments in which the animals can survive, an ecology of priorities and hierarchies different from the evaluations of human life, and the ecologies of action and practice.

Even in the granular lists of preparations and equipment we have seen through this chapter and elsewhere in the book, there can be a bureaucratic kind of care and protection and a hugely empathetic one, even if it resorts to valuations of species life. Yet some of the public narratives around the animal sanctuary's evacuation are quite ahistorical and render out the ecologies of settler colonial management and dispossession from which the sanctuary was born and through which fire has been intensified. It is a species of event that seems rendered out of time with the past ecological, economic, and racial history of the land and the inhabitants of the sanctuary we discussed. What Griffiths saw as the sinister recursions in the "preservation" of life withdrawn to a sanctuary are the strange recursions of the treatment of Indigenous life as less than human within a settler colonial apparatus of control, removal, exclusion, and the transformation of nature that has made more lives vulnerable to bushfires. Compared to the relations of chapter 5, things are almost inverted. The sanctuary of a human population treated as less than human and preserved somehow from their anticipated and premeditated erasure comes to be resembled in a procedural but empathetic protection of animal life made possible by close multispecies relations.

CONCLUSION

THE END

Many authors have consistently used evacuation in science and speculative fiction to signal that the end has come not only for some but for all. We have seen this throughout the book, for example, in the apocalyptic imaginaries rendered on an urban scale in chapter 4 in advance of the atomic bomb and thermonuclear war, and pertinently in the colonial "apocalyptic sublime" portrayal of bushfire conflagrations in William Strutt's imagery of nineteenth-century settler life in Australia in chapter 5, which would anticipate the Black Summer bushfires of 2020. Wildfires are consistently popping up around the world as we endure the extreme weather disturbances of climate change. *Apocalypse* seems the only word possible to describe the ineffable violence of these climatic emergencies and the unstoppable momentum of ecological-carbon feedback loops. As the Coast Guard chief witnessing the evacuation by sea of the residents of the Greek island of Evia in the Southern Peloponnese region in August 2021 exclaimed, "We're talking about the apocalypse, I don't know how to describe it" (quoted in *Journal* 2021).

Evacuation can portend endings. The end of the world. The end of us.

Some end narratives seem to involve evacuation as a vehicle to enable the end, in order to productively reorient attention and thought to some other problem. The end of the world and the evacuation of the Earth's inhabitants have been a way to ruminate on the capacity of the Earth's ecological systems to rejuvenate and move to their own cycles of development without the interference of humans. This leaves apocalyptic wasteland-type planets ripe for the "Edenic reestablishment" (Danowski, de Castro, and Nunes 2016) of the planet once we are gone. In *The World without Us*, Alan Weisman asks the reader to consider mechanisms that might get us to this point, "say . . . Jesus . . . or space aliens rapture us away, either to our heavenly glory or to a zoo somewhere across the galaxy" (2008, 4). Let us linger just for a moment on the idea of Rapture in relation to evacuation.

The Rapture is a 150-year-old Christian evangelical belief that on the eve of the Apocalypse, before Jesus returns to the Earth, a chosen population will be removed to heaven in an instant. It has given rise to a massively successful genre of fiction and the television series *Left Behind*. Rapture is almost a theologically inspired version of the belief of a privileged mobility potential in global emergency. Rapture is the "promise of planetary escape prior to worldly destruction" (Wojcik 1997, 30). As Amy Johnson Frykholm asserts, "The rapture is woven into the fabric of American culture, a part of the culture's hopes, dreams, fears and theology" (2004, 13). Leon Bates, an American aerospace engineer who became an evangelical minister in the 1970s, devised a series of interesting pictograms that borrow heavily and surprisingly from some of the visual logic of evacuation diagrams explored through this book. Presenting his diagrams in the abstract pamphlet titled *Are You Ready?* (1983)—which also apes and precedes emergency agency "readiness" campaigns and slogans seen in the United States and elsewhere (Aradau and Van Munster 2011; Collier and Lakoff 2008)—he explained that the top arrow portrays Jesus descending from the clouds to gather up his "Raptured people." The bottom is the upward-oriented ascension of those saved "being caught up to Meet Him" (Bates 1983, 2) (figure C.1).

Perhaps such an association between Rapture beliefs and the emergency aesthetics of evacuation instructions should not be all that surprising in the collusion of religious faith and emergency proclamations and responses (M. Cooper 2015), and the tendency to atomize evacuation to individualistic and conservative notions of family and household (M. Cooper 2017). The influential Cold War American televangelist and author Hal Lindsey made such an alignment, explaining how "America will survive this perilous situation and endure until He comes to evacuate His people" (quoted in O'Leary 1998, 179). Lindsey suggests the choice is between "extinction [and] evacuation," drawing on the emergent warnings about population growth and global warming of the 1980s. As Stephen O'Leary (1998, 179) contends, Lindsey's rapture was a "divine airlift." Similarly, Daniel Wojcik,

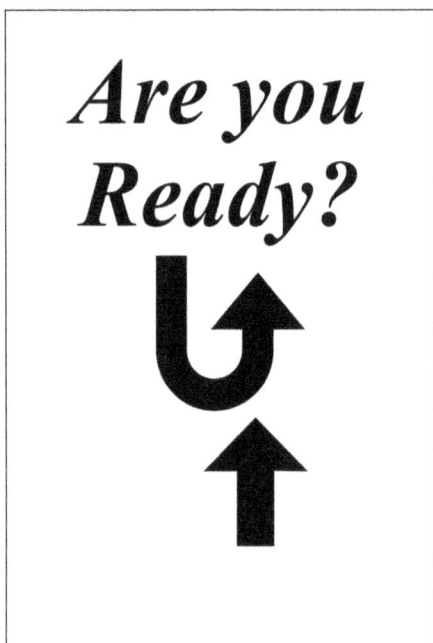

FIGURE C.1. Cover of Leon Bates's pamphlet *Are You Ready?* (Bates 1983).

comparing Lindsey with the influential minister and television evangelist Jerry Falwell, finds that Falwell's rapture belief "typifies this faith in divine *evacuation* prior to Armageddon" (1997, 57). We saw similar juxtapositions in chapter 6 between the Israeli disengagement from Gaza and the (non)evacuations of Hurricane Katrina. New Orleans's disaster was cast as a form of divine retribution: an evacuation for an evacuation.

Interestingly, the Rapture is also embedded within scenarios that circulate around automobility, that symbol of Americanized autonomy and aeromobility in other genres of rapture literature and cinema, and one we have seen in many evacuation practices. As Falwell explained, "You'll be the driver perhaps. You're a Christian. There'll be several people in the automobile with you, maybe someone who is not a Christian. When the trumpet sounds you and the other born-again believers in that automobile will be instantly caught away—you will disappear, leaving behind only your clothes and physical things that cannot inherit eternal life. That unsaved person or persons in the automobile will suddenly be startled to find the car suddenly somewhere crashes" (quoted in Wojcik 1997, 309–10).

The car, in this instance, is another site of evacuation departure, but this time it is not its vehicle. The rupture of the Rapture turns the evacuation into a moment of "mobility violence," represented in a cityscape of mobility

FIGURE C.2. Mobility carnage during the Rapture (Bates [1974] 1991).

carnage (Culver 2018). In one of Bates's paintings, an airplane has crashed into the top of a building, and highway pileups occur. White-clad figures ascend to meet Jesus in the sky (figure C.2). The pamphlet narrates a scene of crashes and wrecks due to the absence of drivers or missing aircrew.

From the Rapture to the vectors of outer space exploration and interplanetary colonization in the face of planetary death, post-Earth life and theologically inspired notions of the end of the world, a rich vein of imagination and aesthetic expressions can be found in stories of planetary evacuation in cinema and fiction, in evacuation boats, subterranean cities, and even the evacuation of the Earth itself into other solar systems and galaxies (Niven 1977; Schmidt 1976, 2000). Evacuation appeals, urging the consideration of longer-term processes of repopulating and potentially reculturing a society in the face of global apocalypse. Thought aesthetically, through different imaginations and beliefs, plans and thought experiments and popular cultural imaginaries, these evacuations of the End help us to reflect critically on evacuation and the recursions of evacuation that flow backward and forward throughout this book.

In the aesthetic and mobility carnage Bates and others have imagined in the Rapture, between an evacuation and those left behind in this kind of selective planetary rescue or escape, this conclusion sifts through some of the twisted meanings, practices, ideas, and concepts in torsion together in the crush. Some also ricochet out. As the book has tried to stay, following Ann Laura Stoler (2016), with the dissonances and discontinuities of evacuation—the "dispersions" as much as the singularities—different versions or forms of evacuation have collided head-on, while others appear much more in passing, only to repeat themselves in other forms much farther down the road, or to be instantiated once more in a kind of memory or specter of evacuation, and often in a kind of aesthetic common sense. Bates's moment of Rapture is a car crash of odd juxtapositions of different end imaginations and aesthetics, including the apocalyptic sublime, planetary evacuation, 9/11 (Culver 2018), and even evacuation signage. These are more of the recursions of evacuation we have traced through the book, or revacuations.

As we saw in chapter 3, *revacuation* was used as a semiofficial terminology and slang (*revac*) to describe the repeat of Japanese American incarceration in an alternative American Cold War in Perry Miyake's (2002) *21st Century Manzanar*. And in Brad Benischek's (2007) graphic novel of the chaos of Hurricane Katrina and its aftermath, revacuation signals a repeatable or more incessant condition of displacement, being evacuated over and over again. Indeed, the Japanese American experience was even contemplated in serious plans for crisis relocation in the 1980s under Ronald Reagan, as we saw at the end of chapter 4. But these recursive revacs are equally swerving. They turn a little like the curving arrow of Bates's imagination, or the lyrebird's remarkable impersonation of an evacuation warning at Taronga Zoo, which captures the Australian accent of the recorded announcer but also the tinniness of its digital and mechanical reproduction. Revacuations are also a kind of response, creatively and ironically sending things up in the manner of the detouring we explored in chapter 3. Or, if you like, they constitute the detouring lines or multiple diagrammings of bodies that resist the evacuation diagrams we have examined. They can be a way to critique, politicize, and deal with evacuation. And they make up evacuation's multiplicities; they detour out to different organizations and assemblages, different versions of evacuation. Thus, planetary evacuations, from the Rapture and beyond, can provide us with another form of what Stoler calls a "recursive analytics"—what we perhaps could think of as "revacuative heuristics"—with which we can reflect on the centralities and dispersions of evacuation as a name, concept, and practice through the past and projected into the future. They are not necessarily about rupture—of course Rapture is a kind of rupture—or continuity or certainly simple repetition but, as Stoler suggests, rather "partial reinscriptions, modified displacements, and amplified recuperations" (2016, 27).

This conclusion explores evacuation through ways that the end of the world has been prophesized, imagined, anticipated, and speculated on, while regrounding its politics and aesthetics to rethinking evacuation back down on Earth. First, we explore the trope of evacuation as an aesthetic and affective narrative device. Second, the chapter considers the negative and empty qualities of evacuation, often signaled in the vacuum-like empty-Earth scenarios common to planetary evacuation. Third, we explore the futural and reproductive logics central to evacuations before returning to evacuation's agonisms.

THE TROPE OF THE EVACUATION DIAGRAM

Evacuation of many kinds and scales has become a common dramatic, aesthetic, and affective trope. It might be compared with what Alfred Hitchcock described as a MacGuffin, a wrapped package, which comes off as a hook for a story. It is a hidden pretext (Žižek 1989), somewhat akin to the "gimmick" as examined by Sianne Ngai (2020). Susan Sontag's famous deconstruction of the science fiction film finds that evacuation is the common dramatic device routed through the scale of the city and obligatory scenes of "panicked crowds stampeding along a highway or a big bridge," directed by "preternaturally calm" authorities: "There is no need to be alarmed" (Sontag 2005, 40). As we saw during the Cold War, even saying the word *evacuation* was a novelty—an aesthetic judgment and utterance of the new to gather interest and excitement (Ngai 2020). As an act of speech, it has absolutely dramatized some moments and also technicalized them too. And perhaps, among other contradictions we have explored between the expressive and affective, it is this seemingly contradictory power of evacuation's aesthetic distributions, the dead but somehow enigmatic technicalities, that lends evacuation the powerful discursive and affective properties discussed throughout this book.

The instructional evacuation diagram has figured as one of the most popular representations of evacuation practices we have examined. The diagram distributes and orders evacuations sensorily and socially, often along normative lines, and literal but diagrammatic lines too: blue lines painted on the road in Wellington, along with their explanation in community messaging; the winding lines of the medical-military evacuation chain for the movement of the wounded home; the flailing lines of falling bodies that broke the woven lines of the escape nets in New York; the polygon lines of the fire emergency app outside Melbourne, which represented areas of fire risk and zones to evacuate from; and the lines of potential of preparation, anxiety, and some form of control.

The evacuation diagram is a diagram of power that is sustained relatively unchanged in how we imagine planetary futures at threat of extinction. Planetary evacuation mobilities play out viscerally in the National Geographic Channel's documentary of 2017, *Evacuate Earth* (Schillinger 2012), which begins in a Hollywood blockbuster–style gravelly movie trailer voice: "Can humankind escape annihilation. Can we evacuate Earth." The first ten minutes or so set up the main threat to Earth's existence through different means of anticipation in expectancy of coming destruction. Comet and meteor showers are first noticed by the public as a harbinger of planetary threat. A rogue traveling neutron star will, in seventy-five years, annihilate the planet, and there is nothing we can do about it and no way to stop it. A digital countdown clock begins running down. Poorly rendered computer graphics depict meteor showers destroying homes. A news report from Agra shows the Taj Mahal destroyed in a fireball. A billion-megaton-like explosion is endlessly replayed throughout the documentary, akin to the overexposed photography of the Bikini Atoll nuclear tests (DeLoughrey 2009). Talking-head experts propose nuclear pulse propulsion as a possible means to travel fast enough to escape the solar system. To emplace a viewer within a media and mediatized spectacle, a rooftop observer witnesses a hail of deadly fiery rain. Such portrayals of evacuation—mobilizing other violent aesthetic styles that accentuate conditions of urgency—happen with certain emergency times, spaces, or conditions. They set the stage for evacuation's need, its potentiality.

The way to resolve this catastrophe in the documentary is to be rational, which is a logical *and* affective relation commonly attached to the diagram of evacuation, as much as other excitable feelings are too. Evacuation seems to be enrolled as an attempt to remove or prevent these excitements, as much as those excitements are the circular causation for some attributions of evacuation: evacuation as flight. Is it possible to move 7 billion people off-planet? How, within seventy-five years, will we develop means to do so, and where will they be moved to? And who should be selected? How do we prevent mass panic, even mass suicides? In part, the emphasis is on technicalities, on rocket engines, on a massive innovation park, on infrastructure. This has also been a key focus of evacuation within this book. Evacuation appears to depend on certain infrastructures and what they promise, just as much as it seems engrained infrastructurally as a habit we are encouraged to resort to in emergency, in some contexts. Evacuation ships, corridors, logistics spaces and airports, staircases, signage, and plans and more plans throughout this book seemingly route and root evacuation in machines and buildings and materials that are meant to prepare and enable bodies to move. In lots of ways, though, they expect far too little of us in impoverished imaginations of who we are as fleshy and social bodies. As we have seen, we diagram differently. The lines of the evacuation diagrams vibrate with something more: other knowings, affects, promises, sensings, dissonances, and discordances.

The possible societal breakdown that is debated and represented in the documentary is not surprising, reflecting the familiar "elite panic" over oncoming emergency (Solnit 2010). Of course, panic is a familiar companion to evacuation, an attached affect we have traced through the book. Its invocation is a frequent and clumsily laden naming of informal or irregular, or inadequate, evacuations *and* emergency affects. Racialized, gendered, and derogatory assumptions with implicit causative properties are tied up in evacuation and panic, which have seen many of those most in need of evacuation blamed for emergency or disaster. We saw this starkly in chapter 1 in vertical evacuation contexts, especially around young, female migrant workers. Panic is often accusatory and entwined in gendered, classed, and raced bodily assumptions. Within the realization of imminent or far-off but absolute threat, the governments in the documentary withhold the truth from their populations in order to stave off panic. This reflects the way authorities treat publics, infantilizing them, imagining them as unreliable and panicky—an approach that even guided preparations for COVID-19 as different countries have admitted, just as it did during the Cold War and preparations for the atomic obliteration of cities in chapter 4. Evacuation practices were a way to try to stave off panic, and their governance and conduct, or conduct of conduct (N. Rose 1999), sought to entrain bodily, automobile, and other movements.

The apparent autonomy of panicky emergency affects in evacuation may be why panic has become such a fetish in popular culture. The American author James Thurber's famous story of evacuation panic in "The Day the Dam Broke" (1933, 45) explores—with some irony—a "shouting, weeping, tangled evacuation of the city" because of a rumor that a dam serving water to Columbus, Ohio, in 1913 had failed and the city was becoming inundated. Animated by socio-psychological notions of the moral and asocial threat of crowds, the affective welding of panic to evacuation is expressive of autonomous, turbulent, exterior forces powerfully moving and moving against authorities who attempt to subdue that panic and movement. It also sets up a peculiar fascination with social structures and orders being upended by a frantic affective life. As we saw in chapter 5, human-animal and sociomaterial and moral relations might be scrambled in a confluence of agencies in part made so visible by the poor recognition of the attachments to other nonhuman companion species and by recognition of their role and rights in emergency. Thurber evocatively describes the "trickles of housewives, children, cripples, servants, dogs, and cats, slipping out of the house past which the main streams flowed, shouting and screaming" (1933, 42). And it is why, perhaps, panic has persisted as an idea that mobilizes state power even when we have seen much research that has questioned panic's occurrence.

In the National Geographic Channel's speculative future, ethics and justice seem to take second fiddle to technopolitical concerns that demand the selection of the fittest or most biologically suitable, or the smartest, discussed later. Some nations simply cannot participate in the international projects that follow—should they even be involved, commentators ask? As Sontag notices in the science fiction tendency of "extreme moral simplification," a "greater range of ethical values is embodied in the décor of these films than in the people" (2005, 43). Tropes of evacuation evoked in the documentary echo the particular temporalities and logics of the future, some of which have underpinned the multiple reproductions tied up in evacuation, discussed further below. They importantly invoke an aesthetic of emergency that is spectacular and that sets the political stakes for evacuation within a common scalar frame of what Bonnie Honig has identified within the "extraordinary, the ruptural, the heroic" (quoted in Rossello 2015, 701), and therefore particular versions of exceptional politics.

And yet what we have explored in this book is how the everyday and the exceptional are often folded much more tightly. Indeed, the spectacular aesthetic of evacuation does much more than draw evacuation into the politics of exception; instead, it conceals the chronic nature of the systematic or structural inequalities and unequal relationships of power that may give rise to emergency in the first place. These may condition a far more everyday sense of emergency in which evacuation is a recurring event within recurrent conditions of precarity, racism, and marginality. Those very spectacular aesthetic divisions may use structural demarcations of the sensible, as we saw in chapter 1 in vertical evacuations, where gendered, raced, and classed proclivities were expressed in representations of fault and victimhood.

EMPTINESS: WORLDS WITHOUT US

Some of the science fiction, thought experiments, and semiserious plans for planetary evacuation tend to assume an end point for a home world that must be left. Disaster, emergency via invasion, depletion of resources, and the poisoning of the environment are all common sources that are today animating a new push for space exploration and multiplanetary futures. Bound up within these imaginations are ideas of what literary theorist Greg Garrard (2012) has called a "disanthropic earth," which is to say that planetary evacuation leaves a strong and common aesthetic of the evacuation zone as the ruins of planetary civilization. Empty, Earth is figured as a far less appropriate medium for human life.

These figurations are especially pertinent within the in medias res openings to imaginations of planetary abandonment and are common to the genres of twenty-first-century ecocide science fiction (Martí 2020). The idea of an empty and evacuated planet (with no one left behind) leaves apparently empty spaces, vacated of life and often with a singular human witness that forgets other beings, who are somehow rendered below the level of human. Once again, evacuation appears to contain ambivalent properties in its negativity. The world has become a semiplayground to a lonesome caretaker-type figure in Joseph Kosinski's *Oblivion* (2013), where Tom Cruise's character, Jack, narrates, in a semigrisly tone, "Everyone's been evacuated . . . nothing human remains." Jack lives out a sort of colonial, frontier, and expansion myth in reverse, a male paradise of sorts lived between a hovering existence on a palatial floating aerial station occupied by his beautiful partner, whom he doesn't trust or seems bored with. Earth is a wasteland covered in the debris of war and environmental failure, as well as a place of natural rebirth or preservation in an area unknown to the surveillance drones.

The lure of an evacuated Earth without us offers a special kind of "peculiar beauty," suggests Garrard (2012), closely associated with the "apocalyptic sublime" we saw earlier in Strutt's painting of the Australian bushfire in chapter 5, reviving colonial imaginations of emptiness, wilderness, and survival even as colonialism moves in the reverse direction in unsettling. Evacuation appears to be able to constitute the settling-expansion practices of colonialism *and* the reversing moves of colonial unsettlement—albeit with remaining and hugely unequal military control—such as in the settler removals from Gaza we saw in chapter 6. In *Oblivion* life is a form of tending and maintenance reminiscent of some sort of pioneering frontier myth chiming with some of the narratives of Japanese American incarceration we saw earlier. For Loraine Haywood (2016), *Oblivion* presents a kind of Garden of Eden theological trope of masculine perfection and neuroticism. The emptiness and affective flatness are peaked by some hugely vertical promontories (L. Hewitt and Graham 2015), whether in the vertical buildup of compacted garbage/rubbish or in dusty morphologies of cities, which become canyon-like ravines where river deltas flow and high-rise skyscrapers are waterfalls. Around these scenes Jack's "bubbleship" swoops and rises with almost omniscient, swiveling vision (Kaplan 2017). Such verticalities are related to the bare synoptic view of the evacuation diagram; the vertical, planar perspectives are also emphasized in the helicopter's shots over Calgary's rural hinterland in the CBC documentary in chapter 4; or the guard tower's and model's-eye view of Manzanar seen between the photography of Ansel Adams and the Manzanar visitor center model. The diagram performs another uninterrupted view from above, with some of the distance and abstraction of vision that vertical perspectives are often criticized for (Adey, Whitehead, and Williams 2013; Dorrian and Poussin 2013).

Such a sense of emptiness from or in the aftermath of evacuation reminds one of different scenes explored in this book: Calgary's apparent emptiness in the Cold War simulations; the desolation of flooded New Orleans; the white spaces in the maps and diagrams of the medical-military chains of evacuation from World War I; the empty ruination caused by Australia's bushfires, leaving burned-out cars and chimney stacks still standing; and the desert isolation of Manzanar's internment or relocation center for Americans of Japanese descent in chapter 3. Haywood (2018) sees in *Oblivion* an expression of the haunting of 9/11, a negative echo of the apparently empty evacuation zones of the twentieth century that have haunted this book. The chemical and nuclear contaminations of Bhopal and Chernobyl have surely added to these imaginaries as expressions of so-called evacuation zones. Yet this emptiness is almost always cast as a human absence, when in fact the world left behind is populated by the all too easily forgotten and abandoned life of abjection (Sinha 2008).

Senses of the abject have recurred frequently within different formations of emptiness we have explored, sometimes conjured in the multiple aesthetics of whiteness in different places through the book. Whiteness tended to express senses of cleanliness—in the ambulance trains of World War I—and the racialized calmness that was perceived to be absent from stereotyped representations of migrant labor in chapter 1 or of Black people in chapter 4. The epigraph in the introduction began with John Furse's extraordinary unpacking of the multiple and specifically clonic meanings of *evacuation*. As waste, and wasted, evacuation has been frequently experienced as and compared to being treated like shit. Evacuation is both an action and a feeling of being scraped off, removed; it is a sense of being used up, wasted away; and it is viscerally expressed in the violent expulsions of chapter 3 or the abject abandonments of chapter 5.

Evacuation is both an aesthetic judgment and an aesthetic experience (Ngai 2020). The deployment of the excremental as a normative rule and an experience of evacuation has been seen in lots of expressions, metaphors, discussions, and literal representations in toilet humor, toilet talk, and narratives, perhaps at their most extreme in the aphasic terminologies we saw in the grotesque legitimizing techniques of evacuation's performance in chapter 3. Evacuation can rely on the construction of communities perceived as worthless and therefore made worthy of punitive evacuations, which should not be called evacuations at all. Or at least if they are, then one of the central points of the book is that a concept of evacuation needs far greater refining in order to capture the multiple and uneven relations to movement, agency, and forms of protection that are held together by it. A clonic conception of evacuation is often tied to the body's rejection of an unwanted substance in order to preserve or protect not the entity ejected but the host. It has also been a powerful expressive and narratological tool, even of resistance. Intra-bodily, clonic evacuations are positioned and juxtaposed alongside the extra-bodily, more traditional senses of the term as geographic or locational displacement. As Perry Miyake's novel on revacuations, *21st Century Manzanar*, showed so effectively in a dystopic alternative United States, the abject resonates through bodies and geographies at multiple scales. In part, these see further recursions of the clonic from a more medieval memory of bodily evacuation. At the same time, the recursive can produce aphasic effects, in the temporal and spatial overlaying of multiple and conflicting meanings and experiences that dispense with neatly defined sequential or linear senses of time (Stoler 2016).

Such a sense of the vacated as abject within planetary evacuations, however, could be read another way. What seems to be assumed as the vacancy of humanity is instead a realization of other agencies and actors, a familiar story in this account of evacuation, and a familiar concealment at the heart of settler colonial jurisprudence. In the sci-fi imaginaries, drones and other robotic life, and of course, subcultures and outcasts and even renewed vegetation, persist and thrive—even if others are intent on destroying them. In many storylines the veil of emptiness lifts at some point to reveal these other inhabitants who were always there, just beneath the surface. Similar unveilings are common to evacuation, taking events like Hurricane Katrina to reveal the importance of human-animal relations, or to demonstrate the racialized disdain and hostility with which the Black communities of New Orleans were treated. The evacuated world without us is in fact very full—of dust, rubbish, refuse, other life-forms—and in this sense we are reminded of just how privileged those who can evacuate really are. In the science fiction worlds, it's the elites, the rich, the most capable, who evacuate. In the real world, this is not entirely the case, but certainly many evacuations we have seen have depended on resources and especially on citizenship, passports, and papers.

Despite the continued sense of the evacuated space as one of abjection, there remains some sort of promise, a providential possibility of reuse, an ethical revaluation of things (Garrard 2012), what Shannon Jackson calls "categorical re-discrimination," in the willingness to decide that the homogenized world that we call "'garbage' deserves to be internally differentiated once again" (2011, 75). Such a rediscrimination has been extremely important in our account of evacuation as more than human. Yes, evacuations are premised on countless objects and infrastructures we have explored, *and* they involve human-animal relations and in many contexts the valorization of different reproductivities. Those treated like garbage may actually perform some of the most promising forms of evacuation and social togetherness. Human-animal relations have been incredibly important as evacuation practices enroll a greater diversity of subjects and objects, caught in taut or looser legal, ethical, and familial relations to each other. Evacuating with or without pets has been particularly contentious, just as those relations have helped us understand the biopolitical and necropolitical logics that sometimes have remapped the boundaries between animal and human life, and between life and death.

FUTURITIES

The relationship between rebirth and reproduction is startlingly central to the ethics of speculation on planetary evacuation (Speier, Lozanski, and Frohlick 2020). Take the movie *Interstellar*'s (Nolan 2014) two methods of planetary evacuation and space colonization. Earth's crops are no longer viable under the "Blight" in a reimagined or recurring dust bowl–era United States of the future. The ideal future for Earth is not in fact evacuation in the form of a colonizing desire to reseed other planets—the so-called plan B—led by the protagonist Cooper's piloted spaceship *Endurance* (a name with its own masculine exploratory lineage). Plan A looks different when Cooper's daughter Murph solves a gravitational equation enabling the Earth's gravity to be suddenly but quietly halted, gently lifting the Earth's population in a giant exodus to space stations. Perhaps this undermines a notion of evacuative space flight, which is usually equated with masculine symbols of reproductivity in the violent eruptions of rockets (Faraci 2014).

In the 2019 cinema version of Chinese author Cixin Liu's (2017) book *The Wandering Earth*, this plays out differently. In the film's spectacular, if not as good or interesting, version, the Earth is evacuating the solar system under threat of the sun exploding. Liu imagines huge eleven-thousand-meter-high Earth Engines constructed to launch the planet to another galaxy—a piloted spaceship Earth. This is about engineering the planet's volume—as fuel—in order to mobilize the earth into an evacuation vessel, even if it breaks the planet's own planetary flows (Szerszynski 2016). Liu follows other science fiction writers and scientists contemplating the planetary "astronomical" engineering necessary to "migrate" the Earth elsewhere (Korycansky, Laughlin, and Adams 2001).

Of course the idea of a spaceship Earth–cum–evacuation vessel is not new. In the emergent international discussions over the realization of environmental vulnerability (Ward and Dubos 1972), the idea of a spaceship Earth orbited around concerns over population growth, environmental degradation, and resource depletion—some of the very same concerns that fed Rapture theology. As the Earth is driven away in *The Wandering Earth*, internal evacuations are necessary. The planet loses its atmosphere, and its mass is so disrupted in feeding the vast engines that earthquakes shake the underground cities. Magma flows overrun them while emergency laws determine that the eldest members of society go to the end of the line to wait for the elevators to evacuate them to safety. The couple at the center of the story actually get married, and in that they are, by law, permitted to procreate. The ruling Coalition has decided that only one out of three couples is permitted a chance of a baby, determined by lottery. Given the unreal conditions, the protagonists' chance of procreation is made entirely bittersweet.

At stake here are the "reproductive futurities" that we started with through the oddly conjoined evacuation/abortion debate in the book's introduction. For Lee Edelman (2004), the insistence on the figure of the child within politics is inherently problematic if we are to challenge and refigure the heterosexual norms at the heart of our social order. Edelman worries about the "child as the culturally pervasive emblem of the motivating end, albeit endlessly postponed, of every political vision as a *vision of futurity*" (1998, 22). Cixin's characters' unease at the world their children will be born into is a common concern in other cultural criticisms of wartime, as Paul Saint-Amour (2015) discovers in Virginia Woolf's meditation that other kinds of futures might be possible. Woolf imagines "a curb on reproduction . . . as an emancipatory feature . . . a means by which women might develop other openings for their creative power" (Saint-Amour 2015, 127).

This was a concern Jeffner Allen (1996) saw in evacuation's protective and affirmative possibilities, which we saw reproduced in the neocolonial adoption debates in Haiti in chapter 7—against the "no future" voiced by Nicolas Sarkozy—or the so-called Katrina babies of New Orleans. Of course a great deal of emphasis within the evacuation "disengagement" contestation in Israel hinged around the apparent fertilities of the land the settlers were being evacuated from too. Woolf and Edelman warn of the futurities symbolized by children and the reproductive social orders that we could find in many planetary and nonplanetary evacuation tropes.

In thinking through the politics and ethics of evacuation, even of the planet, should fertility matter, or, for that matter, measures of age, fitness, genetic adaptability or intelligence or value, education, class, or wealth? Why should the future of the human be so selective, and if it is, is it worth it? Of course some wonder if we're better off dying out as a species rather than betraying some of our values in order to continue surviving on elsewhere, resting on futures from which so many are excluded. What isn't posed is, What if people simply don't want to go? What if we decide on a slow and deliberate end to humanity?

In many respects not evacuating—and ending life—is impossible to imagine or realize. Death might seem a more unimaginable scenario than the scenario itself. In science fiction, evacuation tends to offer the promise of a beginning or the beginning of a climax, the genesis of a dramatic narrative, perhaps an origin myth for the birth of a people, a sort of recirculation perhaps. Primarily these kinds of imaginations render evacuations and emergency away from the ordinary or everyday and in this way can resort to an exceptional bio- and even selective necropolitics, as we have seen in the multiple abuses of the term and in the different violences evacuation might perform and disguise. But is it better that we, forewarned of death, decide not to evacuate—as if growing old might be a more romantic way of meeting the future and maybe not meeting it at all? Yet how can we imagine a future that children may be born into but in which they will not have a chance of their own future?

The Wandering Earth sees society reassessing law, ethics, and moral values to the extent that it has done away with its superfluous professions and academic disciplines, in favor of fields such as engineering and mathematics. Within these kinds of futures, evacuation's dominant technical register seems the ally of narrow kinds of societal imaginations, a kind of pathway of extrasolar transportation and mobility without any recognition of how society might prepare itself for the journey through arts and culture, through

FIGURE C.3. The Cooper farmstead in *Interstellar* (Nolan 2014).

discussions of philosophy, ethics, or belief. The social regresses even if we advance technologically. If we go back to the National Geographic Channel documentary, outside of building the right vehicle for our evacuation, genetic engineering is turned to as the only way for survival through genetic diversity, even to rule out people genetically susceptible to disease. Those with diabetes should not be allowed to go. People mail in their own genetic material in an "evacuation eligibility kit." What kind of society would we be if we did this in order to live on, if we closed off the ethical and political decisions about who and how we evacuate so that the fittest or most adaptable or reproductively fertile or competent are those who get to leave?

Curiously, one of the last scenes in *Interstellar* returns to a different kind of reproductive fertility, familiar more to the agricultural practices and "arborescent metaphors" (Malkki 1992) tied up in evacuation, and especially its colonial orientations of settling practices, as we saw in the recursive narratives around Japanese American detention or the Gaza settlers miraculously settling the apparently infertile soil. A farmstead ends up being preserved on the cylinder/torus-shaped Cooper Station, orbiting Saturn (figure C.3). It is Cooper's old farmstead as a museum exhibit—which has been laboriously reconstructed on the space station amid verdant cornfields. A "heritage futures" project has criticized this mode of historic remembering in the film. Why, they argue, should this be what heritage looks like in the future, where kids still play baseball on the greenest baseball diamond in a replica of small-town America? Why is "a somewhat dull period house of our own present" what is remembered, "as if heritage cannot have a visionary future of its own and cannot be rescued from its own problems on Earth"? Why should heritage mean "faithful material reconstruction"? they ask (Holtorf 2015). The recursions of evacuation can mean the same things are reproduced with little possibility to remember the past, or imagine a future, which could be otherwise.

AGONISMS

To be sure, the multiple evacuations explored in this book elude easy categorization. We began our consideration of Adi Ophir's (2007) providential/catastrophic dichotomy at the start of the book, and we have seen numerous moments where those logics break down, existing in far wider senses of degrees or intensities, sometimes colliding, contradicting, and coexisting in different forms. Perhaps the most extreme example of this was the Nazis' duplicitous use of evacuation as *Evakuierung*, a terminology of extreme ejection, violence, and death, used as a way to manipulate and disguise in order to perversely care for a biologically purified notion of the nation-state and territory and an ethnically valued civilian population under Allied air raids. In its recursions and revacuations, we see multiple and contradictory notions of evacuation holding together, sometimes very loosely, their meaning tethered to and conjoined tentatively with notions of protection. And consider the confusion of these, for instance, in the muddling of evacuation and adoption in Haiti in chapter 7, or the use of terms such as *preservation*, *refuge*, and *sanctuary* in the treatment of Japanese Americans, or Indigenous First Nations people, or animals in chapters 3 and 8. Terms, concepts, and practices overlap and flow into one another. They fold and refold old and new, revealing "new surfaces, and new planes" (Stoler 2016, 26) in a far more complicated and contradictory relation to life outside of Ophir's binary. To return to Bonnie Honig's (2009) "agonistic" politics of emergency, Honig seeks productive tensions, agonistic contentions, and partnerships generative of critique, democratic impulses, and new sites of power. Perhaps these animate Ophir's binary of the "catastrophic" and "providential" states as opposite tendencies into more contentious dialogue and exchange?

Part of the politics of this is legitimation. Evacuation seems to demand legitimate and illegitimate grounds, performances, and subjects. Even the emergency or crisis being evacuated from is legitimated in some way by an evacuation, just as authorities, actors, and states gain or lose some of their legitimacy from evacuations. Of course, the events and evacuations we have seen recently in Ukraine were preceded by the United States and its coalition partners, and other countries, evacuating their nationals from Kabul airport as the Taliban retook control of Afghanistan following the withdrawal in 2021. Evacuations flood the news. The hastily performed evacuations from Kabul, apparently long in the planning, were shocking. They illustrate that who is deemed a legitimate evacuee is precariously given: someone with the right piece of paper, passport, ID, look, ethnicity. Even the human-animal evacuation debates we explored in chapter 5 surface once more, as a former British marine, Paul "Pen" Farthing, made headlines by pursuing and raising funding for a chartered transport evacuation plane for rescued cats and dogs by the Nowzad animal welfare charity that he had set up in Afghanistan. The parity of evacuation between human and animal was questioned when it appeared that the sanctuary's some 170 cats and dogs were evacuated with British military assistance within the airport, while the local Afghan staff the charity employed were turned away from the southern airport circle gate by the Taliban. Operation Ark—as it was named—raised $450,000 from private donations and celebrity support. The event raised familiar but difficult equivalences between human and animal, demonstrating the apparent disdain with which some human lives are regarded.

In the midst of the debacle, the event powerfully ruptured assumptions about sovereignty, rights, care, and control. Some were worried about government officials whose time was being wasted or taken up by the former marine's political influence. Many of the cats and dogs actually belonged to UK embassy and United Nations staff and contractors who were not allowed to bring them aboard their own evacuation/repatriation flights and who, by one media outlet, were judged as having "dumped" their animals at Nowzad. For others the experiential modality of the animals' mobility was important, their bodies imagined as taking up valuable passenger seats, precious landing slots at the airport as the timeline of withdrawal counted down, while others worried about the heat stress endured by the animals waiting to be allowed into the airport. In this human-animal politics of evacuation mobility, the relative stoppages, moves, and streams of bodies caught in relational ties of autonomy and constraint matter greatly. Moments like this, in the tension or contention of those ties, help realize multispecies vulnerability to be judged as worthy for evacuation, yet found wanting. Some commentators and British government officials—even defense ministers—sought to reassure that the pet evacuation was not being blocked but that a hierarchy of mobility ordering would stay. The pet evacuation would not cut in line. An animal rights campaigner even highlighted the privatized nature of the operation in a "Dunkirk spirit" (Dyer, quoted in BBC 2021) as former and mythologized evacuations recur. Indexing evacuations as moral and ethical through the evacuation, or not, of animals and nonhuman subjects is a common but difficult practice, yet it lays bare challenges to the unequal structures that guide evacuation and its experience, the conflicts among more state-led, privatized, and collective kinds of evacuation.

As we saw in chapter 7's overlay of events in Libya and Haiti, when thought logistically and viapolitically, the modes of evacuation's production are the objects around which vehicular modalities can politicize and contest mobility. Similarly, it is the privileged forms of mobility we call evacuations, by military and commercial or private aircraft, that have formed many understandings of and reports on Kabul's evacuation. Some are given privileged rights to remain in the countries they enter, while the millions of people displaced by the Taliban who seek refuge in much slower and more arduous movements have not been seen in the same way. A Toyota car was pictured taking up space on a cargo plane as hundreds of people sat alongside it holding on to the cargo rope. The car became an object of concern for its excessiveness—inside the Kabul cargo plane—strapped down in almost the same way as the people who were lucky enough to be treated like cargo. Cars have been a key modality of evacuation mobility. With other vehicles they too are in agonistic tension. Between their privatizing division of care and comfort, autonomy, and resources, provided to or withdrawn from evacuating and nonevacuating bodies, they can become sites of visibility, political remonstration, and deliberation.

The body has been a crucial battleground of contention in this book. Of course, the violence on those who were not deemed legitimate in the Afghanistan evacuations was clear to see, as comparisons were sought in a recursive triangulation made by the media, including the historian William Dalrymple. Connections were cast between the desperate struggle of people clinging onto departing aircraft leaving Kabul, whose bodies fell to the ground, and the leaping falling bodies of 9/11—of course harking back to earlier scenes in the same city's high-rise factory fires—and similar scenes from the evacuation of the United States from Saigon in 1975. Evacuation and falling bodies once again recur and overlay one another, sometimes surfacing as multiple and overlapping traumas, perhaps made most clear in the Israeli evacuation of Jewish settlers in chapter 6. The disengagement evacuations were so traumatic in part because they were perceived as a recurrence of other persecutions, violences, removals, and forced displacements of Jewish people. The fact that the evacuations were being performed by the "people's army" on themselves was ungrounding.

During the COVID-19 pandemic, the evacuating body also continued to resonate across rights, vehicles, borders, and bodies. Stigmatization was common for those repatriated through evacuation flights, as was the case in Ukraine well before the Russian invasion, as well as those who were not, such as many African nationals in China. India's Vande Bharat program of evacuation flights repatriated millions of Indian citizens from abroad and from across India from May 2020, following political pressure from the Indian diaspora on the Hindu-nationalist government. Many complained at the exorbitant costs of a ticket on these flights, pricing out the poor from accessing them. Others examined the disproportionate way the evacuations excluded women's bodies. Pujarinee Mitra (2021) assesses the difficulties experienced by pregnant women, for whom unequal attachments to male partners may impede their decision-making but who are subject to additional forms of "reproductive surveillance" on their bodies. The contracted airlines enforced extremely strict and patriarchal rules on pregnant women traveling on Air India aircraft, which meant that pregnant women rarely met the complex legal requirements to provide the documentation, medical advice, or medical chaperoning that was demanded. For Mitra, echoing Deborah Lupton's writings on normative social ideals of a "contained body," and the pregnant body as a "container" whose expressions and leakages may challenge those norms, the pregnant body cannot seem to live up to the image of an "economically stable body" assumed within the structures of evacuative repatriation.

With this in mind, we have routinely examined the embedding of particular imaginations of the bodies evacuation pretends to move. So-called fitness for evacuation often works as a standard against which "unfit" and "maladaptive" bodies and behaviors are named and judged. The *we* becomes remarkably—and at other moments more subtly—circumscribed by particular imaginations and models of what a human being is, of who *we* are. Maladaptive, wrong, unfit, panicking, too social, bodies appeared not to be streamlined enough, or right, hardy, or strong enough, to evacuate, or they *should* be evacuated because they are too weak or feminine. The wrong gender or sexuality. Raced. Too promiscuous. Diagramming differently. The "us," then, of who is left behind helps us challenge the notion of who or what is evacuated or doesn't get to. And this gets at a key concern of this volume. Evacuations may put a group of strangers, other units, families, or organizations together, sometimes in close and uncomfortable proximities. Evacuation tears apart earthly assemblies of relations—social and familial—classed structures and hierarchies too, although new ones are soon forged. In lots of ways, the planetary evacuation narratives discussed in this chapter actually animate the difficult and problematic relations of evacuation experienced on the ground, and they invite challenges for the critical ethical and political questions of how or why we might experience and plan for or live evacuations better. While wildly imaginative in some ways, the planetary evacuation tropes are troublingly conservative and derivative in their assumptions.

There are some promising and hopeful possibilities, however. In evacuation's many agonisms, there are optimisms. We saw how the excessive promiscuities of evacuation were celebrated as ways to survive, where bodily differences and apparently ill or maladapted bodies saw people coming together. Even in war, intimate bodily connections were used to move people to medical care, becoming animalistic and mechanistic to do so in sometimes homosocial relations. Hands and arms grip one another. Shoes are shared. Soldiers and citizens grieve and console one another in disengagement. The most promising moments of evacuation seem to found in intimate bodily sympathies, conspiracies, and solidarities, which, even if the source of much maligning, constitute generous yet vulnerable encounters across human, animal, and social divides.

More recent approaches to interstellar evacuation and travel, such as the Project Persephone, have begun engineering and philosophical approaches to imagine some sort of starship or "worldship" capable of lifting populations off the Earth in the far future in a living spaceship (Armstrong 2016). The project leader, Rachel Armstrong, has imagined a more symbiotic relationship between the "worldship" and the life that sustains and depends on it, imagining organically grown architectures, a liveliness and unpredictability to the life that will inhabit the ship. Persephone imagines a flatter, more symbiotic, and companionable sort of evacuation compared to the nihilistic and hierarchical relations observed in other evacuations involving nonhuman life. In tune with efforts to develop a "planetary [or interplanetary] social thought" (N. Clark and Szerszynski 2020), Project Persephone imagines more open understandings of the ship's relationship to the energies and radiation penetrating it, perhaps in distinction from the "unsustainable" and "closed systems" approaches redolent of a "Spaceship Earth" (Armstrong 2013).

Such projects diagram other evacuation possibilities, gestured within more open and inclusive planetary evacuation ships, and other more solidaristic and collective evacuations throughout this book. I think there are useful notes of caution in this. Evacuation is not simply destined to recur, to reproduce the representational drawings and diagrams of evacuation at face value, as if what Derek Gregory (2019b) called the "linear geometry" of the chain-like or treelike structures discussed throughout should become the order and direction through which we continue to go on evacuating, reproducing ourselves and our social relations and structures. Even if they resemble phylogenetic ladder representations, it is too easy to assume that they will produce hierarchical, hereditary, teleological, and species-singular notions of evacuation.

As others have asked with regard to the lineage of these diagrammatic structures within biology, inspired in part by Lyn Margulis and coevolutionary theories and thinking around endosymbiosis (Hejnol 2017), are there other possibilities that are less "discouraging of curiosity" (100), leaving possible multidirectional, nonteleological branches, multispecies and companion evacuations that are more plural and shared? We have seen these promises in the book even in the flailing lines that trace the confusing outlines of falling and reforming (in multiple senses of political action against working conditions via embodied solidarities) women's bodies portrayed around high-rise emergencies, or in the detouring word plays around evacuation—to *scat*, for example. Thought palimpsestically, evacuation's multiplicities have worked with structures of violence and control, especially in its polysemiotic meanings, but they have also been overlain and rewired by other social and civil practices, aesthetic and affective transitions. Humor has been a strong interruption to evacuation's aesthetic utterances.

So these odd futures of planetary and space evacuations, born from the fragilities of life in the Anthropocene, help us think in another direction. The imagination of evacuating the Earth can be a critical vehicle with which to reengage ethically and practically with life, human, nonhuman, posthuman, right here. In shifting efforts away, as Woolf suggests, from evacuation or procreation and their logic of futurity, we can come to terms with more creative and affirmative reproductions. We should ask difficult questions of planetary evacuations in order to ask harder questions of terrestrial evacuation. We should be finding out, as films and fictions tend to dramatize and interrogate, who, in fact, gets to go, who doesn't, why and why not. In some cases, evacuation may be for the best but not perhaps for everyone. How to include those who do not want to go? How to take care of and account for (im)mobile bodies in evacuation, who are more than the mobile atoms or singular lines that represent them, but bodies coming together in sympathy and solidarity?

I hope this book has not provided us with a get-out clause for evacuation in its demonstration of the often emptiness of the term, its vacuity, its powers of erasure—even of itself—but rather a way to call out when we plan and call for an evacuation, about what we mean, about who or what is being protected, and for whose benefit. It means calling out those events we call evacuations and the way we historicize them and find relevance in them in the present. It means being careful with what we call them, and in what context an evacuation is uttered and how and by whom it is muttered or declared. It means to consider the multiple arts, technologies, and practices of evacuation—with attention to its names, concepts, and practices—as aesthetic experiences and judgments of how we move, are expected to move, are moved, or may not move.

Ultimately, it means reclaiming evacuation from multiple points of view and rehabilitating the term: to not evacuate evacuation perhaps but to fill it with its dispersions and singularities, its recurring revacuating contexts, histories, and conditions, and the multiple words, practices, feelings, and meanings that repeat in, attach to, and cohere around it.

INTRODUCTION

1   "Darwin Disaster—Situation Report," January 1975, folder 34, Red Cross Archives, Melbourne University Archives.

2   Director of Social Work, Red Cross Australia, Victoria Division, "Report of Work Undertaken and Impressions Gained While in Darwin from 3rd–16th January, 1975," folder 34, Red Cross Archives, Melbourne University Archives.

3   Ruary Bucknall, 1990, Oral History Interview, TS No. 599, Northern Territory Archives, Darwin, NT.

4   "*Emergency Exit*: Ilya and Emilia Kabakov," Fine Art Biblio, 1993, https://fineartbiblio.com/artworks/ilya-and-emilia-kabakov/934/emergency-exit.

2. MOBILE MEDICAL-MILITARY MACHINES

1   68 Parl. Deb. H.C. (1914) col. 1407.

2   Copy of a Letter from Colonel Arthur Lee, M.P., to Lord Kitchener, Paris, October 12, 1914, p. 3.

3   Report of the Meeting Held at the Medical Board Room, January 15, 1917, p. 2, RAMC/365/4, Wellcome Trust Library, London.

4   Report of the Meeting Held at the Medical Board Room, January 15, 1917, p. 2.

5   Almroth Wright, Memorandum on the Necessity of Creating at the War Office a Medical Intelligence and Investigation Department to Get the Best Possible Treatment for the Wounded; Diminish Invaliding; and Return the Men to the Ranks in Shortest Time, January 15, 1917, RAMC/365/4, Wellcome Trust Library, London.

6   Arthur Sloggett, Memorandum, January 15, 1917, pp. 10–11, RAMC/365/4, Wellcome Trust Library, London.

7   Arthur Sloggett to Almroth Wright, January 15, 1917, RAMC/365/4, Wellcome Trust Library.

8   Almroth Wright to Arthur Sloggett, January 17, 1917, 2, RAMC/365/4, Wellcome Trust Library.

9   Jaromir Baron von Mundy was the founder of the Vienna Ambulance Service (Figl and Pelinka 2005).

10 Thomas Longmore, "On the Geneva Convention of August the 22nd 1864," lecture delivered to the Royal United Services Institute, August 22, 1864, p. 11, Wellcome Trust Library.

11 Longmore, "On the Geneva Convention," p. 11.

12 The Royal Commission on South African Hospitals in 1901 termed Deelfontein "a palace of luxury" (Crichton-Harris 2009, 74).

13 91 Parl. Deb. H.C. (1901) cols. 517–56.

14 68 Parl. Deb. H.C. (1914) col. 1410 .

15 68 Parl. Deb. H.C. (1914) col. 1411.

16 An Act to Establish a Uniform System of Ambulances in the Armies of the United States, 1864, 38th Cong., 1st sess., ch. 27, p. 20.

17 See the papers of Colonel John H. Plumridge, RAMC/1752/2/3–4, Wellcome Trust Library.

18 Longmore, "On the Geneva Convention," 15.

19 Longmore, "On the Geneva Convention," 15.

20 Longmore, "On the Geneva Convention," 15.

21 I'm grateful to curators at the archives of the National Railway Museum in York for allowing me to access their ambulance train archives, which included plans and brochures.

3. EVACUATION AND EUPHEMISM

1   I am very grateful to Hartmut Behr for pushing me on this point in a previous version of this argument.

2   Irving v. Penguin Books Limited, Deborah E. Lipstadt, [2000] EWHC QB 115, 5.192, accessed March 14, 2024, https://www.hdot.org/judge/.

3   *Irving*, 6.107, https://www.hdot.org/judge/.

4   Korematsu v. United States, 323 U.S. at 224 (1944).

5   *Korematsu*, 323 U.S. at 230.

6   *Korematsu*, 323 U.S. at 232.

7   *Korematsu*, 323 U.S. at 233.

8   To avoid any misreading, my fundamental point here is not to critique Lipstadt's devastating work on Holocaust denial but rather to point out that lexical complexities, distortions, and power to distort can infect the writing of even those who are the most careful to expose it.

9   Interview of Dillon S. Meyer on the Relocation of Japanese-Americans, ca. 1943, Records of the War Relocation Authority, National Archives Identifier: 2284719, US National Archives, https://catalog.archives.gov /id/2284719.

10 Commission on Wartime Relocation and Internment of Civilians, *Hearings*, Congress Subcommittee on the Judiciary, 96th Cong., 2d sess., on H.R. 5499, 1980, p. 70.

11 Speech of Hon. John Collier, Commissioner of Indian Affairs, June 27, 1942, 4–5, Japanese American Evacuation and Resettlement Records, Bancroft Library, University of California, Berkeley.

12 Testimony of Ralph M. Gelvin, Assistant Project Manager Colorado River War Relocation Project, Poston, Arizona, "Congressional Committee on Un-American Activities," *Hearings*, Congress Special Committee on Un-American Activities 1938–44, vol. 15, 1943, p. 8858.

13 Alexander Leighton, "General Plans for Poston," February 22, 1943, Sociological Research Project, Poston, Arizona, Japanese American Evacuation and Resettlement Records, BANC MSS 72/233, Bancroft Library.

14 Leighton, "General Plans for Poston," 5.

15 Memo to Wade Head, "Memo on WRA Policy," Bureau of Sociological Affairs, August 9, 1942, 1, Japanese American Evacuation and Resettlement Records, Bancroft Library.

16 Memo to Wade Head, "Memo on WRA Policy."

17 Danny Iwanaga, "So Interned!," *Poston Notes and Activities*, spring 1943, 46, Poston Relocation Center, Online Archive of California, BANC MSS 67/14 c, folder J9.01, Bancroft Library.

18 W. Wado Hoad, untitled message, "Messages," *Press Bulletin*, 1942, 4, Online Archive of California, BANC MSS 67/14 c, folder J2.94, Bancroft Library.

19 Norris E. James, untitled message, "Messages," *Press Bulletin*, 1942, 4.

20 R. H. Rupkey, untitled message, "Messages," *Press Bulletin*, 1942, 5.

21 William S. Sharp, untitled message, "Messages," *Press Bulletin*, 1942, 5.

22 Frank Mizusawa, "Poston Agrarians," *Press Bulletin*, 1942, Ag-1.

23   "Ban on Japanese Literature Placed," *Racemaker* 1, no. 24, July 8, 1942, WRA Archives, Melbourne University Library.

24   *Manzanar Free Press*, April 9, 1944.

4. "THE CITY IS TO BE EVACUATED"

1   Statement of General Lucius D. Clay, Chairman, President's Advisory Committee on a National Highway Program, March 11, 1955, Subcommittee on Public Roads, Committee on Public Works, 84th Cong., pp. 395–96.

2   President Eisenhower, Special Message to the Congress Regarding a National Highway Program, February 22, 1955, White House Office, Office of the Press Secretary to the President, Box 4, Press Releases Feb. 8–March 14, 1955, NAID #16857605, https://www.eisenhowerlibrary .gov/sites/default/files/research/online-documents/interstate-highway -system/1955-02-22-message-to-congress.pdf.

3   Statement of Robert Moses, City Construction Co-ordinator of New York City, presented by Honorable David L. Lawrence, Mayor of Pittsburgh, Pennsylvania, Advisory Committee on a National Highway Program, General Lucius D. Clay, Chair, October 7, 1954, Washington, DC, pp. 48–50.

4   Statement of Burton N. Behling, Economist, Association of American Railroads, Transportation building, Washington, DC, Advisory Committee on a National Highway Program, General Lucius D. Clay, Chair, October 7, 1954, Washington, DC, p. 183.

5   Statement of Burton N. Behling, p. 187.

6   Statement of A. D. Condon, General Counsel for the Independent Advisory Committee to the Trucking Industry Inc., Advisory Committee on a National Highway Program, General Lucius D. Clay, Chair, October 7, 1954, Washington, DC, p. 45.

7   *Federal Civil Defense Act of 1950: Hearings*, Congress Committee on Armed Services, 81st Cong., 2d sess., vol. 2, p. 181.

8   Val Peterson, testimony to "Civil Defense for National Survival," 1956, Subcommittee of the Committee on Government Operations House of Representatives, 84th Cong., 2d sess., p. 1389.

9   Civil Defense Program: statement of Val Peterson, February 26, 1955, Hearings before the Subcommittee on Civil Defense of the Committee of Armed Services, US Senate, 84th Cong., 1st sess., p. 81.

10   Landstreet, quoted in the symposium report, *Studies of Preattack Evacuation*, based on the symposium held November 5–6, 1954, Battle Creek, Michigan, compiled and edited by Bertrand Klass, Committee

on Disaster Studies, National Academy of Sciences–National Research Council, Archives of the National Academy of Sciences, p. 19.

11   Landstreet, quoted in Klass, *Studies in Preattack Evacuation*, 18.

12   "Proposed Time-Space Car Tracking Plan for Bremerton, Washington. Operation Rideout on June 24, 1954," prepared by John Mathewson, in *Operation Rideout*, by William Garrison, Harry Matthewson, William D. White, and Bertrand Klass, Committee on Disaster Studies, Archives of the National Academy of Sciences, p. 13.

13   Williams, quoted in Klass, *Studies in Preattack Evacuation*, 1.

14   Questionnaire, Operation Rideout, director, Kitsap County Civil Defense, Bremerton, Washington, June 18, 1954, Committee on Disaster Studies, Archives of the National Academy of Sciences.

15   Garrison et al., *Operation Rideout*, 2.

16   Garrison et al., *Operation Rideout*, 4.

17   Robert Livingston to Lawrence Livingston Jr., April 12, 1954, Committee on Disaster Studies, Archives of the National Academy of Sciences.

18   Garrison et al., *Operation Rideout*, 2.

19   Garrison et al., *Operation Rideout*, 4.

20   Garrison et al., *Operation Rideout*, 4.

21   Garrison et al., *Operation Rideout*, 8.

22   Garrison et al., *Operation Rideout*, 7.

23   Garrison et al., *Operation Rideout*, 9.

24   Garrison et al., *Operation Rideout*, 9.

25   Garrison et al., *Operation Rideout*, 10.

26   L. Livingstone, Bertrand Klass, and John Rohrer, *Operations Walkout, Rideout, and Scat: Studies of Civil Defense Dispersal Test Exercises in Spokane, Bremerton, and Mobile* 1955, Organised by the Committee on Disaster Studies, for the Federal Civil Defense Administration, Archives of the National Academy of Sciences, p. 2.

27   Mr. Harry Williams, National Research Council, to Dr. John Rohrer, May 20, 1954, Committee on Disaster Studies, Archives of the National Academy of Sciences.

28   "A Report on Operation Scat: A Preliminary Report to the Committee on Disaster Studies, National Research Council," by Dr. John Rohrer, Director, Institute of Urban Studies, Tulane University, Committee on Disaster Studies, Archives of the National Academy of Sciences.

29   Participant Observer J, in Rohrer, "Report on Operation Scat," 4.

30   Research Observer B, in Rohrer, "Report on Operation Scat," 5.

31   Research Observer W, in Rohrer, "Report on Operation Scat," 2.

32   Research Observer W, in Rohrer, "Report on Operation Scat," 9.

33   Research Observer W, in Rohrer, "Report on Operation Scat," 9.

34   Research Observer B, in Rohrer, "Report on Operation Scat," 8.

35   Research Observer S, in Rohrer, "Report on Operation Scat," 6.

36   Research Observer W, in Rohrer, "Report on Operation Scat," 9.

37   Research Observer D, in Rohrer, "Report on Operation Scat," 7.

38   Research Observer D, in Rohrer, "Report on Operation Scat," 3.

39   Research Observer D, in Rohrer, "Report on Operation Scat," 8.

40   "Operation Lifesaver," Instructions for Evacuation, City of Calgary, Civil Defence, September 1955. Author's collection.

41   Statement of Val Peterson, February 26, 1955, Civil Defense Subcommittee of the Committee of Armed Services, 84th Cong., 1st sess., p. 82.

42   Statement of W. L. Shaffer, deputy director, Weld County, Civil Defense Agency, Colorado, April 5, 1955, Civil Defense Subcommittee of the Committee of Armed Services, 84th Cong., 1st sess.

43   *4 Wheels to Survival: Civil Defense and Your Car*, Federal Civil Defense Authority, 1955, leaflet and envelope addressed to residents of the Savannah-Chatham city area, Savannah Municipal Archives, Georgia.

44   *4 Wheels to Survival*, 4.

45   "150,000 School Children Evacuated," *Georgia Alert* 7, no. 4 (July 1957), Georgia State Archives.

46   "State of Georgia Survival Plan Project Proposals for Atlanta and Its Support Area," Civil Defense Division, Office of the State Director, Atlanta, 1956, Georgia State Archives.

5. COMPANION EVACUATIONS AT THE BOUNDARIES OF LIFE

1   Toshio Yatsushiro would write about his experiences of evacuation at the Poston Colorado River camp (Yatsushiro, Ishino, and Matsumoto 1944).

2   Sir Gordon Johnson to Mr. Shelling, October 15, 1942, Horseferry House, HO 186/1224, National Archives, UK.

3   John Anderson to Robert Morgan, MP, July 4, 1940, HO 186/1224, National Archives, UK.

4   D. J. Lidbury, Ministry of Home Security, to Chief Constables (Evacuation Areas), July 30, 1940, HO 186/1224, National Archives, UK.

5   Office of Regional Commissioner, South East Regional Office, August 22, 1942, HO 186/1224, National Archives, UK.

6   "Disposal of Cats and Dogs in Evacuation Areas," November 6, 1942, HO 186/1224, National Archives, UK.

7   "Draft: Domestic Pets in Evacuation Towns," 1942, HO 186/224; and Draft Circular for Issue to Chief Constables by Regional Commissioners in Nos 4 and 12 Regions, 1942, HO 186/1224, National Archives, UK.

8   Draft Circular for Issue to Chief Constables.

9   R. H. Franklin, Ministry of Agriculture and Fisheries, letter, August 8, 1940, HO 186/1224, National Archives, UK.

10   Donald Ferguson, Ministry of Agriculture and Fisheries, "Circular Letter to County War Agricultural Executive Committees in Negland and Wales: Evacuation of Livestock from Coastal Areas," July 12, 1940, HO 186/1224, National Archives, UK.

11   Circular Letter to County War Agricultural Executive Committees in England and Wales, September 1940, HO 186/224, National Archives, UK.

12   Report on the Service for Urban Areas, NARPAC, April 9, 1942, p. 1, HO 186/2075, National Archives, UK.

13   Adam S. Windle to the Home Office, March 24, 1932, HO 186/1224, National Archives, UK.

14   In 2007 Jerry Sneed again reported to *Time*, "To a lot of people, these animals are their children" (McCulley 2007).

15   The Centers for Disease Control and Prevention (n.d.) of the US Department of Health and Human Services have produced their own "Pet Disaster Check List."

16   2006 Louisiana Laws RS 14:102—Definitions; cruelty to animals.

6. A DISENGAGEMENT

1   As I noted in my acknowledgments, given that the final version of the book went into production in late 2022–early 2023, it is grossly inadequate in its anticipation of or application to the terrible and asymmetric violence of the Hamas attacks of October 7, 2023, and Israel's bombing and invasion of the Gaza Strip. Great caution and care should be taken in trying to apply this book's discussion to those events.

2   Appendix No. 130/A of the Disengagement Plan Implementation Bill, 5755-2005, which became law on February 18, 2005. I'm grateful to Craig Jones for his help in sourcing the implementation bill papers.

3   The report cites Pnina Shukrun-Nagar (2008, 341). I'm very grateful to Pnina and Mor Shilon for their thoughts, suggestions, and linguistic help with Hebrew.

4   Shukrun-Nagar's fascinating paper draws on Moshe Arens and other writers collected within Ari Shavit's (2005) *A Land Divided*, which features over thirty texts on the disengagement by Israeli writers and commentators.

5   Getty Images, caption text for image 53348677MDL049, titled "Israeli Troops Train to Forcibly Evacuate Settlers," Marco Di Lauro / Stringer, 2005, https://www.gettyimages.co.uk/detail/news-photo/israeli -soldier-playing-the-role-of-extremist-jewish-news-photo/53353356 ?adppopup=true.

6   "Israelis at Home—Gush Katif, Summer 2005," accessed February 21, 2024, http://www.gushkatif.net.

7. SEEING EVACUATION LOGISTICALLY

1   Under the item "Orphans Getting Great Care at Emb. PaP," in email from Department of State to Oscar Flores, aide to Secretary of State Hillary Clinton, "Fw: Haiti Major Developments Report (MDR)— 12:00," January 24, 2010, Hillary Clinton Email Archive, WikiLeaks, p. 1, https://wikileaks.org/clinton-emails/emailid/25786.

2   "Orphans Getting Great Care," 1.

3   "Visit to Haiti—Statements Made by Nicolas Sarkozy, President of the Republic, during His Joint Press Conference with René Préval, President of the Republic of Haiti," February 17, 2010, France in the United Kingdom: French Embassy in London, accessed February 1, 2020, https://uk .ambafrance.org/President-Sarkozy-s-press,16874.

4   "Visit to Haiti—Statements Made by Nicolas Sarkozy."

5   "Temporary Exclusion of the Assessment of Overflight Fees for Humanitarian Flights Related to the January 12, 2010, Earthquake in Haiti," notice by the Federal Aviation Administration, *Federal Register*, May 11, 2010.

6   American Embassy Cable, Port-au-Prince, Haiti, TFHA0I: EMBASSY PORT AU PRINCE EARTHQUAKE SITREP AS OF 1800 DAY 4, January 16, 2010, WikiLeaks, https://search.wikileaks.org/plusd/cables /10PORTAUPRINCE50_a.html.

7   "Turmoil in Libya," Red24, accessed February 1, 2020, https://www .red24.com/uploads/Casestudy_Libyaunrest.pdf.

1   State Attorney General Rob Hulls, session of the Victorian Parliament 2008–2009, February 24, 2009, https://www.parliament.vic.gov.au /downloadhansard/pdf/Assembly/Feb-Jun%202009/Assembly%20 Extract%2024%20February%202009%20from%20Book%202.pdf.

2   There are strange parallels to and inversions of William Knight's famous *Collins Street* 1839 watercolor (Edmonds 2010, 47), which contains echoes of removal and an existence "out of time."

3   John Brumby and Lynne Kosky, transcript of conversation on Friday, February 6, 2009, Evidence Compiled by the Victorian Bushfires Royal Commission, http://royalcommission.vic.gov.au/getdoc/e63bf956-f328 -45d2-b87b-05f8482a3571/WIT.005.001.2383.pdf.

4   "Transcript of Proceedings," 2009 Victorian Bushfires Royal Commission, June 16, 2009, http://royalcommission.vic.gov.au/getdoc/32d93df1 -8563-4f9e-b984-34a1fad25667/Transcript-VBRC_Day-024_16-Jun -2009.PDF. Karen Kissane (2010) also draws on the resident's testimony.

5   "Transcript of Proceedings," 2009 Victorian Bushfires Royal Commission, June 16, 2009, 150.

6   "Transcript of Proceedings," 2009 Victorian Bushfires Royal Commission, June 16, 2009, 176.

7   "Evacuation during Wildfire," Country Fire Association internal briefing, 2009 Victorian Bushfires Royal Commission, evidence submission, 1995, WIT.3004.003.0049, p. 3, http://royalcommission.vic.gov.au/getdoc /3e26a2d2-86b4-4528-84e7-18a68fe49d3a/WIT.3004.003.0049.PDF.

8   "Transcript of Proceedings," 2009 Victorian Bushfires Royal Commission, June 16, 2009, 177.

9   "Transcript of Proceedings," 2009 Victorian Bushfires Royal Commission, June 3, 2009, 73, http://royalcommission.vic.gov.au/getdoc /5564dc6a-34af-4331-819e-3259f04c50c9/Transcript-VBRC_Day-017 _03-Jun-2009.PDF.

10   "Transcript of Proceedings," 2009 Victorian Bushfires Royal Commission, June 3, 2009, 76.

11   Firefoxes Australia is a women's group and network that emerged from communities located in the Kinglake Ranges as they recovered from the Black Saturday disaster fires of 2009. See https://firefoxes.au/.

12   I'm hugely grateful to Healesville Sanctuary and Zoos Victoria for talking me through the events of Black Saturday and the sanctuary's experience of the evacuation.

ABC News. 2009. "An Internment Camp within an Internment Camp." February 9, 2009. https://abcnews.go.com/US/story?id=4310157&page=1.

Abel, Elizabeth. 2008. "Double Take: Photography, Cinema, and the Segregated Theater." *Critical Inquiry* 34 (S2): S2–20.

Adams, Ansel. 1944. *Born Free and Equal.* New York: US Camera.

Adams, Guy. 2010. "Adoption Agencies Warned Off Haiti's Orphans." *Independent*, January 19, 2010. https://www.independent.co.uk/news/world/americas/adoption-agencies-warned-off-haitis-orphans-1871972.html.

Adey, Peter. 2010. *Aerial Life: Spaces, Mobilities, Affects.* Oxford: Wiley Blackwell.

Adey, Peter. 2016. "Emergency Mobilities." *Mobilities* 11 (1): 32–48.

Adey, Peter. 2020. "Evacuated to Death: The Lexicon, Concept, and Practice of Mobility in the Nazi Deportation and Killing Machine." *Annals of the American Association of Geographers* 110 (3): 808–26.

Adey, Peter, and Ben Anderson. 2011. "Event and Anticipation: UK Civil Contingencies and the Space-Times of Decision." *Environment and Planning A: Economy and Space* 43 (12): 2878–99.

Adey, Peter, Mark Whitehead, and Alison Williams, eds. 2013. *From Above: War, Violence and Verticality.* London: Hurst.

Adler, H. G. 2017. *Theresienstadt, 1941–1945: The Face of a Coerced Community.* Cambridge: Cambridge University Press.

Aguiar, Marian. 2011. *Tracking Modernity: India's Railway and the Culture of Mobility.* Minneapolis: University of Minnesota Press.

Aguirre, Benigno E. 1983. "Evacuation as Population Mobility." *International Journal of Mass Emergencies and Disasters* 1 (3): 415–37.

Ahmed, Sara, Claudia Castada, Anne-Marie Fortier, and Mimi Sheller, eds. 2003. *Uprootings/Regroundings: Questions of Home and Migration.* Oxford: Berg.

AlarmPhone. 2015. "Ferries Not Frontex! 10 Points to Really End the Deaths of Migrants at Sea." *openDemocracy*, June 4, 2015. https://www.opendemocracy.net/en/beyond-trafficking-and-slavery/ferries-not-frontex-10-points-to-really-end-deaths-of-migrants-at-sea/.

Allen, Jeffner. 1996. *Sinuosities, Lesbian Poetic Politics*. Bloomington: Indiana University Press.

Alloun, Esther. 2018. "'That's the Beauty of It, It's Very Simple!': Animal Rights and Settler Colonialism in Palestine–Israel." *Settler Colonial Studies* 8 (4): 559–74.

Alush, Zvi. 2005. "Rabbi: Hurricane Punishment for Pullout." *ynetnews*, September 7, 2005. https://www.ynetnews.com/articles/0,7340,L-3138779,00.html.

Amar, Paul. 2012. "Global South to the Rescue: Emerging Humanitarian Superpowers and Globalizing Rescue Industries." *Globalizations* 9 (1): 1–13.

Anderson, Ben. 2014. *Encountering Affect: Capacities, Apparatuses, Conditions*. Aldershot, UK: Ashgate.

Anderson, Ben. 2017. "Emergency Futures: Exception, Urgency, Interval, Hope." *Sociological Review* 65 (3): 463–77.

Anderson, Ben, and Peter Adey. 2011. "Affect and Security: Exercising Emergency in 'UK Civil Contingencies.'" *Environment and Planning D: Society and Space* 29 (6): 1092–109.

Andrews, Maggie. 2019. *Women and Evacuation in the Second World War: Femininity, Domesticity and Motherhood*. London: Bloomsbury.

Aradau, Claudia, and Rens Van Munster. 2011. *Politics of Catastrophe: Genealogies of the Unknown*. London: Routledge.

Arendt, Hannah. 1963. *Eichmann in Jerusalem*. London: Penguin.

Argus. 1883. "Black Thursday." March 16, 1883.

Armstrong, Rachel. 2013. "Project Persephone." *Centauri Dreams*, September 6, 2013. https://www.centauri-dreams.org/2013/09/06/project-persephone/.

Arutz Sheva / Israel National News TV. 2011. "South Carolina Senator Visits Gush Katif Memorial Museum." IsraelNationalNews.com, December 12, 2011. https://www.youtube.com/watch?v=SmdnsF3WyUc.

Ashkenzi, Eli. 2005. "Ovadia Yosef: Katrina Is God's Punishment for Disengagement." *Haaretz*, September 8, 2005. https://www.haaretz.com/1.4940323.

Attewell, Wesley. 2018. "'From Factory to Field': USAID and the Logistics of Foreign Aid in Soviet-Occupied Afghanistan." *Environment and Planning D: Society and Space* 36 (4): 719–38.

Averill, Jason D., Dennis S. Mileti, Richard, D. Peacock, Erica D. Kuligowski, Norman Groner, Guylene Proulx, Paul A. Reneke, and Harold E. Nelson. 2005. *Occupant Behavior, Egress, and Emergency Communications: Federal Building and Fire Safety Investigation of the World Trade Center Disaster*. Washington, DC: National Institute of Standards and Technology (NIST). https://permanent.fdlp.gov/lps59595/lps59595.pdf.

Azoulay, Ariella. 2011. *From Palestine to Israel: A Photographic Record of Destruction and State Formation, 1947–1950*. Vol. 4. London: Pluto.

Azoulay, Ariella. 2012. *Different Ways Not to Say Deportation*. Vancouver: Fillip Editions.

Bahr, D. 2007. *The Unquiet Nisei: An Oral History of the Life of Sue Kunitomi Embrey*. New York: Palgrave Macmillan.

Baigorria, Osvaldo. 2009. "La guerra rizomática." *Página/12*, February 8, 2009. https://www.pagina12.com.ar/diario/suplementos/radar/9-5100-2009 -02-08.html.

Bailey, James. 2011. "'Repetition, Boredom, Despair': Muriel Spark and the Eichmann Trial." *Holocaust Studies* 17 (2–3): 185–206.

Baird, John, R. K. Betts, Frederik Graff, C. H. Banes, Strickland Kneass, William D. Marks, Henry G. Morris, J. B. Lippincott, and Isaac Norris. 1881. "Report of Committee of the Franklin Institute on Fire-Escapes and Elevators." *Journal of the Franklin Institute* 112 (6): 408–14.

Baker, William K., and Susan D. Chan, eds. 2007. "Special Issue on Crisis Management in Zoos." *Animal Keepers Forum* 34 (11–12): 455–588.

Barber, Daniel A. 2019. "Emergency Exit." *e-flux Architecture*, September 2019. https://www.e-flux.com/architecture/overgrowth/284030/emergency-exit/.

Barker-Devine, Jenny. 2006. "'Mightier than Missiles': The Rhetoric of Civil Defense for Rural American Families, 1950–1970." *Agricultural History* 80 (4): 415–35.

Barnes, Trevor J. 2001. "Lives Lived and Lives Told: Biographies of Geography's Quantitative Revolution." *Environment and Planning D: Society and Space* 19 (4): 409–29.

Barnes, Trevor J. 2012. "Reopke Lecture in Economic Geography: Notes from the Underground: Why the History of Economic Geography Matters: The Case of Central Place Theory." *Economic Geography* 88 (1): 1–26.

Barnes, Trevor J., and Matthew Farish. 2006. "Between Regions: Science, Militarism, and American Geography from World War to Cold War." *Annals of the Association of American Geographers* 96 (4): 807–26.

Barnes, Trevor J., and Claudio Minca. 2012. "Nazi Spatial Theory: The Dark Geographies of Carl Schmitt and Walter Christaller." *Annals of the Association of American Geographers* 103 (3): 669–87.

Barry, Kaya. 2017. "The Aesthetics of Aircraft Safety Cards: Spatial Negotiations and Affective Mobilities in Diagrammatic Instructions." *Mobilities* 12 (3): 365–83.

Barry, Kaya. 2020. "Diagramming the Uncertainties of COVID-19: Scales, Spatialities, and Aesthetics." *Dialogues in Human Geography* 10 (2): 282–86.

Barton, Gregory. 2000. "Keepers of the Jungle: Environmental Management in British India, 1855–1900." *Historian* 62 (3): 557–74.

Bates, Leon. (1974) 1991. *A Tribulation Map*. Sherman, TX: Bible Believers' Evangelistic Association. https://www.bbea.org/languageFiles /TribulationMapNEWcombined.pdf.

Bates, Leon. 1983. *Are You Ready?* Sherman, TX: Bible Believers' Evangelistic Association. https://www.bbea.org/images/pdfs/doc_tracts/Are%20 You%20Ready%20Tract.pdf.

Baum, Marsha L. 2016. "'Room on the Ark?': The Symbolic Nature of US Pet Evacuation Statutes for Nonhuman Animals." In *Considering Animals*, edited by Carol Freeman, Elizabeth Leane, and Yvette Watt, 105–18. New York: Routledge.

Bayor, Ronald H. 1988. "Roads to Racial Segregation: Atlanta in the Twentieth Century." *Journal of Urban History* 15 (1): 3–21.

BBC. 2019. "Grenfell Survivor 'Had to Trust' Firefighters' Advice." *BBC News*, November 5, 2019. https://www.bbc.co.uk/news/av/uk-50309512/grenfell -survivor-had-to-trust-firefighters-advice.

BBC. 2021. "Afghanistan: Operation Ark Luton Evacuation Plane Is Swapped." *BBC News*, August 26, 2021. https://www.bbc.co.uk/news/uk-england-beds -bucks-herts-58340272.

Beattie, James. 2011. *Empire and Environmental Anxiety: Health, Science, Art and Conservation in South Asia and Australasia, 1800–1920*. Berlin: Springer.

Beevor, W. C. 1914. "The Removal of Sick and Wounded in Motor-Lorries: A Warning and a Counter Proposal." *Journal of the Royal Army Medical Corps*, July 1, 1914, 66–68.

Ben, Eyal. 2006. "Gaza Pullout: A Year On." *ynetnews*, August 17, 2006. https:// www.ynetnews.com/articles/0,7340,L-3292340,00.html.

Benischek, Brad. 2007. *Revacuation*. New Orleans: Press Books.

Beres, Louis René. 1983. "Subways to Armageddon." *Society* 20 (6): 7–10.

Berg, Nicolas. 2014. *Luftmenschen: Zur Geschichte einer Metapher*. Göttingen, Germany: Vandenhoeck und Ruprecht.

Bergquist, Kathleen Ja Sook. 2009. "Operation Babylift or Babyabduction? Implications of the Hague Convention on the Humanitarian Evacuation and 'Rescue' of Children." *International Social Work* 52 (5): 621–33.

Beriwal, Madhu. 2006. "Preparing for a Catastrophe: The Hurricane Pam Exercise." Statement before the Senate Homeland Security and Governmental Affairs Committee, January 24, 2006. https://www.hsgac.senate.gov/imo /media/doc/012406Beriwal.pdf.

Berlant, Lauren Gail. 2011. *Cruel Optimism*. Durham, NC: Duke University Press.

Bernard, Andreas. 2014. *Lifted: A Cultural History of the Elevator*. New York: NYU Press.

Bernardie-Tahir, Nathalie, and Camille Schmoll. 2014. "Opening Up the Island: A 'Counter-Islandness' Approach to Migration in Malta." *Island Studies Journal* 9 (1): 43–56.

Beyes, Timon, and Christian De Cock. 2017. "Adorno's Grey, Taussig's Blue: Colour, Organization and Critical Affect." *Organization* 24 (1): 59–78.

Bhattacharya, Sumangala. 2015. "'Those Two Thin Strips of Iron': The Uncanny Mobilities of Railways in British India." *Nineteenth-Century Contexts* 37 (5): 411–30.

Bhattacharya, Tithi. 2017. *Social Reproduction Theory: Remapping Class, Recentering Oppression*. London: Pluto.

Bissell, David. 2016. "Micropolitics of Mobility: Public Transport Commuting and Everyday Encounters with Forces of Enablement and Constraint." *Annals of the American Association of Geographers* 106 (2): 394–403.

Bissell, David. 2022. "The Anaesthetic Politics of Being Unaffected: Embodying Insecure Digital Platform Labour." *Antipode* 54 (1): 86–105.

Bissell, David, Maria Hynes, and Scott Sharpe. 2012. "Unveiling Seductions beyond Societies of Control: Affect, Security, and Humour in Spaces of Aeromobility." *Environment and Planning D: Society and Space* 30 (4): 694–710.

Blair, Jennifer. 2008. "Fire Escape." *Victorian Review* 34 (1): 52–56.

Blanchard, B. Wayne. 1985. *American Civil Defense 1945–1984: The Evolution of Programs and Policies*. Emmitsburg, MD: National Emergency Training Center, Federal Emergency Management Agency.

Blatman, Daniel. 2011. *The Death Marches: The Final Phase of Nazi Genocide.* Cambridge, MA: Harvard University Press.

Bonilla, Yarimar. 2020. "The Coloniality of Disaster: Race, Empire, and the Temporal Logics of Emergency in Puerto Rico, USA." *Political Geography* 78 (April): 102181. https://doi.org/10.1016/j.polgeo.2020.102181.

Bonner, Hugh, and Lawrence Veiller. 1900. *Special Report on Fire Escapes in New York and Brooklyn: Prepared for the Tenement House Commission of 1900*. New York: Evening Post Job Printing House.

Boucher, Leigh, and Lynette Russell. 2015. "Introduction: Colonial History, Postcolonial Theory and the 'Aboriginal Problem' in Colonial Victoria." In *Settler Colonial Governance in Nineteenth-Century Victoria*, edited by Leigh Boucher and Lynette Russell, 1–26. Canberra: ANU Press.

Bowditch, Henry. 1863. *A Brief Plea for an Ambulance System for the Army of the United States*. Boston: Ticknor and Fields.

Boyer, Paul. 2005. *By the Bomb's Early Light: American Thought and Culture at the Dawn of the Atomic Age*. Chapel Hill: University of North Carolina Press.

Braedley, Susan, and Meg Luxton. 2015. Foreword to *Precarious Worlds: Contested Geographies of Social Reproduction*, edited by Katie Meehan and Kendra Stauss, vii–xv. Atlanta: University of Georgia Press.

Braun, Stuart. 2008. "Last Refuge: Remembering Coranderrk Aboriginal Station." *Hindsight*, ABC, May 11, 2011. https://www.abc.net.au/listen/programs/hindsight/last-refuge-remembering-coranderrk-aboriginal/3261602.

Braverman, Irus. 2017. "Captive: Zoometric Operations in Gaza." *Public Culture* 29 (1) (81): 191–215.

Bristow, Tom, and Andrea Witcomb. 2016. "Melancholy and the Continent of Fire." In *A Cultural History of Climate Change*, edited by Tom Bristow and Thomas Ford, 72–86. London: Taylor and Francis.

British Medical Journal. 1903. "Report of the Royal Commission on the South African War." 2 (2226) (August 29): 484–87.

British Medical Journal. 1917. "The Royal Army Medical Corps and Its Work." 2 (2955) (August 18): 217–24.

Broshear, Nathan. 2010. "Airmen Fly Predator in Controlled Airspace over Haiti." US Air Force News, January 29, 2010. https://www.af.mil/News/Article-Display/Article/117786/airmen-fly-predator-in-controlled-airspace-over-haiti/.

Brown, Holly Cade. 2017. "Figuring Giorgio Agamben's 'Bare Life' in the Post-Katrina Works of Jesmyn Ward and Kara Walker." *Journal of American Studies* 51 (1): 1–19.

Buist, H. Massac. 1914. "Motor Ambulances in War Services." *British Medical Journal* 2 (September 20): 544–46.

Buist, H. Massac. 1916. "Motor Ambulance Development." *British Medical Journal* 1 (January 1): 18.

Burtch, Andrew P. 2012. *Give Me Shelter: The Failure of Canada's Cold War Civil Defence.* Vancouver: UBC Press.

Burton, Jeffery F., and Mary M. Farrell. 2013. "'Life in Manzanar Where There Is a Spring Breeze': Graffiti at a World War II Japanese American Internment Camp." In *Prisoners of War*, edited by Harold Mytum and Gilly Carr, 239–69. Berlin: Springer.

Burton, Robert. 1628. *The Anatomy of Melancholy: What It Is, with All the Kinds, Causes, Symptomes, Prognostickes, & Seuerall Cures of It in Three Partitions, with Their Severall Sections, Members & Subsections, Philosophically, Medicinally, Historically, Opened & Cut Up.* Oxford: Henry Cripps.

Butler, Judith. 2004. *Precarious Life: The Powers of Violence and Mourning.* London: Verso.

Bywater, H. E. 1940. "Air Raid Precautions for Animals." *Veterinary Journal* 96 (6): 221–31.

Cahill, Caitlin, and Rachel Pain. 2019. "Representing Slow Violence and Resistance." *ACME: An International Journal for Critical Geographies* 18 (5): 1054–65.

Campbell, Clare, and Christy Campbell. 2015. *Dogs of Courage: When Britain's Pets Went to War, 1939–45.* London: Little, Brown.

Cao, Yinghui, Bryan J. Boruff, and Ilona M. McNeill. 2017. "The Smoke Is Rising but Where Is the Fire? Exploring Effective Online Map Design for Wildfire Warnings." *Natural Hazards* 88 (3): 1473–501.

Carden-Coyne, Ana. 2014. *The Politics of Wounds: Military Patients and Medical Power in the First World War.* Oxford: Oxford University Press.

Carpio, Genevieve. 2019. *Collisions at the Crossroads: How Place and Mobility Make Race.* Berkeley: University of California Press.

Carrera, Elena. 2013. *Emotions and Health, 1200–1700.* Leiden: Brill.

Carroll, Rory, and Daniel Nasaw. 2010. "US Accused of Annexing Airport as Squabbling Hinders Aid Effort in Haiti." *Guardian*, January 17, 2010. https://www.theguardian.com/world/2010/jan/17/us-accused-aid-effort-haiti.

Carter, Paul. 1996. "Turning the Tables—or, Grounding Post-colonialism." In *Text, Theory, Space: Land, Literature and History in South Africa and Australia*, edited by Kate Darian-Smith, Liz Gunner, and Sarah Nuttall, 23–35. London: Routledge.

Catton, Theodore, and Diane L. Krahe. 2018. *The Sands of Manzanar: Japanese American Confinement, Public Memory, and the National Park Service.* Washington, DC: National Park Service, US Department of the Interior. https://www.oah.org/site/assets/files/10217/manz_admin_history.pdf.

Centers for Disease Control and Prevention. 2018. "Pets in Evacuation Centers." Last reviewed November 7, 2018. https://www.cdc.gov/healthypets/emergencies/pets-in-evacuation-centers.html.

Centers for Disease Control and Prevention. n.d. "Pet Disaster Kit: Easy as 1-2-3." Accessed March 18, 2024. https://www.cdc.gov/healthypets/resources/disaster-prep-pet-emergency-checklist.pdf.

Cesarani, David. 2016. *Final Solution: The Fate of the Jews, 1933–1949.* London: Pan Macmillan.

Chapman, Alix. 2017. "Katrina Babies: Reproducing Deviance in the Future Unknown." In *Navigating Souths: Transdisciplinary Explorations of a US Region*, edited by Michelle Grigsby Coffey and Jodi Skipper, 72–85. Athens: University of Georgia Press.

Chatham-Savannah Defense Council. 1955. *Escape from the H-bomb.* Leaflet. Savannah, GA: Chatham-Savannah Defense Council.

Cheney, Kristen. 2014. "'Giving Children a Better Life?': Reconsidering Social Reproduction, Humanitarianism and Development in Intercountry Adoption." *European Journal of Development Research* 26 (2): 247–63.

Chivers, C. J. 2011. "On Ship of Evacuees from Libya, Harrowing Tales." *New York Times*, April 11, 2011. https://www.nytimes.com/2011/04/19/world/africa/19evacuees.html.

Christian, Jenna Marie, and Lorraine Dowler. 2019. "Slow and Fast Violence." *ACME: An International Journal for Critical Geographies* 18 (5): 1066–75.

Chua, Charmaine, Martin Danyluk, Deborah Cowen, and Laleh Khalili. 2018. "Introduction: Turbulent Circulation: Building a Critical Engagement with Logistics." *Environment and Planning D: Society and Space* 36 (4): 617–29.

Clair, William S., and John S. Clair. 2004. *The Road to St. Julien: The Letters of a Stretcher-Bearer of the Great War.* Cheltenham: Pen and Sword Books.

Clark, Ian. 2015. *"A Peep at the Blacks": A History of Tourism at Coranderrk Aboriginal Station, 1863–1924.* Berlin: De Gruyter.

Clark, Nigel, and Bronislaw Szerszynski. 2020. *Planetary Social Thought: The Anthropocene Challenge to the Social Sciences.* Cambridge, UK: Polity.

Clarke, David B., Marcus A. Doel, and Francis X. McDonough. 1996. "Holocaust Topologies: Singularity, Politics, Space." *Political Geography* 15 (6–7): 457–89.

Clarsen, Georgine. 2017. "'Australia—Drive It Like You Stole It': Automobility as a Medium of Communication in Settler Colonial Australia." *Mobilities* 12 (4): 520–33.

CNN. 2005. "American Morning: Hurricane Katrina's Aftermath." *CNN Transcripts*, September 3, 2005. http://edition.cnn.com/TRANSCRIPTS/0509 /03/ltm.05.html.

Cole, Tim. 2011. *Traces of the Holocaust: Journeying in and out of the Ghettos.* London: Continuum.

Cole, Tim. 2016. *Holocaust Landscapes*. London: Bloomsbury.

Collier, Stephen J. 2009. "Topologies of Power: Foucault's Analysis of Political Government beyond 'Governmentality.'" *Theory, Culture and Society* 26 (6): 78–108.

Collier, Stephen J. 2014. "Neoliberalism and Natural Disaster: Insurance as Political Technology of Catastrophe." *Journal of Cultural Economy* 7 (3): 273–90.

Collier, Stephen J., and Andrew Lakoff. 2008. "Distributed Preparedness: The Spatial Logic of Domestic Security in the United States." *Environment and Planning D: Society and Space* 26 (1): 7–28.

Colls, Rachel, and Bethan Evans. 2014. "Making Space for Fat Bodies? A Critical Account of 'the Obesogenic Environment.'" *Progress in Human Geography* 38 (6): 733–53.

Connerly, Charles E. 2002. "From Racial Zoning to Community Empowerment: The Interstate Highway System and the African American Community in Birmingham, Alabama." *Journal of Planning Education and Research* 22 (2): 99–114.

Cook, Nancy, and David Butz. 2016. "Mobility Justice in the Context of Disaster." *Mobilities* 11 (3): 400–419.

Cooper, Melinda. 2015. "The Theology of Emergency: Welfare Reform, US Foreign Aid and the Faith-Based Initiative." *Theory, Culture and Society* 32 (2): 53–77.

Cooper, Melinda. 2017. *Family Values: Between Neoliberalism and the New Social Conservatism.* Cambridge, MA: MIT Press.

Cooper, Robyn. 1995. "The Fireman: Immaculate Manhood." *Journal of Popular Culture* 28 (4): 139–70.

Cooter, Roger, Mark Harrison, and Steve Sturdy. 1998. "Introduction: Of War, Medicine and Modernity." In *War, Medicine and Modernity*, edited by Roger Cooter, Mark Harrison, and Steve Sturdy, 1–21. Midsomer Norton, UK: Sutton.

Corbett, Glenn. 2018. "How the Design of the World Trade Center Claimed Lives on 9/11." *History Channel*, September 10, 2018. https://www.history .com/news/world-trade-center-stairwell-design-9-11.

Corvi, Steven J. 1998. "Men of Mercy: The Evolution of the Royal Army Veterinary Corps and the Soldier-Horse Bond during the Great War." *Journal of the Society for Army Historical Research* 76 (308): 272–84.

Cowen, Deb. 2014. *The Deadly Life of Logistics: Mapping Violence in Global Trade*. Minneapolis: University of Minnesota Press.

Cox, Helen. 1998. "Women in Bushfire Territory." In *The Gendered Terrain of Disaster: Through Women's Eyes*, edited by Elaine Enarson and Betty Hearn Morrow, 133–42. Westport, CT: Praeger.

Crary, Jonathan. 2001. *Suspensions of Perception: Attention, Spectacle, and Modern Culture*. Cambridge, MA: MIT Press.

Cresswell, Tim. 2001. *The Tramp in America*. London: Reaktion Books.

Cresswell, Tim. 2006. *On the Move: Mobility in the Modern Western World*. New York: Routledge.

Crichton-Harris, Ann. 2009. *Poison in Small Measure: Dr. Christopherson and the Cure for Bilharzia*. Leiden: Brill.

Crimp, Douglas. 2004. *Melancholia and Moralism: Essays on AIDS and Queer Politics*. Cambridge, MA: MIT Press.

Culver, Gregg. 2018. "Death and the Car: On (Auto) Mobility, Violence, and Injustice." *ACME: An International Journal for Critical Geographies* 17 (1): 144–70.

Cunningham, Sophie. 2012. *Melbourne*. Sydney: UNSW Press.

Cunningham, Sophie. 2014. *Warning: The Story of Cyclone Tracy*. Melbourne: Text.

Cutter, Susan, and Kent Barnes. 1982. "Evacuation Behavior and Three Mile Island." *Disasters* 6 (2): 116–24.

Czamanski-Cohen, J. 2010. "'Oh! Now I Remember': The Use of a Studio Approach to Art Therapy with Internally Displaced People." *Arts in Psychotherapy* 37 (5): 407–13.

Dahlenburg, Virginia, and Hamish Curry. 2011. "Virginia Dahlenburg on 'Black Thursday, February 6th, 1851.'" Transcript, State Library Victoria, Melbourne, February 11, 2011. https://www.slv.vic.gov.au/asset/video/1394.

Daily Mail. 1903. "Gallant Fire Rescues: Panic-Stricken Families in the East End." March 24, 1903, 3.

Daily Mail Brisbane. 1921. "A Vanishing Heritage." July 9, 1921, 6.

Daniels, Roger. 2005. "Words Do Matter: A Note on Inappropriate Terminology and the Incarceration of the Japanese Americans." In *Nikkei in the Pacific Northwest: Japanese Americans and Japanese Canadians in the Twentieth Century*, edited by Louis Fiset and Gail Nomura, 183–207. Seattle: University of Washington Press.

Daniels, Stephen. 1993. *Fields of Vision: Landscape Imagery and National Identity in England and the United States*. London: Polity.

Danowski, Deborah, and Eduardo V. de Castro. 2016. *The Ends of the World*. Oxford: Wiley.

Das, Santanu. 2006. *Touch and Intimacy in First World War Literature*. Cambridge: Cambridge University Press.

Davis, Tracy C. 2002. "Between History and Event: Rehearsing Nuclear War Survival." *TDR: The Drama Review* 46 (4): 11–45.

Davis, Tracy C. 2007. *Stages of Emergency: Cold War Nuclear Civil Defense.* Durham, NC: Duke University Press.

Day, Bill. 1975. "Darwin after Tracy: Showing the Ugly Side of 'Private Property' Syndrome." *Sydney Tribune*, February 18, 1975, 5.

Dekel, Rachel. 2010. "Mental Health Practitioners' Experiences during the Shared Trauma of the Forced Relocation from Gush Katif." *Clinical Social Work Journal* 38 (4): 388–96.

Delaney, Janice, Mary Jane Lupton, and Emily Toth. 1988. *The Curse: A Cultural History of Menstruation.* Urbana: University of Illinois Press.

Deleuze, Gilles. 1998. *Gilles Deleuze: Essays Critical and Clinical.* London: Verso.

Deleuze, Gilles. 2006. *Foucault.* London: Continuum.

DeLoughrey, Elizabeth. 2009. "Radiation Ecologies and the Wars of Light." *MFS: Modern Fiction Studies* 55 (3): 468–98.

Department of Buildings. 1968. *Building Laws of the City of New York: 1938 Building Code. Edited and Amended to December 6, 1968.* New York: City of New York.

Department of Defense (DOD). 2011. *Military and Security Developments Involving the People's Republic of China 2011.* Annual Report to Congress. https://dod.defense.gov/Portals/1/Documents/pubs/2011_CMPR_Final.pdf.

DeWitt, John L. 1943. *Final Report, Japanese Evacuation from the West Coast, 1942.* Washington, DC: U.S. Government Printing Office.

Díaz, Eva. 2014. *The Experimenters: Chance and Design at Black Mountain College.* Chicago: University of Chicago Press.

Diffrient, David Scott. 2018. "Elevator to the Shallows: Spatial Verticality and the Questionable Depth of Social Relations in *Mad Men*." *New Review of Film and Television Studies* 16 (3): 324–52. https://doi.org/10.1080/17400309.2018.1479181.

Doan, Laura. 2006. "Primum Mobile: Women and Auto/Mobility in the Era of the Great War." *Women: A Cultural Review* 17 (1): 26–41.

Donini, Antonio. 2010. "The Far Side: The Meta Functions of Humanitarianism in a Globalised World." *Disasters* 34 (s2): s220–37.

Dorling, Danny, Ben Wheeler, Mary Shaw, and Richard Mitchell. 2007. "Counting the 21st Century Children of Britain: The Extent of Advantage and Disadvantage." *Twenty-First Century Society* 2 (2): 173–89.

Dorrian, Mark, and Frédéric Pousin, eds. 2013. *Seeing from Above: The Aerial View in Visual Culture.* London: I. B. Tauris.

Dowler, Kevin. 2019. "Restoring the 'Appropriate Relation.'" *Politics and Animals* 5:1–16.

Downing, Johnette. 2019. *Take Your Pets with You.* Wiggle Worm Records LLC, in conjunction with the Louisiana Department of Agriculture and Forestry. https://youtu.be/Zoq5GgRLPMY.

Dwyer, Jim, and Kevin Flynn. 2011. *102 Minutes: The Unforgettable Story of the Fight to Survive inside the Twin Towers.* New York: Macmillan.

Edelman, Lee. 1998. "The Future Is Kid Stuff: Queer Theory, Disidentification, and the Death Drive." *Narrative* 6 (1): 18–30.

Edelman, Lee. 2004. *No Future: Queer Theory and the Death Drive*. Durham, NC: Duke University Press.

Edmonds, Penelope. 2010. *Urbanizing Frontiers: Indigenous Peoples and Settlers in 19th-Century Pacific Rim Cities*. Vancouver: UBC Press.

Elad-Strenger, Julia, Zvi Fajerman, Moran Schiller, Avi Besser, and Golan Shahar. 2013. "Risk-Resilience Dynamics of Ideological Factors in Distress after the Evacuation from Gush Katif." *International Journal of Stress Management* 20 (1): 57–75.

Elden, Stuart. 2013. *The Birth of Territory*. Chicago: University of Chicago Press.

Elden, Stuart, and Alison J. Williams. 2009. "The Territorial Integrity of Iraq, 2003–2007: Invocation, Violation, Viability." *Geoforum* 40 (3): 407–17.

Eriksen, Christine. 2014a. *Gender and Wildfire: Landscapes of Uncertainty*. New York: Routledge.

Eriksen, Christine. 2014b. "Gendered Risk Engagement: Challenging the Embedded Vulnerability, Social Norms and Power Relations in Conventional Australian Bushfire Education." *Geographical Research* 52 (1): 23–33.

Eriksen, Christine, and Don L. Hankins. 2014. "The Retention, Revival, and Subjugation of Indigenous Fire Knowledge through Agency Fire Fighting in Eastern Australia and California." *Society and Natural Resources* 27 (12): 1288–303.

Esh, Shaul. 1963. "Words and Their Meanings." *Yad Vashem Studies* 5:133–67.

Eugenios, Jillian. 2015. "Katrina Changed Animal Evacuation Laws." CNN, August 28, 2015. https://edition.cnn.com/2015/08/14/news/hurricane-katrina-dog-rescue/index.html.

Evans, Richard J. 2002. *Telling Lies about Hitler: The Holocaust, History and the David Irving Trial*. London: Verso.

Evatt, G. H. J. 1884. *Ambulance Organization, Equipment, and Transport*. International Health Exhibition Handbooks. London: William Clowes and Sons.

Faraci, D. 2014. "INTERSTELLAR and the Death of the Penis." *Birth.Movies.Death*, November 9, 2014. Accessed October 1, 2020. https://birthmoviesdeath.com/2014/11/09/interstellar-and-the-death-of-the-penis.

Faria, Caroline, Jovah Katushabe, Catherine Kyotowadde, and Dominica Whitesell. 2021. "'You Rise Up . . . They Burn You Again': Market Fires and the Urban Intimacies of Disaster Colonialism." *Transactions of the Institute of British Geographers* 46 (1): 87–101.

Farish, Matthew. 2010. *The Contours of America's Cold War*. Minneapolis: University of Minnesota Press.

Fauzi, Mohd Hashairi, Alzamani Mohd Idrose, Abu Hassan Asaari Abdullah, Jahlelawati Zul, and Nur Haslinda Mohd Nordin. 2014. "The Pattern of Injuries or Medical Emergencies during High-Rise Evacuation Drill." *Journal of Pioneering Medical Sciences* 4 (2): 81–84.

Favret, Mary. 2009. *War at a Distance: Romanticism and the Making of Wartime.* Princeton, NJ: Princeton University Press.

Federal Civil Defense Administration. 1951. *Rural Family Defense.* Washington, DC: U.S. Government Printing Office.

Federal Civil Defense Administration. 1954. *A Report on the Washington Conference of National Women's Advisory Committee, Federal Civil Defense Administration, October 26 and 27, 1954.* Washington, DC.

Federal Civil Defense Administration. 1957. *Annual Statistical Report, June 30, 1957.* National Headquarters, Battle Creek Michigan.

Feeny-Hart, Alison. 2013. "The Little-Told Story of the Massive WWII Pet Cull." BBC *Magazine*, October 2013.

Feige, Michael. 2009. *Settling in the Hearts: Jewish Fundamentalism in the Occupied Territories.* Detroit, MI: Wayne State University Press.

Feinberg, Alexander. 1952. "Civil Defense Tests Disaster Technique as City Is 'Bombed.'" *New York Times*, October 1, 1952.

Feldman, Jeffrey David. 2006. "Contact Points: Museums and the Lost Body Problem." In *Sensible Objects: Colonialism, Museums and Material Culture*, edited by Elizabeth Edwards, Chris Gosden, and Ruth B. Phillips, 245–68. London: Bloomsbury.

Feldman, Yotam. 2007. "Dr. Naveh, or, How I Learned to Stop Worrying and Walk through Walls." *Haaretz*, October 25, 2007. https://www.haaretz .com/1.4990742.

Felman, Shoshana. 2001a. "A Ghost in the House of Justice: Death and the Language of the Law." *Yale Journal of Law and the Humanities* 13 (1): 241–82.

Felman, Shoshana. 2001b. "Theaters of Justice: Arendt in Jerusalem, the Eichmann Trial, and the Redefinition of Legal Meaning in the Wake of the Holocaust." *Critical Inquiry* 27 (2): 201–38.

FEMA. 2013. "Standing Rock Sioux Tribe—Severe Storms and Flooding." FEMA-4123-DR. June 25, 2013. Washington, DC: Federal Emergency Management Administration. https://www.fema.gov/sites/default/files/2020 -09/PDA_Report_FEMA-4123-DR-SRST.pdf.

Fendel, Hillel. 2005. "14-Year-Old Girl in Prison for Talking Freshly." Israel National News, July 18, 2005. http://www.israelnationalnews.com/News /News.aspx/86013.

Fendel, Hillel. 2006. "Gush Katif Fundraiser Provides Faith, Hope and Laughs." Israel National News, December 6, 2006. http://www.israelnationalnews .com/News/News.aspx/105264.

Fendel, Hillel. 2010. "Gush Katif: Past and Future." Israel National News, July 15, 2010. http://www.israelnationalnews.com/News/News.aspx /138612#.U_uWBGJdXDg.

Ferguson, K. V. M. 1947. "Non-professional Forestry Training in Victoria." *Australian Forestry* 11 (1): 52–56.

Figl, Markus, and Linda E. Pelinka. 2005. "Jaromir Baron von Mundy— Founder of the Vienna Ambulance Service." *Resuscitation* 66 (2): 121–25.

Fink, Sheri. 2013. *Five Days at Memorial: Life and Death in a Storm-Ravaged Hospital.* London: Atlantic Books.

Fitch, James Marston, John Templer, and Paul Corcoran. 1974. "The Dimensions of Stairs." *Scientific American* 231 (4): 82–91.

Fleming, Victor, dir. 1939. *Gone with the Wind.* MGM.

Foreign and Commonwealth Office (FCO). 2012. *Review of FCO Consular Evacuation.* Corporate Report. https://www.gov.uk/government/publications /review-of-consular-evacuation-procedures.

Forschungsgesellschaft Flucht und Migration. 2020. "Flugzeug nach Lesbos: 'Let's Bring Them Here.'" FFM-ONLINE, October 6, 2020. https://ffm -online.org/flugzeug-nach-lesbos-lets-bring-them-here/.

Foucault, Michel. 1977. *Discipline and Punish: The Birth of the Prison.* Translated by Alan Sheridan. New York: Vintage Books.

Four Paws. 2019. "Shut Down Rafah Zoo in Gaza." Accessed February 21, 2024. https://help.four-paws.org/en-GB/shut-down-rafah-zoo-gaza.

Fraser, Douglas M., and Wendell S. Hertzelle. 2010. "Haiti Relief: An International Effort Enabled through Air, Space, and Cyberspace." *Air and Space Power Journal* 24 (4): 5–12.

Fraser, Peg. 2018. *Black Saturday: Not the End of the Story.* Melbourne: Monash University Publishing.

Fraser, Peg. 2019. "Messy History." *Cultural Studies Review* 25 (2): 262–64.

Freeman, Carla. 2000. *High Tech and High Heels in the Global Economy: Women, Work, and Pink-Collar Identities in the Caribbean.* Durham, NC: Duke University Press.

Freeman, Cordelia. 2020. "Viapolitics and the Emancipatory Possibilities of Abortion Mobilities." *Mobilities* 15 (6): 896–910.

Freud, Sigmund. 1908. "On the Sexual Theories of Children." In *The Standard Edition of the Complete Psychological Works of Sigmund Freud*, 9:209–26. London: Hogarth.

Fritz, Charles E., and Eli S. Marks. 1954. "The NORC Studies of Human Behavior in Disaster." *Journal of Social Issues* 10 (3): 26–41.

Fritz, Charles E., and Harry B. Williams. 1957. "The Human Being in Disasters: A Research Perspective." *Annals of the American Academy of Political and Social Science* 309 (1): 42–51.

Fry, Erika. 2014. "The Escape Business." *Fortune*, November 13, 2014. https:// fortune.com/2014/11/13/global-hotspot-evacuation-planners/.

Frykholm, Amy Johnson. 2004. *Rapture Culture: Left Behind in Evangelical America.* Oxford: Oxford University Press.

Fujitsuka, Satoshi. 2001. "Designing Safety Signs for Evacuation Centers: Mr. Yukio Ota/ President, Ota Yukio Design Associates." Great Hanshin-Awaji Earthquake + Creative Timeline Project, March 2001. https://tm19950117.jp/en/interview/2931/.

Fuller, Gillian. 2014. "Queue." In *The Routledge Handbook of Mobilities*, edited by Peter Adey, David Bissell, Kevin Hannam, Peter Merriman, and Mimi Sheller, 205–13. London: Routledge.

Fuller, Gillian, and Ross Harley. 2004. *Aviopolis: A Book about Airports*. London: Black Dog.

Gabriel, Richard A. 2013. *Between Flesh and Steel: A History of Military Medicine from the Middle Ages to the War in Afghanistan*. Sterling, VA: Potomac Books.

Galea, Edwin R., Lynn Hulse, Rachel Day, Asim Siddiqui, and Gary Sharp. 2012. "The UK WTC 9/11 Evacuation Study: An Overview of Findings Derived from First-Hand Interview Data and Computer Modelling." *Fire Mater* 36:501–21. https://doi.org/10.1002/fam.1070 521.

Gallie, F. S. 1915. "Transportation of the Sick and Wounded." *War Medicine* 2 (2): 184–88.

Garrard, Greg. 2012. "Worlds without Us: Some Types of Disanthropy." *SubStance* 41 (1): 40–60.

Garrett, Bradley. 2020. *Bunker: Building for the End Times*. New York: Simon and Schuster.

Garrison, Dee. 2006. *Bracing for Armageddon: Why Civil Defense Never Worked*. Oxford: Oxford University Press.

Georgia, Civil Defense Division. 1952. "Civil Defense Manual for Georgia Schools." In co-operation with the State Department of Education. Atlanta, GA: Civil Defense Division.

Gershon, Robyn R. M., Kristine A. Qureshi, Marcie S. Rubin, and Victoria H. Raveis. 2007. "Factors Associated with High-Rise Evacuation: Qualitative Results from the World Trade Center Evacuation Study." *Prehospital and Disaster Medicine* 22 (3): 165–73.

Ghamari-Tabrizi, Sharon. 2009. *The Worlds of Herman Kahn: The Intuitive Science of Thermonuclear War*. Cambridge, MA: Harvard University Press.

Ghamari-Tabrizi, Sharon. 2013. "Death and Resurrection in the Early Cold War: The Grand Analogy of the Disaster Researchers." In *Aufbruch ins Unversicherbare*, edited by Leon Hempel, Marie Bartels, and Thomas Markwart, 335–78. Berlin: Transcript.

Ghertner, D. Asher, Hudson McFann, and Daniel M. Goldstein, 2020. "Introduction: Security of and beyond the Biopolitical." In *Futureproof: Security Aesthetics and the Management of Life*, edited by D. Asher Ghertner, Hudson McFann, and Daniel M. Goldstein, 1–32. Durham, NC: Duke University Press.

Giaccaria, Paolo, and Claudio Minca. 2011. "Nazi Biopolitics and the Dark Geographies of the Selva." *Journal of Genocide Research* 13 (1–2): 67–84.

Giaccaria, Paola, and Claudio Minca. 2016. "Life in Space, Space in Life: Nazi Topographies, Geographical Imaginations, and *Lebensraum*." *Holocaust Studies* 22 (2–3): 151–71.

Gigliotti, Simone. 2009. *The Train Journey: Transit, Captivity, and Witnessing in the Holocaust*. New York: Berghahn.

Ginsberg, Mitch. 2014. "After Evacuating Gaza, a Lonely General of Faith Struggles for Israel's Salvation." *Times of Israel*, August 4, 2014. https://www.timesofisrael.com/after-evacuating-gaza-a-lonely-general-of-faith-struggles-for-israels-salvation/.

Giroux, Henry A. 2015. *Stormy Weather: Katrina and the Politics of Disposability*. New York: Routledge.

Gleason, Arthur, and Helen H. Gleason. 1916. *Golden Lads*. Toronto: McClelland, Goodchild and Stewart.

Goddard, Jonathan C. 2016. "Reginald Harrison: Liverpool's First Urologist." *Journal of Clinical Urology* 9 (1): 4–8.

Goldberg, Ori. 2018. *Faith and Politics in Iran, Israel and the Islamic State: Theologies of the Real*. Cambridge: Cambridge University Press.

Goldberger, Paul. 1976. "City Fire Prevention: A Maze for Experts." *New York Times*, January 20, 1976, 22.

Gómez Reus, Teresa. 2012. "Fighting for Fame: The 'Heroines of Pervyse' and the Disputed Construction of a Public Image." *Women: A Cultural Review* 23 (3): 300–322.

Gordon, Linda. 2006. "Dorothea Lange Photographs the Japanese American Internment." In *Impounded: Dorothea Lange and the Censored Images of Japanese American Internment*, edited by Linda Gordon and Gary Y. Okihiro, 5–46. New York: W. W. Norton.

Gordon, Neve, and Nicola Perugini. 2016. "The Politics of Human Shielding: On the Resignification of Space and the Constitution of Civilians as Shields in Liberal Wars." *Environment and Planning D: Society and Space* 34 (1): 168–87.

Gough-Brady, C. 2012. "Fire Words: Developing a Bushfire Narrative." *Griffiths Review* 35 (Autumn): 233–37.

Government Accountability Office. 2018. *Emergency Management: Implementation of the Major Disaster Declaration Process for Federally Recognized Tribes*. May 23, 2018. GAO 18-443. Washington, DC: U.S. Government Accountability Office.

Graham, Stephen. 2002. "Bulldozers and Bombs: The Latest Palestinian–Israeli Conflict as Asymmetric Urbicide." *Antipode* 34 (4): 642–49.

Graham, Stephen. 2005. "Cities under Siege: Katrina and the Politics of Metropolitan America." *Items: Insights from the Social Sciences*, June 11, 2006. https://items.ssrc.org/understanding-katrina/cities-under-siege-katrina-and-the-politics-of-metropolitan-america/.

Graham, Stephen. 2011. *Cities under Siege: The New Military Urbanism*. London: Verso.

Graham, Stephen. 2018. "Elite Avenues: Flyovers, Freeways and the Politics of Urban Mobility." *City* 22 (4): 527–50.

Graham, Stephen, and Simon Marvin. 2002. *Splintering Urbanism: Networked Infrastructures, Technological Mobilities and the Urban Condition*. London: Routledge.

Grappi, Giorgio. 2018. "Asia's Era of Infrastructure and the Politics of Corridors: Decoding the Language of Logistical Governance." In *Logistical Asia*, edited by Brett Neilson, Ned Rossiter, and Ranabir Samaddar, 175–98. Berlin: Springer.

Grayzel, Susan R. 2012. *At Home and under Fire: Air Raids and Culture in Britain from the Great War to the Blitz*. Cambridge: Cambridge University Press.

Greenberg, Hanan. 2005. "Evacuation Notice: Please Follow Me to Bus." *ynetnews*, August 14, 2005. https://www.ynetnews.com/articles/0,7340,L-3127212,00.html.

Gregory, Derek. 2011. "The Everywhere War." *Geographical Journal* 177 (3): 238–50.

Gregory, Derek. 2012. "From a View to a Kill: Drones and Late Modern War." *Theory, Culture and Society* 28 (7–8): 188–215.

Gregory, Derek. 2013. "Medical Military Machines and Casualties of War, 1914–2014." *Geographical Imaginations*, October 16, 2013. https://geographicalimaginations.com/2013/10/16/medical-military-machines-and-the-casualties-of-war-1914-2014/.

Gregory, Derek. 2015a. "Gabriel's Map: Cartography and Corpography in Modern War." In *Geographies of Knowledge and Power*, edited by Peter Meusburger, Derek Gregory, and Laura Suarsana, 89–121. Dordrecht: Springer.

Gregory, Derek. 2015b. "The Natures of War." *Antipode* 48 (1): 3–56.

Gregory, Derek. 2019a. "Being Wounded." *Geographical Imaginations*, June 25, 2019. https://geographicalimaginations.com/2019/06/25/being-wounded/.

Gregory, Derek. 2019b. "The Leaden Hours." *Geographical Imaginations*, January 31, 2019. https://geographicalimaginations.com/2019/01/31/the-leaden-hours/.

Gregory, Derek. 2019c. "More-than-Human Casualties." *Geographical Imaginations*, May 3, 2019. https://geographicalimaginations.com/2019/05/03/more-than-human-casualties/.

Grenfell Action Group. 2017. "Grenfell Tower—the KCTMO Culture of Negligence." June 19, 2017. https://grenfellactiongroup.wordpress.com/2017/06/19/grenfell-tower-the-kctmo-culture-of-negligence/.

Griffiths, Tom. 2001. *Forests of Ash: An Environmental History*. Cambridge: Cambridge University Press.

Griffiths, Tom. 2012. "The Language of Catastrophe: Forgetting, Blaming and Bursting into Colour." *Griffith Review*, no. 35, 46–58.

Grimal, Francis, and Graham Melling. 2011. "British Action in Libya 2011: The Lawful Protection of Nationals Abroad." *Denning Law Journal* 23:165–77.

Grimm, David. 2015. *Citizen Canine: Our Evolving Relationship with Cats and Dogs*. New York: PublicAffairs.

Grossman, David, and Dan Newling. 2017. "Dramatic Grenfell Baby Story Probably Never Happened." BBC *News*, October 9, 2017. https://www.bbc .co.uk/news/uk-41550836.

Gudis, Catherine. 2013. "I Thought California Would Be Different: Defining California through Visual Culture." In *A Companion to California History*, edited by William Deverell and David Igler, 40–74. Oxford: Wiley-Blackwell.

Guha-Sapir, Deborati, Thomas Kirsch, Shayna Dooling, and Adam Sirois, with Mark DerSarkissian. 2011. *Independent Review of the US Government Response to the Haiti Earthquake*. Funded by the United States Administration for International Development (USAID), Macfadden & Assoc., Incorporated. https://pdf.usaid.gov/pdf_docs/pdacr222.pdf.

Haaretz. 2005. "IDF Aims to Complete Gaza Evacuation within 10 Days." English edition, August 16, 2005.

Hacohen, Gershon. 2018. "The Israeli Security Concept: Wandering through a Maze." *Israel National News*, November 15, 2018. http://www .israelnationalnews.com/Articles/Article.aspx/23004.

Hacohen, Gershon. 2019. "14 Years since the Destruction of Gush Katif." Begin-Sadat Center for Strategic Studies, August 28, 2019. https://besacenter.org /destruction-gush-katif/.

Hague Conference on Private International Law (HCCH). 2010. "Haiti Earthquake and Intercountry Adoption of Children." Information Note to States and Central Authorities. https://www.hcch.net/en/news-archive/details/ ?varevent=183.

Hakol Chai. 2005. "Hakol Chai Enters Gaza to Rescue Abandoned Animals." Rescuing Abandoned Animals in Gaza, August 24, 2005. http://www .chai-online.org/en/campaigns/settlements/campaigns_settlements.htm.

Hall, Timothy. 1980. *Darwin 1942: Australia's Darkest Hour*. Sydney: Methuen Australia.

Hammack, Phillip L. 2018. "Disjunctive: Social Justice, Black Identity, and the Normality of Black People." In *The Oxford Handbook of Social Psychology and Social Justice*, edited by Phillip L. Hammack, 129–40. Oxford: Oxford University Press.

Hammond, Elizabeth E., and Dan Maloney. 2007. "Hurricane Preparedness: Lessons Learned from Hurricane Katrina." *Animal Keepers Forum* 34 (11–12): 546–77.

Handmer, John, Saffron O'Neil, and Damien Killalea. 2010. *Review of Fatalities in the February 7, 2009, Bushfires*. Report prepared for the Victorian

Bushfires Royal Commission, April 13, 2010. Melbourne: Centre for Risk and Community Safety, RMIT University, and Bushfire CRC.

Hanley, Lynsey. 2017. "Look at Grenfell Tower and See the Terrible Price of Britain's Inequality." *Guardian*, June 16, 2017. https://www.theguardian .com/commentisfree/2017/jun/16/grenfell-tower-price-britain-inequality -high-rise.

Hansen, Christine, and Tom Griffiths. 2012. *Living with Fire: People, Nature and History in Steels Creek*. Clayton, VIC: CSIRO Publishing.

Harcourt, Bernard E. 2009. *Illusion of Order: The False Promise of Broken Windows Policing*. Cambridge, MA: Harvard University Press.

Hardwick, M. J. 2015. *Mall Maker: Victor Gruen, Architect of an American Dream*. Philadelphia: University of Pennsylvania Press.

Harel, Amos. 2017. "Settlements Do Not Serve Israel's Security Needs, Say Former Generals." *Haaretz*, June 5, 2017. https://www.haaretz.com/israel -news/.premium-israeli-ex-generals-settlements-do-not-serve-security -needs-1.5480013.

Harel, Amos, and Baruch Shary. 2005. "Evacuators Too Violent, Say Soldiers Playing Role of Settlers." *Haaretz*, August 11, 2005. https://www.haaretz .com/1.4930681.

Harris, Andrew. 2015. "Vertical Urbanisms: Opening Up Geographies of the Three-Dimensional City." *Progress in Human Geography* 39 (5): 601–20.

Harrison, Mark. 2010. *The Medical War: British Military Medicine in the First World War*. Oxford: Oxford University Press.

Harrison, Rebecca. 2015. "Writing History on the Page and Screen: Mediating Conflict through Britain's First World War Ambulance Trains." *Historical Journal of Film, Radio and Television* 35 (4): 559–78.

Harris-Perry, Melissa. 2010. "Michael Vick, Racial History and Animal Rights." *Nation*, December 30, 2010. https://www.thenation.com/article/archive /michael-vick-racial-history-and-animal-rights/.

Hasson, Nir. 2005. "Without the Soldiers, There Would Have Been No Netzarim." *Haaretz*, August 23, 2005. https://www.haaretz.com/1.4935207.

Hata, Donald Teruo, and Nadine Ishitani Hata. 2011. *Japanese Americans and World War II: Mass Removal, Imprisonment, and Redress*. Arlington Heights, IL: Harlan Davidson.

Haywood, Loraine. 2016. "Nostalgia for Eden: Joseph Kosinski's *Oblivion*, an Apocalyptic Genesis." *Seachanges: Journal for Women Scholars in Religion and Theology* 7:1–19.

Haywood, Loraine. 2018. "Reflecting Absence, Mediating 'the Real': *Oblivion* as a Requiem for 9/11." *Performance of the Real Working Papers* 1 (2): 26–46.

Healesville and Yarra Glen Guardian. 1927. "Coranderrk Reserve: Wanted for Settlement. Minister Favorable." September 24, 1927, 2.

Healesville and Yarra Glen Guardian. 1936. "Sanctuary Committee: Healesville Shire Council." December 12, 1936, 1.

Hejnol, Andreas. 2017. "Ladders, Trees, Complexity, and Other Metaphors in Evolutionary Thinking." In *Arts of Living on a Damaged Planet: Ghosts and Monsters of the Anthropocene*, edited by Anna Lowenhaupt Tsing, Heather Anne Swanson, Elaine Gan, and Nils Bubandt, 87–102. Minneapolis: University of Minnesota Press.

Helbing, Dirk, Illés J. Farkas, Peter Molnar, and Tamás Vicsek. 2002. "Simulation of Pedestrian Crowds in Normal and Evacuation Situations." In *Pedestrian and Evacuation Dynamics*, edited by K. Knoflacher, P. Gattermann, N. Waldau, and M. Schreckenberg, 21–58. Berlin: Springer.

Helbing, Dirk, Illés Farkas, and Tamás Vicsek. 2000. "Simulating Dynamical Features of Escape Panic." *Nature* 407 (6803): 487–90.

Heller, Charles. 2020. "De-confine Borders: Towards a Politics of Freedom of Movement in the Time of the Pandemic." Working Paper No. 147, Centre on Migration, Policy and Society, University of Oxford. https://www.compas.ox.ac.uk/wp-content/uploads/WP-2020-147-Heller_De-confine_Borders_Towards_Politics_Freedom_Movement_Time_Pandemic.pdf.

Heller, Charles, and Lorenzo Pezzani. 2014. "Liquid Traces: The Left-to-Die Boat Case." In *Forensis: The Architecture of Public Truth*, edited by Forensic Architecture, 657–83. Berlin: Sternberg.

Herzig-Yoshinaga, Aiko. 2009. "Words Can Lie or Clarify: Terminology of the World War II Incarceration of Japanese Americans." *Manzanar Committee*. Accessed May 29, 2024. https://manzanarcommittee.org/wp-content/uploads/2010/03/wordscanlieorclarify-ahy.pdf.

Hetherington, Kregg. 2019. Introduction to *Infrastructure, Environment, and Life in the Anthropocene*, edited by Kregg Hetherington, 1–16. Durham, NC: Duke University Press.

Hewitt, Kenneth. 1980. "*The Environment as Hazard* by Ian Burton, Robert W. Kates and Gilbert F. White." *Annals of the Association of American Geographers* 70 (2): 306–11.

Hewitt, Lucy, and Stephen Graham. 2015. "Vertical Cities: Representations of Urban Verticality in 20th-Century Science Fiction Literature." *Urban Studies* 52 (5): 923–37.

Hinted, Harold B. 1950. "City Won't Be Evacuated in a Raid, Mayor Testifies, Asking U.S. Funds." *New York Times*, December 16, 1950.

Hirabayashi, Lane Ryo, and Kenichiro Shimada. 2009. *Japanese American Resettlement through the Lens: Hikaru Iwasaki and the WRA's Photographic Section, 1943–1945*. Boulder: University Press of Colorado.

Hoare, Liam. 2013. "Gush Katif Nostalgia Is Based upon Illusion." The Blogs, *Jewish News*, August 19, 2013. https://blogs.timesofisrael.com/gush-katif-nostalgia-is-based-upon-illusion/.

Hoare, Philip. 2001. *Spike Island: The Memory of a Military Hospital*. London: 4th Estate.

Hoare, Philip. 2018. "The Animal Victims of the First World War Are a Stain on Our Conscience." *Guardian*, November 7, 2018. https://www.theguardian.com/commentisfree/2018/nov/07/animal-victims-first-world-war.

Hochberg, Gil Z. 2015. *Visual Occupations: Violence and Visibility in a Conflict Zone*. Durham, NC: Duke University Press.

Hodgetts, Timothy, and Jamie Lorimer. 2020. "Animals' Mobilities." *Progress in Human Geography* 44 (1): 4–26.

Holmes, Frederic Morell. 1899. *Firemen and Their Exploits: With Some Account of the Rise and Development of Fire-Brigades, of Various Appliances for Saving Life at Fires and Extinguishing the Flames*. London: S. W. Partridge.

Holtorf, Cornelius. 2015. "Heritage Futures in *Interstellar*." Heritage Futures. Accessed April 20, 2019. https://heritage-futures.org/heritage-futures-interstellar/.

Home Office. 1939. "Air Raid Precautions for Animals." Air Raid Precautions Handbook No. 12. Air Raid Precautions Department. London: Her Majesty's Stationery Office.

Honig, Bonnie. 2009. *Emergency Politics: Paradox, Law, Democracy*. Princeton, NJ: Princeton University Press.

Honig, Bonnie. 2014. "Three Models of Emergency Politics." *Boundary 2* 41 (2): 45–70.

Honig, Bonnie. 2021. *A Feminist Theory of Refusal*. Cambridge, MA: Harvard University Press.

Hooper, Chloe. 2018. *The Arsonist: A Mind on Fire*. Sydney: Penguin.

Hooper, Colette. 2014. *Railways of the Great War*. London: Bantam.

Hopkins, Mary Alden. 1911. "1910 Newark Factory Fire." *McClure's Magazine* 36 (6): 663–65.

Horowitz, Sara R. 1997. *Voicing the Void: Muteness and Memory in Holocaust Fiction*. Albany: State University of New York Press.

Houtum, Henk van, and Rodrigo Bueno Lacy. 2019. "The Migration Map Trap: On the Invasion Arrows in the Cartography of Migration." *Mobilities* 15 (2): 196–219.

Howell, Philip, and Hilda Kean. 2018. "The Dogs That Didn't Bark in the Blitz: Transpecies and Transpersonal Emotional Geographies on the British Home Front." *Journal of Historical Geography* 61 (July): 44–52.

Humane Society of the United States. 2008. *DISASTER Preparedness for Pets*. Washington, DC: Humane Society of the United States.

Human Rights Watch. 2011. "Libya: Stranded Foreign Workers Need Urgent Evacuation." March 2, 2011. https://www.hrw.org/news/2011/03/02/libya-stranded-foreign.

Hunner, Jon. 2014. *Inventing Los Alamos: The Growth of an Atomic Community*. Norman: University of Oklahoma Press.

Hunt, Nancy Rose. 2019. "Aphasia, History, and Duress." *History and Theory* 58 (3): 437–50.

Hutchinson, John F. 2018. *Champions of Charity: War and the Rise of the Red Cross*. London: Taylor and Francis.

Hutt City Council. 2017. "Spotted! The First Tsunami Blue Lines in Lower Hutt Have Started to Be Painted." Facebook, July 3, 2017. https://www.facebook .com/huttcitycouncil/posts/pfbid068N4QmZkTf2hndgCf2jrcCLhvrjDk YEceYeqzqfAHtxB2ghbBzAJi2YSFtJ9iFZpl.

Hyland, Adrian. 2015. *Kinglake-350*. Melbourne: Text.

Hyndman, Jennifer. 2012. "The Geopolitics of Migration and Mobility." *Geopolitics* 17 (2): 243–55.

Iklé, Fred, and Harry V. Kincaid. 1956. *Social Aspects of Wartime Evacuation of American Cities: With Particular Emphasis on Long-Term Housing and Reemployment*. Disaster Study No. 4. Washington, DC: National Academy of Sciences, National Research Council.

Ingraham, Joseph. 1950a. "Huge Garage Plan Backed by Mayor." *New York Times*, December 29, 1950.

Ingraham, Joseph. 1950b. "More One-Way Avenues Set as City Speeds Defense Plan." *New York Times*, December 30, 1950.

Ishizuka, Karen L. 2005. "Coming to Terms: America's Concentration Camps." In *Common Ground: The Japanese American National Museum and the Culture of Collaborations*, edited by Akemi Kikumura-Yano, Lane Ryo Hirabayashi, and James A. Hirabayashi, 101–22. Boulder: University Press of Colorado.

Israel State Commission of Inquiry. 2010. "Inquiry into the Handling of the Evacuees from Gush Katif and Northern Samaria by the 'Official Authorities.'" Jerusalem. Chair: Justice Eliahu Mazza.

Jackson, A. Marshall. "Tells Development of Ambulance Service." *New York Times*, February 28, 1915, 12.

Jackson, Lucas. 2015. "Artifacts from the 9/11 Museum." Reuters—the Wider Image, September 15, 2015. https://widerimage.reuters.com/story/artefacts -from-the-911-museum.

Jackson, Shannon. 2011. *Social Works: Performing Art, Supporting Publics*. London: Routledge.

Japanese American Citizens League. 2020. *Power of Words Handbook: A Guide to Language about Japanese Americans in World War II*. San Francisco: National JACL Power of Words II Committee.

James, H. E. R, and C. E. Pollock. 1910. "Notes on the Conveyance of the Sick and Wounded by Rail, with Special Reference to Improvised Methods." *Journal of the Royal Army Medical Corps* 15 (3): 276–85.

Janis, I. 1950. "Psychological Problems of A-Bomb Defense." *Bulletin of Atomic Scientists* 6 (8–9): 256–62.

Jewish Telegraphic Agency. 2005. "The Disengagement Summer 'Disengagement Dogs' and Cats Seek New Owners after Withdrawal." *Jewish Telegraph*, August 30, 2005. https://www.jta.org/2005/08/30/archive/the-disengagement-summer-disengagement-dogs-and-cats-seek-new-owners-after-withdrawal.

Jin, Jongsoon, and Geoboo Song. 2017. "Bureaucratic Accountability and Disaster Response: Why Did the Korea Coast Guard Fail in Its Rescue Mission during the Sewol Ferry Accident?" *Risk, Hazards, and Crisis in Public Policy* 8 (3): 220–43.

Johnson, B. S., ed. 1968. *The Evacuees*. London: Victor Gollancz.

Jonassen, Wendi. 2012. "7 Years after Katrina, New Orleans Is Overrun by Wild Dogs." *Atlantic*, August 2012. https://www.theatlantic.com/national/archive/2012/08/7-years-after-katrina-new-orleans-is-overrun-by-wild-dogs/.

Jones, Edgar, and Simon Wessely. 2001. "The Origins of British Military Psychiatry before the First World War." *War and Society* 19 (2): 91–108.

Jones, Owen. 1926. "Forestry in Victoria." *Empire Forestry* 5 (1): 87–101.

Journal (Dublin). 2021. "'We're Talking about the Apocalypse': Thousands Flee Wildfires Burning Out of Control in Greece." August 6, 2021. https://www.thejournal.ie/thousands-flee-wildfires-in-greece-5517174-Aug2021/.

Joyce, Katherine. 2013. *The Child Catchers: Rescue, Trafficking, and the New Gospel of Adoption*. New York: PublicAffairs.

Kaczor, Christopher. 2005. *The Edge of Life: Human Dignity and Contemporary Bioethics*. Dordrecht: Springer.

Kallis, Aristotle. 2008. *Genocide and Fascism: The Eliminationist Drive in Fascist Europe*. London: Routledge.

Kaplan, Caren. 2017. *Aerial Aftermaths: Wartime from Above*. Durham, NC: Duke University Press.

Kaszynski, William. 2000. *The American Highway: The History and Culture of Roads in the United States*. Jefferson, NC: McFarland.

Kates, Robert W. 1971. "Natural Hazard in Human Ecological Perspective: Hypotheses and Models." *Economic Geography* 47 (3): 438–51.

Kates, Robert W., and Ian Burton. 2008. "Gilbert F. White, 1911–2006: Local Legacies, National Achievements, and Global Visions." *Annals of the Association of American Geographers* 98 (2): 479–86.

Katz, Cindi. 2008. "Bad Elements: Katrina and the Scoured Landscape of Social Reproduction." *Gender, Place and Culture* 15 (1): 15–29.

Kean, Hilda. 2015. "The Dog and Cat Massacre of September 1939 and the People's War." *European Review of History / Revue Européenne d'histoire* 22 (5): 741–56.

Keane, Thomas H., and Lee Hamilton. 2004. *The 9/11 Commission Report: Final Report of the National Commission on Terrorist Attacks upon the United*

States. Washington, DC: National Commission on Terrorist Attacks upon the United States.

Kehoe, Isobel, dir. 1957. *Operation Lifesaver*. National Film Board of Canada, Canadian Broadcasting Corporation.

Kelman, Ari. 2003. *A River and Its City: The Nature of Landscape in New Orleans*. Berkeley: University of California Press.

Kendall, Tim. 2006. *Modern English War Poetry*. Oxford: Oxford University Press.

Kerry, Stephen Craig. 2017. "Transgender People in Australia's Northern Territory." *International Journal of Transgenderism* 18 (2): 129–39.

Khan, Yasmin. 2007. *The Great Partition: The Making of India and Pakistan*. New Haven, CT: Yale University Press.

Kidd, Rosalind. 1997. *The Way We Civilise: Aboriginal Affairs, the Untold Story*. Brisbane: University of Queensland Press.

Kim, Claire Jean. 2015. *Dangerous Crossings: Race, Species, and Nature in a Multicultural Age*. Cambridge: Cambridge University Press.

Kipling, Rudyard. 1900. "With Number Three." Kipling Society. Accessed March 13, 2024. https://www.kiplingsociety.co.uk/tale/numberthree.htm.

Kissane, Karen. 2010. *Worst of Days: Inside the Black Saturday Firestorm*. Sydney: Hachette Australia.

Klein, Aaron. 2005. "Did God Send Katrina as Judgment for Gaza?" *WND*, September 7, 2005. https://www.wnd.com/2005/09/32196/.

Klein, Kelly R., and Nanci E. Nagel. 2007. "Mass Medical Evacuation: Hurricane Katrina and Nursing Experiences at the New Orleans Airport." *Disaster Management and Response* 5 (2): 56–61.

Klinke, Ian, and Mark Bassin. 2018. "Introduction: *Lebensraum* and Its Discontents." *Journal of Historical Geography* 61 (July): 53–58.

Kock, Gerhard. 1997. *Der Führer sorgt für unsere Kinder: Die Kinderlandverschickung im Zweiten Weltkrieg*. Paderborn: F. Schöningh.

Kohl, Katrin. 2014. "Bearing Witness: The Poetics of H. G. Adler and W. G. Sebald." In *Witnessing, Memory, Poetics: H. G. Adler and W. G. Sebald*, edited by Helen Finch and Lynn L. Wolff, 81–111. Norwich: Boydell and Brewer.

Kolker, Robert. 2011. "Stairwell A." *New York Magazine*, August 27, 2011. https://nymag.com/news/9-11/10th-anniversary/stairwell-a/.

Kory, Ronen, Alon Carney, and Sody Naimer. 2013. "Health Ramifications of the Gush Katif Evacuation." *Israel Medical Association Journal* 15 (3): 137–42.

Korycansky, D. G., Gregory Laughlin, and Fred C. Adams. 2001. "Astronomical Engineering: A Strategy for Modifying Planetary Orbits." *Astrophysics and Space Science* 275 (4): 349–66.

Kosinski, Joseph, dir. 2003. *Oblivion*. Universal Pictures.

Kotef, Hagar. 2015. *Movement and the Ordering of Freedom: On Liberal Governances of Mobility*. Durham, NC: Duke University Press.

Kra-Oz, Tal. 2013. "Revisiting Israel's Disengagement from the Gaza Strip Eight Years Later." *Tablet Magazine*, August 15, 2013. https://www.tabletmag.com /scroll/141376/a-tour-of-the-gush-katif-museum.

Kristeva, Julia. 1982. *Powers of Horror: An Essay on Abjection*. New York: Columbia University Press.

Kulczuga, Aleksander. 2010. "Haiti Helps Pentagon Dissipate the Fog of War." *Daily Caller*, February 18, 2010. https://dailycaller.com/2010/02/18/haiti -helps-pentagon-dissipate%20-the-fog-of-war/.

Lacy, Michael G., and Kathleen C. Haspel. 2011. "Apocalypse: The Media's Framing of Black Looters, Shooters, and Brutes in Hurricane Katrina's Aftermath." In *Critical Rhetorics of Race*, edited by Michael G. Lacy and Kent A. Ono, 21–46. New York: NYU Press.

Lancet. 1914. "The War: The Organisation of Medical Services in the Field." 184 (4753): 873.

Landstreet, Barent. 1955. "Evacuation and Dispersal" (1954). In *A Report on the Washington Conference of National Women's Advisory Committee, Federal Civil Defense Administration, October 26 and 27, 1954, Washington, DC*, 13–23. Washington, DC: Federal Civil Defense Administration.

Langmaid, Aaron. 2009. "With This Ring of Fire I Thee Wed, Then Get Out of Here." *Daily Telegraph*, February 12, 2009. https://www.dailytelegraph.com .au/with-this-ring-of-fire-i-thee-wed-then-get-out-of-here/news-story/eeee9 457b6040e75dcb8c79aef61edof?sv=14843d6733dc4a459e0babb4a29ee75e.

Larkin, Brian. 2018. "Promising Forms: The Political Aesthetics of Infrastructure." In *The Promise of Infrastructure*, edited by Nikhil Anand, Akhil Gupta, and Hannah Appel, 175–202. Durham, NC: Duke University Press.

Larrey, Dominique Jean. 1814. *Memoirs of Military Surgery, and Campaigns of the French Armies, on the Rhine, in Corsica, Catalonia, Egypt, and Syria; at Boulogne, Ulm, and Austerlitz; in Saxony, Prussia, Poland, Spain, and Austria*. London: Joseph Cushing.

Larsen, Elizabeth Foy. 2010. "Haiti's Adoption Free for All." *Daily Beast*, January 25, 2010. https://www.thedailybeast.com/haitis-adoption-free-for-all.

Lauder, S. 2009. "Heatwave Saves Worst for Last" [Transcript]. *PM*, ABC News. Accessed January 5, 2018. http://www.abc.net.au/pm/content/2008 /s2484719.htm.

Lee, Trymaine. 2006. "New Orleans Disaster Practice Takes to the Water." *EMS World*, September 20, 2006. https://www.emsworld.com/news/10410536 /new-orleans-disaster-practice-takes-water.

Leib, Jonathan, and Thomas Chapman. 2011. "Jim Crow, Civil Defense, and the Hydrogen Bomb: Race, Evacuation Planning, and the Geopolitics of Fear in 1950s Savannah, Georgia." *Southeastern Geographer* 51 (4): 578–95.

Lemaire, Léa. 2019. "The European Dispositif of Border Control in Malta: Migrants' Experiences of a Securitized Borderland." *Journal of Borderlands Studies* 34 (5): 717–32.

Levinas, Emmanuel. 1991. *Totality and Infinity: An Essay on Exteriority.* Dordrecht: Kluwer.

Levy, Gideon. 2003. "End the Fake Evacuations." Electronic Intifada, June 24, 2003. https://electronicintifada.net/content/end-fake-evacuations/4647.

Levy, Yagil. 2007. "The Embedded Military: Why Did the IDF Perform Effectively in Executing the Disengagement Plan?" *Security Studies* 16 (3): 382–408.

Levy, Yagil. 2010. "The Clash between Feminism and Religion in the Israeli Military: A Multilayered Analysis." *Social Politics* 17 (2): 185–209.

Lewin, Avrohom Shmuel. 2019. "Disbanding Gush Katif a Failed Experiment, Says General Who Helped Carry Out Evacuation." *Jewish News Syndicate,* August 20, 2019. https://www.jns.org/disbanding-gush-katif-a-failed-experiment-says-general-who-helped-carry-out-evacuation/.

Lewin, Eyal. 2015. "The Disengagement from Gaza: Understanding the Ideological Background." *Jewish Political Studies Review* 27 (1–2): 15–32.

Li, Darryl. 2006. "The Gaza Strip as Laboratory: Notes in the Wake of Disengagement." *Journal of Palestine Studies* 35 (2): 38–55.

Li, Darryl. 2008. "From Prison to Zoo: Israel's 'Humanitarian' Control of Gaza." *Adalah's Newsletter* 44 (January): 1–3. https://www.adalah.org/uploads/oldfiles/Public/files/English/Publications/Articles/Prison-Zoo-Israel-Humanitarian-Control-Gaza-Darryl-Li.pdf.

Libération. 2010. "Adoption en Haïti: Les parents français toujours mobilisés." January 23, 2010. https://www.liberation.fr/societe/2010/01/23/adoption-en-haiti-les-parents-francais-toujours-mobilises_606043/.

Lifton, Robert Jay. 1986. *The Nazi Doctors: Medical Killing and the Psychology of Genocide.* New York: Basic Books.

Light, Jennifer S. 2003. *From Warfare to Welfare: Defense Intellectuals and Urban Problems in Cold War America.* Baltimore, MD: Johns Hopkins University Press.

Lillquist, Karl. 2010. "Farming the Desert: Agriculture in the World War II–Era Japanese-American Relocation Centers." *Agricultural History* 81 (1): 74–104.

Lipstadt, Deborah E. 1994. *Denying the Holocaust: The Growing Assault on Truth and Memory.* London: Penguin.

Lisle, Debbie. 2016. *Holidays in the Danger Zone: Entanglements of War and Tourism.* Minneapolis: University of Minnesota Press.

Literary Digest. 1912. "147 Dead, Nobody Guilty." January 6, 1912. http://trianglefire.ilr.cornell.edu/primary/newspapersMagazines/ld_010612.html.

Liu, Cixin. 2017. *The Wandering Earth: And Other Short Stories.* London: Head of Zeus.

Llewellyn, Chris. 1987. *Fragments from the Fire: The Triangle Shirtwaist Company Fire of March 25, 1911: Poems.* New York: Puffin Books.

Lobo-Guerrero, Luis. 2007. "Biopolitics of Specialized Risk: An Analysis of Kidnap and Ransom Insurance." *Security Dialogue* 38 (3): 315–34.

London News. 1898. "A Wonder in London: Lives Saved at a Fire for the First Time by a 'Horsed-Escape.'" November 2, 1898.

London School of Hygiene and Tropical Medicine. 1909. *Report of the Ambulance Committee*. London: Her Majesty's Stationery Office.

Longmore, Thomas. 1869a. *A Treatise on the Transport of Sick and Wounded Troops*. London: Her Majesty's Stationery Office.

Longmore, Thomas. 1869b. *Thomas Longmore's Scrapbook of Illustrations of Appliances for the Transport of Wounded Soldiers*. RAMC 275. Wellcome Collection, London.

Lord, Amnon. 2015. "The 'Etrog': The Media, the Courts and Prime Minister Sharon during the Disengagement." *Jewish Political Studies Review* 27 (1–2): 65–78.

Lozowick, Yaacov. 1999. "Malice in Action." *Yad Vashem Studies* 27:287–330.

Lozowick, Yaacov. 2005. *Hitler's Bureaucrats: The Nazi Security Police and the Banality of Evil*. Edinburgh: A&C Black.

Lye, Colleen. 2009. *America's Asia: Racial Form and American Literature, 1893–1945*. Princeton, NJ: Princeton University Press.

Lyon, David. 2006. "Airport Screening, Surveillance, and Social Sorting: Canadian Responses to 9/11 in Context." *Canadian Journal of Criminology and Criminal Justice* 48 (3): 397–411.

Lyotard, Jean-François. 1988. *Le Différend*. Minneapolis: University of Minnesota Press.

Macdonald, Neil, David Chester, Heather Sangster, Beverly Todd, and Janet Hooke. 2012. "The Significance of Gilbert F. White's 1945 Paper 'Human Adjustment to Floods' in the Development of Risk and Hazard Management." *Progress in Physical Geography* 36 (1): 125–33.

Mackay, R. 2011. "'Going Backwards in Time to Talk about the Present': *Man on Wire* and Verticality after 9/11." *Comparative American Studies* 9 (1): 3–20.

Mackay, Ruth. 2016. *Waiting for the Sky to Fall: The Age of Verticality in American Narrative*. Columbus: Ohio State University Press.

Maleta, Yulia. 2009. "Playing with Fire: Gender at Work and the Australian Female Cultural Experience within Rural Fire Fighting." *Journal of Sociology* 45 (3): 291–306.

Malkki, Liisa. 1992. "National Geographic: The Rooting of Peoples and the Territorialization of National Identity among Scholars and Refugees." *Cultural Anthropology* 7 (1): 24–44.

Manchester Guardian. 1915a. "Ambulance Train on View in Manchester." December 17, 1915, 5.

Manchester Guardian. 1915b. "L and Y Ambulance Train for Use in France." July 10, 1915, 6.

Manchester Guardian. 1915c. "New Ambulance Train." February 11, 1915, 3.

Manchester Guardian. 1915d. "New Ambulance Train." December 17, 1915, 3.

Manchester Guardian. 1916a. "The Journey of the Wounded: From Hospital Ship to Manchester." August 31, 1916, 10.

Manchester Guardian. 1916b. "Railway Ambulance Train." January 28, 1916, 3.

Maravillas, Francis. 2012. "Un/Settled Geographies: Vertigo and the Predicament of Australia's Postcoloniality." In *New Voices, New Visions: Challenging Australian Identities and Legacies,* edited by Catriona Elder and Keith Moore, 17–37. Cambridge: Cambridge Scholars.

Margolies, Eleanor. 2003. "Were Those Boots Made Just for Walking? Shoes as Performing Objects in Everyday Life and in the Theatre." *Visual Communication* 2 (2): 169–88.

Marshall, Clarence J. 1917. "The Conservation of Horses in War." In *University Lectures Delivered by Members of the Faculty in the Free Public Lecture Course, 1916–1917,* vol. 4, 89–98. Philadelphia: University of Pennsylvania.

Martí, Mónica. 2020. "Ecocritical Archaeologies of Global Ecocide in Twenty-First-Century Post-apocalyptic Films." In *Avenging Nature: The Role of Nature in Modern and Contemporary Art and Literature,* edited by Eduardo Valls Oyarzun, Rebeca Gualberto Valverde, Noelia Malla Garcia, María Colom Jiménez, and Rebeca Cordero Sánchez, 179–94. Lexington, KY: Lexington Books.

Mas, Erick, Anawat Suppasri, Fumihiko Imamura, and Shunichi Koshimura. 2012. "Agent-Based Simulation of the 2011 Great East Japan Earthquake/Tsunami Evacuation: An Integrated Model of Tsunami Inundation and Evacuation." *Journal of Natural Disaster Science* 34 (1): 41–57.

Masco, Joseph. 2014. *The Theater of Operations: National Security Affect from the Cold War to the War on Terror.* Durham, NC: Duke University Press.

Massumi, Brian. 2002. *Parables for the Virtual: Movement, Affect, Sensation.* Durham, NC: Duke University Press.

Massumi, Brian. 2011. *Semblance and Event: Activist Philosophy and the Occurrent Arts.* Cambridge, MA: MIT Press.

Mayhew, Emily. 2013. *Wounded: From Battlefield to Blighty, 1914–1918.* London: Random House.

McCormack, Derek P. 2004. "Drawing Out the Lines of the Event." *Cultural Geographies* 11 (2): 211–20.

McCormack, Derek P. 2008. "Engineering Affective Atmospheres on the Moving Geographies of the 1897 Andree Expedition." *Cultural Geographies* 15 (4): 413–30.

McCormack, Derek P. 2010. "Remotely Sensing Affective Afterlives: The Spectral Geographies of Material Remains." *Annals of the Association of American Geographers* 100 (3): 640–54.

McCormack, Derek P. 2018. *Atmospheric Things: On the Allure of Elemental Envelopment.* Durham, NC: Duke University Press.

McCulley, Russell. 2007. "Saving Pets from Another Katrina." *Time Magazine,* June 6, 2007.

McEnaney, Laura. 2000. *Civil Defense Begins at Home: Militarization Meets Everyday Life in the Fifties*. Princeton, NJ: Princeton University Press.

McEvoy, Arthur F. 1995. "The Triangle Shirtwaist Factory Fire of 1911: Social Change, Industrial Accidents, and the Evolution of Common-Sense Causality." *Law and Social Enquiry* 20 (2): 621–51.

McFann, Hudson, and Alexander Hinton. 2018. "Impassable Visions: The Cambodia to Come, the Detritus in Its Wake." In *A Companion to the Anthropology of Death*, edited by Antonius C. G. M. Robben, 223–35. Oxford: Wiley.

McGreal, Chris. 2005. "Teenage Girls Stay in Jail as Israel Cracks Down on Settler Protests." *Guardian*, August 2, 2005. https://www.theguardian.com /world/2005/aug/02/israel.

McHugh, Susan. 2013. "'A Flash Point in Inuit Memories': Endangered Knowledges in the Mountie Sled Dog Massacre." *ESC: English Studies in Canada* 39 (1): 149–75.

McKeon, Peter Joseph. 1910. "Fire as a Social Destroyer." *Survey* 25 (December 3, 1910): 343–44.

McKeon, Peter Joseph. 1911. "Fires, Factories and Prevention: The Newark Casualty—the New York Dangers." *Survey* 25 (January 7): 532–46.

McNabb, Megan. 2007. "Pets in the Eye of the Storm: Hurricane Katrina Floods the Courts with Pet Custody Disputes." *Animal Law* 14 (1): 71–108.

McNeill, William Hardy. 1997. *Keeping Together in Time: Dance and Drill in Human History*. Cambridge, MA: Harvard University Press.

A Medical Officer on the Ambulance Ship "St. David." 1915. "The Overseas Transport of the Wounded: Life and Procedure on Board an Ambulance Ship." *Hospital*, April 17, 1915, 57–58.

Mennel, Timothy. 2004. "Victor Gruen and the Construction of Cold War Utopias." *Journal of Planning History* 3 (2): 116–50.

Meotti, Giulio. 2012. "Gush Katif: Was It Right to Turn Grass into Sand?" *Israel National News*, August 18, 2012. http://www.israelnationalnews.com /News/News.aspx/159015.

Meyer, Jessica. 2019. *An Equal Burden: The Men of the Royal Army Medical Corps in the First World War*. Oxford: Oxford University Press.

Milner, Elya Lucy. 2020. "Devaluation, Erasure and Replacement: Urban Frontiers and the Reproduction of Settler Colonial Urbanism in Tel Aviv." *Environment and Planning D: Society and Space* 38 (2): 267–86.

Ministry of Information. 1918. *The Care of Our Wounded*. Topical Film Company. Imperial War Museum, IWM 162. https://www.iwm.org.uk /collections/item/object/1060022666.

Mitchell, Margaret. 1936. *Gone with the Wind*. New York: Avon Books.

Mitra, Pujarinee. 2021. "The Vande Bharat Scam: Women, Social Standing, and Evacuation Flights to India under Covid-19." *Journal of Comparative Literature and Aesthetics* 44 (1): 79–91.

Miyake, Perry. 2002. *21st Century Manzanar: A Novel*. Los Angeles: Really Great Books.

Mizuno, Takeya. 2003. "Government Suppression of the Japanese Language in World War II Assembly Camps." *Journalism and Mass Communication Quarterly* 80 (4): 849–65.

Molotch, Harvey. 2014. *Against Security: How We Go Wrong at Airports, Subways, and Other Sites of Ambiguous Danger*. Princeton, NJ: Princeton University Press.

Moore, John. 1920. *Army Veterinary Service in War*. London: H. and W. Brown.

Moore, Sarah A. 2012. "Garbage Matters: Concepts in New Geographies of Waste." *Progress in Human Geography* 36 (6): 780–99.

Morris, D. 2005. "Lema'an Achai (RBS)—Emergency Campaign for Gush Katif." *Maof*, November 30, 2005. http://maof.rjews.net/english/37-english/8993-lemaan-achai-rbs-emergency-campaign-for-gush-katif.

Morshed, Adnan. 2015. *Impossible Heights: Skyscrapers, Flight, and the Master Builder*. Minneapolis: University of Minnesota Press.

Moses, Robert. 1940. "New York Opens Bottlenecks: City's New Parkways Ease Traffic in Many . . ." *New York Times*, October 13, 1940.

Moses, Robert. 1942. "What Happened to Haussmann." *Architectural Forum* 77 (1): 57–66.

Moses, Robert. 1951. "The Traffic Menace, in Both Peace and War." *New York Times*, April 29, 1951.

Moses, Robert. 1957. "Civil Defense Fiasco." *Harper's Magazine*, November 1957, 29–34.

Mrázek, Rudolf. 2002. *Engineers of Happy Land: Technology and Nationalism in a Colony*. Princeton, NJ: Princeton University Press.

Nagel, D. 1935a. "Das Räumungsproblem im zivilen Luftschutz: Räumungsproblem in Kriegsgeschichte." *Gasschutz und Luftschutz* 5 (5): 113–18.

Nagel, D. 1935b. "Das Räumungsproblem im zivilen Luftschutz: Räumungsproblem in Kriegsgeschichte." *Gasschutz und Luftschutz* 5 (6): 143–47.

Nagel, D. 1935c. "Das Räumungsproblem im zivilen Luftschutz 2: Räumung und Unterkunft." *Gasschutz und Luftschutz* 5 (10): 249–54.

Nanni, Giordano, and Andrea James. 2013. *Coranderrk: We Will Show the Country*. Canberra: Aboriginal Studies Press.

National Bureau of Standards. 1935. *Design and Construction of Building Exits*. Washington, DC: US Department of Commerce.

National Railway Museum. 2019. "Ambulance Trains: Bringing the First World War Back Home." RailwayMuseum, February 1, 2019. https://www.railwaymuseum.org.uk/objects-and-stories/ambulance-trains-bringing-first-world-war-home.

NBC News. 2020. "Haiti Quake Creates Thousands of New Orphans." January 19, 2020. http://www.nbcnews.com/id/34934553/ns/world_news-haiti/t/haiti-quake-creates-thousands-new-orphans/#.XaTIbS2ZN-o.

Neale, Timothy. 2018a. "'Are We Wasting Our Time?': Bushfire Practitioners and Flammable Futures in Northern Australia." *Social and Cultural Geography* 19 (4): 473–95.

Neale, Timothy. 2018b. "Digging for Fire: Finding Control on the Australian Continent." *Journal of Contemporary Archaeology* 5 (1): 79–90.

Neale, Timothy, Rodney Carter, Trent Nelson, and Mick Bourke. 2019. "Walking Together: A Decolonising Experiment in Bushfire Management on Dja Dja Wurrung Country." *Cultural Geographies* 26 (3): 341–59.

Neumann, Boaz. 2011. *Land and Desire in Early Zionism*. Waltham, MA: Brandeis University Press.

Newark Evening News. 1910. "Fifteen Girls Die in High St. Factory Fire." November 26, 1910, 1–2.

New York Times. 1910. "The Newark Disaster." November 28, 1910, 8.

New York Times. 1937. "New Building Code Passes Alderman after 4 Year Wait." July 21, 1937, 1, 9.

New York Times. 1940. "German-Type Road Is Urged for US." June 28, 1940, 21.

New York Times. 1973. "Department Store Fire in Japan Kills 100." November 30, 1973.

New York Tribune. 1911. "More Than 140 Die as Factory Flames Sweep through Three Stories of Factory Building on Washington Place." March 26, 1911, 1.

Ngai, Sianne. 2012. *Our Aesthetic Categories: Zany, Cute, Interesting*. Cambridge, MA: Harvard University Press.

Ngai, Sianne. 2020. *Theory of the Gimmick: Aesthetic Judgment and Capitalist Form*. Cambridge, MA: Harvard University Press.

9/11 Memorial and Museum. 2010. "Shoes Symbolize Acts of Compassion during 9/11 Evacuations." Accessed November 10, 2017. https://www.911memorial.org/connect/blog/shoes-symbolize-acts-compassion-during-911-evacuations.

9/11 Memorial and Museum. 2015. "Museum Guide for Visitors with Children." Accessed November 2019. https://911memorial.org/learn.

9News Sydney. 2022. "A Lyrebird Has Been Caught Mimicking the 'Evacuate Now' Siren at Sydney's Taronga Zoo." X, November 13, 2022. https://twitter.com/9NewsSyd/status/1591682390534942721.

Niven, Larry. 1977. *A World Out of Time*. New York: Del Rey.

Nixon, Rob. 2011. *Slow Violence and the Environmentalism of the Poor*. Cambridge, MA: Harvard University Press.

Noah, Timothy. 2005. "The Puppies and Kittens Act of 2005." *Slate*, September 22, 2005. https://slate.com/news-and-politics/2005/09/the-puppies-and-kittens-act-of-2005.html.

Nolan, Christopher, dir. 2014. *Interstellar*. Warner Bros.

Nolen, R. Scott. 2005. "Katrina's Other Victims: Animals' Plight Prompts Outcry for Change." American Veterinary Medical Association, October 15,

2005. https://www.avma.org/javma-news/2005-10-15/katrinas-other -victims-october-15-2005.

Northern Herald. 1926. "Victorian Settlers Have Narrow Escapes." March 3, 1926, 15.

Oakes, Guy. 1995. *The Imaginary War: Civil Defense and American Cold War Culture.* Oxford: Oxford University Press.

Observer. 1915. "The Queen's Surprise Visit." March 28, 1915, 13.

O'Connor, Lee Thomas. 2010. "Take Cover, Spokane: A History of Backyard Bunkers, Basement Hideaways, and Public Fallout Shelters of the Cold War." Master's thesis, Washington State University.

An Officer of the R.A.M.C. at the Front. 1915a. "Impressions from a Field Ambulance: Tires, Equipment and the Health of the Army." *Hospital,* January 16, 1915, 351.

An Officer of the R.A.M.C. at the Front. 1915b. "The Work of the in R.A.M.C. in War: Evacuation Difficulties: The Triumph of the Motor Ambulance." *Hospital,* January 2, 1915, 307–8.

Off Our Backs: A Women's Newsjournal. 1970. "Triangle Shirtwaist Fire." March 19, 1970, 9.

O'Grady, Nathaniel. 2018. *Governing Future Emergencies: Lived Relations to Risk in the UK Fire and Rescue Service.* London: Palgrave.

O'Hagan, John T. 1977. *High Rise—Fire and Life Safety.* New York: Dun-Donnelley.

Okamura, Raymond Y. 1982. "The American Concentration Camps: A Cover-Up through Euphemistic Terminology." *Journal of Ethnic Studies* 10 (3): 95–109.

O'Leary, Stephen D. 1998. *Arguing the Apocalypse: A Theory of Millennial Rhetoric.* Oxford: Oxford University Press.

Ophir, Adi. 2007. "The Two-State Solution: Providence and Catastrophe." *Theoretical Inquiries in Law* 8 (1): 117–60.

O'Reilly, Conor. 2011. "'From Kidnaps to Contagious Diseases': Elite Rescue and the Strategic Expansion of the Transnational Security Consultancy Industry." *International Political Sociology* 5 (2): 178–97.

Oriolus. 1939. "Nature Notes: How Nature Has Conquered Drought." *The Age,* January 10, 1939, 2.

Overall, Christine. 2015. "Rethinking Abortion, Ectogenesis, and Fetal Death." *Journal of Social Philosophy* 46 (1): 126–40.

Overy, Richard. 2013. *The Bombing War: Europe, 1939–1945.* London: Penguin.

Paley, Morton D., and Northrop Frye. 1986. *The Apocalyptic Sublime.* New Haven, CT: Yale University Press.

Pallister-Wilkins, Polly. 2015. "The Humanitarian Politics of European Border Policing: Frontex and Border Police in Evros." *International Political Sociology* 9 (1): 53–69.

Pallister-Wilkins, Polly. 2020. "Hotspots and the Geographies of Humanitarianism." *Environment and Planning D: Society and Space* 38 (6): 991–1008.

Pallott, Peter. 2006. "Flee the Danger Zone with Evacuation Cover." *Telegraph*, August 24, 2006. https://www.telegraph.co.uk/news/health/expat-health /4201144/Flee-the-danger-zone-with-evacuation-cover.html.

Park, Robert E., and Ernest W. Burgess. 1925. *The City.* Chicago: University of Chicago Press.

Parker, Dorothy R. 1994. *Singing an Indian Song: A Biography of D'Arcy McNickle.* Lincoln: University of Nebraska Press.

Parsons, Brinckerhoff, Quade & Douglas, Ammann & Whitney, 1955. *General Plan: Milwaukee County Expressway System.* Milwaukee, WI: Milwaukee County Expressway Commission.

Pauls, Jake, dir. 1979. *The Stair Event.* 18 mins. National Research Council of Canada, Ottawa.

Pauls, Jake. 1984. "The Movement of People in Buildings and Design Solutions for Means of Egress." *Fire Technology* 20 (1): 27–47.

Pauls, Jake L., John J. Fruin, and J. M. Zupan. 2007. "Minimum Stair Width for Evacuation, Overtaking Movement and Counterflow—Technical Bases and Suggestions for the Past, Present and Future." In *Pedestrian and Evacuation Dynamics 2005,* edited by Nathalie Waldau, Peter Gattermann, Hermann Knoflacher, and Michael Schreckenberg, 57–69. Berlin: Springer.

Peiss, Kathy. 1986. *Cheap Amusements: Working Women and Leisure in Turn-of-the-Century New York.* Philadelphia: Temple University Press.

Pence, Patricia Lanier, Paula Phillips Carson, Kerry D. Carson, J. Brooke Hamilton, and Betty Birkenmeier. 2003. "And All Who Jumped Died: The Triangle Shirtwaist Factory Fire." *Management Decision* 41 (4): 407–21.

Perry, Ronald W. 1979. "Evacuation Decision-Making in Natural Disasters." *Mass Emergencies* 4 (1): 25–38.

Perugini, Nicola. 2014. "The Moral Economy of Settler Colonialism: Israel and the 'Evacuation Trauma.'" *History of the Present* 4 (1): 49–74.

Perugini, Nicola. 2019. "Settler Colonial Inversions: Israel's 'Disengagement' and the Gush Katif 'Museum of Expulsion' in Jerusalem." *Settler Colonial Studies* 9 (1): 41–58.

Peter, Marc, Jr. 1950. "Lessons from the Last War." *Bulletin of Atomic Scientists,* August–September 1950, 252–55.

Peterson, Val. 1955. Interview in *U.S. News and World Report,* April 8, 1955, 74–78.

Pezzani, Lorenzo, and Charles Heller. 2013. "A Disobedient Gaze: Strategic Interventions in the Knowledge(s) of Maritime Borders." *Postcolonial Studies* 16 (3): 289–98.

Philo, Chris, and Chris Wilbert. 2000. "Animal Spaces, Beastly Places: An Introduction." In *Animal Spaces, Beastly Places: New Geographies of Human-*

Animal Relations, edited by Chris Philo and Chris Wilbert, 1–34. London: Routledge.

Phu, Thy. 2008. "The Spaces of Human Confinement: Manzanar Photography and Landscape Ideology." *Journal of Asian American Studies* 11 (3): 337–71.

Pierce, Bert. 1948. "Highways Urged as Defense Aids." *New York Times*, March 4, 1948.

Plant, Bob. 2011. "Welcoming Dogs: Levinas and 'the Animal' Question." *Philosophy and Social Criticism* 37 (1): 49–71.

Powell, J. M. 1991. *An Historical Geography of Modern Australia: The Restive Fringe*. Cambridge: Cambridge University Press.

Preston, John. 2007. *Whiteness and Class in Education*. Dordrecht: Springer.

Preston, John. 2018. *Grenfell Tower: Preparedness, Race and Disaster Capitalism*. London: Palgrave.

Prieto, Sara. 2015. "'Without Methods': Three Female Authors Visiting the Western Front." *First World War Studies* 6 (2): 171–85.

Protevi, John. 2009. *Political Affect: Connecting the Social and the Somatic*. Minneapolis: University of Minnesota Press.

Qikiqtani Inuit Association. 2013. *Qikiqtani Truth Commission: Thematic Reports and Special Studies 1950–1975: Analysis of the RCMP Sled Report*. Nunavut: Inhabit Media Inc. (Iqaluit).

Quarantelli, Enrico Louis. 1980. *Evacuation Behavior and Problems: Findings and Implications from the Research Literature*. Columbus: Ohio State University Disaster Research Center.

Rabinbach, Anson. 1992. *The Human Motor: Energy, Fatigue, and the Origins of Modernity*. Los Angeles: University of California Press.

Rancière, Jacques. 2004. *The Politics of Aesthetics*. London: Bloomsbury.

Raymonds, Jack. 1947. "Moses Is in Berlin on Secret Mission: Clay Says Only That He Will Study Military Government and Make Recommendations." *New York Times*, June 28, 1947.

Read, Peter. 1996. *Returning to Nothing: The Meaning of Lost Places*. Cambridge: Cambridge University Press.

Ready.gov. 2018. "Disaster Tips: If you evacuate, take your pets with you." Facebook, September 7, 2018. https://www.facebook.com/watch/?v=293975461203392.

Redfield, Peter. 2008. "Vital Mobility and the Humanitarian Kit." In *Biosecurity Interventions: Global Health and Security in Question*, edited by Andrew Lakoff and Stephen J. Collier, 147–71. New York: Columbia University Press.

Reiff, R. 1975. "Office Stairs Take Novel Twist." *New York Times*, January 5, 1975, 6.

Reilly, Frances. 2008. "Operation 'Lifesaver': Canadian Atomic Culture and Cold War Civil Defence." *Past Imperfect* 14:46–85.

Richards, Gary. 2010. "Queering Katrina: Gay Discourses of the Disaster in New Orleans." *Journal of American Studies* 44 (3): 519–34.

Rickards, Lauren, Tim Neale, and Matt Kearnes. 2017. "Australia's National Climate: Learning to Adapt?" *Geographical Research* 55 (4): 469–76.

Roberts, Patrick S. 2013. *Disasters and the American State: How Politicians, Bureaucrats, and the Public Prepare for the Unexpected.* Cambridge: Cambridge University Press.

Rohrer, John H., Seymour H. Baron, E. L. Hoffman, and D. V. Swander. 1954. "The Stability of Autokinetic Judgments." *Journal of Abnormal and Social Psychology* 49 (4, pt. 1): 595–97.

Rose, Mark H., and Raymond A. Mohl. 2012. *Interstate: Highway Politics and Policy since 1939.* Knoxville: University of Tennessee Press.

Rose, Nikolas. 1999. *Powers of Freedom: Reframing Political Thought.* Cambridge: Cambridge University Press.

Rose, Sonya O. 2003. *Which People's War? National Identity and Citizenship in Wartime Britain, 1939–1945.* Oxford: Oxford University Press.

Rosen, Gillad, and Igal Charney. 2016. "Divided We Rise: Politics, Architecture and Vertical Cityscapes at Opposite Ends of Jerusalem." *Transactions of the Institute of British Geographers* 41 (2): 163–74.

Rosenbaum, Alan S. 2019. *Is the Holocaust Unique? Perspectives on Comparative Genocide.* London: Taylor and Francis.

Rosler, Martha. 1988. *In the Place of the Public: Airport Series.* Cantz: Ostfildern-Ruit.

Rossello, Diego. 2015. "Ordinary Emergences in Democratic Theory: An Interview with Bonnie Honig." *Philosophy Today* 59 (4): 699–710.

Royal Commission. 2009. *2009 Victorian Bushfires Royal Commission: Interim Report.* Bernard Teague, Chair; Ronald Mcleod, Commissioner; Susan Pascoe, Commissioner. Melbourne: Government Printer for the State of Victoria.

Royal Commission on South African Hospitals. 1901. *Report of the Royal Commission Appointed to Consider and Report upon the Care and Treatment of the Sick and Wounded during the South African Campaign: Presented to Both Houses of Parliament by Command of Her Majesty / Royal Commission on South African Hospitals.* London: Her Majesty's Stationery Office.

Saint-Amour, Paul K. 2011. "Applied Modernism: Military and Civilian Uses of the Aerial Photomosaic." *Theory, Culture and Society* 28 (7–8): 241–69.

Saint-Amour, Paul K. 2015. *Tense Future: Modernism, Total War, Encyclopedic Form.* Oxford: Oxford University Press.

Salter, Mark B. 2007. "Governmentalities of an Airport: Heterotopia and Confession." *International Political Sociology* 1 (1): 49–66.

Sanders, Kerry. 2005. "Horrible Scenes at New Orleans Airport." *NBC News,* September 2, 2005.

Sanford, Christopher, Jonathan Jui, Helen C. Miller, and Kathleen A. Jobe. 2007. "Medical Treatment at Louis Armstrong New Orleans International

Airport after Hurricane Katrina: The Experience of Disaster Medical Assistance Teams WA-1 and OR-2." *Travel Medicine and Infectious Disease* 5 (4): 230–35.

Scanlan, John. 2005. *On Garbage*. London: Reaktion Books.

Schaberg, Christopher. 2015. *The End of Airports*. London: Bloomsbury.

Schiffrin, Deborah. 2001. "Language and Public Memorial: America's Concentration Camps." *Discourse and Society* 12 (4): 505–34.

Schillinger, Ted, dir. 2012. *Evacuate Earth*. National Geographic Channel.

Schivelbusch, Wolfgang. 1986. *The Railway Journey: The Industrialization and Perception of Time and Space*. Berkeley: University of California Press.

Schmidt, Stanley. 1976. "How to Move the Earth." *Analog* 96 (5): 59–72.

Schmidt, Stanley. 2000. *Lifeboat Earth*. New York: Berkley Pub Corp.

Schmitz-Berning, C. 1998. *Vokabular des Nationalsozialismus*. Berlin: W. de Gruyter.

Schraubi, Globarius. 1942. "Yule Tide, Greetings Friends!" *Trek*, December 1942, 12–16.

Schweik, Susan. 1989. "The 'Pre-poetics' of Internment: The Example of Toyo Suyemoto." *American Literary History* 1 (1): 89–109.

Scott, James C. 1999. *Seeing Like a State: How Certain Schemes to Improve the Human Condition Have Failed*. New Haven, CT: Yale University Press.

Sebald, W. G. 2001. *Austerlitz*. London: Hamish Hamilton.

Segal, Rafi, David Tartakover, and Eyal Weizman. 2003. *A Civilian Occupation: The Politics of Israeli Architecture*. London: Verso.

Seiler, Cotten. 2009. *Republic of Drivers: A Cultural History of Automobility in America*. Chicago: University of Chicago Press.

Selman, Peter. 2011. "Intercountry Adoption after the Haiti Earthquake: Rescue or Robbery?" *Adoption and Fostering* 35 (4): 41–49.

Shani, Liron. 2011. "Nationalism between Land and Environment: The Conflict between 'Oranges' and 'Greens' over Settling in the East Lakhish Area." Research Paper No. 6. College Park, MD: Joseph and Alma Gildenhorn Institute of Israel Studies, University of Maryland.

Shavit, Ari. 2005. *A Land Divided: Israelis Think about Disengagement*. Tel Aviv: Keter.

Sheleg, Yair. 2007. "The Political and Social Ramifications of Evacuating Settlements in Judea, Samaria, and the Gaza Strip: Disengagement 2005 as a Test Case." Policy Paper No. 42. Jerusalem: Israel Democracy Institute.

Sheleg, Yair. 2008. "The Disengagement of 2005: An Interview with Yair Sheleg." Israel Democracy Institute, May 18, 2008. https://en.idi.org.il /articles/6982.

Sheller, Mimi. 2013. "The Islanding Effect: Post-disaster Mobility Systems and Humanitarian Logistics in Haiti." *Cultural Geographies* 20 (2): 185–204.

Shragai, Nadav. 2005. "Gush Katif Surfer Teens Threaten Group Suicide on the Waves." *Haaretz*, August 3, 2005. https://www.haaretz.com/1.4927805.

Shukrun-Nagar, Pnina. 2008. "Disengagement Terminologies." *Studies in Israeli and Modern Jewish Society* 18:341–61.

Simmons, Kristen. 2017. "Settler Atmospherics." Member Voices, *Fieldsights*, November 20, 2017. https://culanth.org/fieldsights/settler -atmospherics.

Sinha, Indra. 2008. *Animal's People: A Novel.* New York: Simon and Schuster.

Skandalakis, Panagiotis N., Panagiotis Lainas, Odyseas Zoras, John E. Skandalakis, and Petros Mirilas. 2006. "'To Afford the Wounded Speedy Assistance': Dominique Jean Larrey and Napoleon." *World Journal of Surgery* 30:1392–99.

Smith, F. 1918. "The Work of the British Army Veterinary Corps at the Fronts." *Journal of the Royal Society of Arts* 67 (3449): 80–92.

Smith, Neil. 2006. "There's No Such Thing as a Natural Disaster." *Items: Insights from the Social Sciences*, June 11, 2006. https://items.ssrc.org/understanding -katrina/theres-no-such-thing-as-a-natural-disaster/.

Sobo, Elisa J. 1993. "Bodies, Kin, and Flow: Family Planning in Rural Jamaica." *Medical Anthropology Quarterly* 7 (1): 50–73.

Sodero, Stephanie. 2019. "Vital Mobilities: Circulating Blood via Fictionalized Vignettes." *Cultural Geographies* 26 (1): 109–25.

Sofsky, Wolfgang, and William Templer. 2013. *The Order of Terror: The Concentration Camp.* Princeton, NJ: Princeton University Press.

Solnit, Rebecca. 2004. *River of Shadows: Eadweard Muybridge and the Technological Wild West.* London: Penguin.

Solnit, Rebecca. 2010. *A Paradise Built in Hell: The Extraordinary Communities That Arise in Disaster.* New York: Penguin.

Somfalvi, Atilla. 2005. "Large-Scale Pullout Drill Ends." *ynetnews*, August 10, 2005. https://www.ynetnews.com/articles/0,7340,L-3125315,00.html.

Sontag, Susan. 2005. "The Imagination of Disaster." In *Liquid Metal: The Science Fiction Film Reader*, edited by Sean Redmond, 40–47. New York: Columbia University Press.

Sopko, Andrew. 2015. "An (Im)Balance of Expectations: Civil Defence in Ottawa, 1951–1962." Master's thesis, Carleton University, Ottawa.

Sorin, Gretchen Sullivan. 2009. "'Keep Going': African Americans on the Road in the Era of Jim Crow." PhD diss., State University of New York at Albany.

A South African Campaigner. 1899. "The Medical Aspects of the Boer War." *British Medical Journal* 2 (2030) (November 25): 1485–87.

Speier, Amy, Kristin Lozanski, and Susan Frohlick. 2020. "Reproductive Mobilities." *Mobilities* 15 (2): 107–19.

Spillett, Richard. 2014. "Goodbye to All That: Poignant Archive Pictures Show the Ambulance Trains That Transported Soldiers Wounded on the Western Front Back to Hospitals across Britain during the First World War." *Daily Mail*, April 16, 2014. https://www.dailymail.co.uk/news

/article-2605886/Poignant-archive-pictures-ambulance-trains-transported
-soldiers-wounded-First-World-War-hospitals-Britain.html.

Srinivasan, Krithika. 2019. "Remaking More-Than-Human Society: Thought Experiments on Street Dogs as 'Nature.'" *Transactions of the Institute of British Geographers* 44 (2): 376–91.

Star, Susan Leigh. 1999. "The Ethnography of Infrastructure." *American Behavioral Scientist* 43 (3): 377–91.

Stargardt, Nicholas. 2011. "Beyond 'Consent' or 'Terror': Wartime Crises in Nazi Germany." *History Workshop Journal* 72 (1): 190–204.

Stein, Leon. 2011. *The Triangle Fire*. Ithaca, NY: Cornell University Press.

Steinbacher, Sybille. 2005. *Auschwitz: A History*. London: Penguin.

Steneck, Nicholas J. 2005. "Everybody Has a Chance: Civil Defense and the Creation of Cold War West German Identity, 1950–1968." PhD diss., Ohio State University.

Stephen, Andrew. 2010. "To Haiti with Hate, from the US Right." *New Statesman*, January 21, 2010. Accessed January 5, 2019. https://www .newstatesman.com/international-politics/2010/01/haiti-aid-american -afghanistan.

Stephens, Marguerita. 2003. "White without Soap: Philanthropy, Caste and Exclusion in Colonial Victoria, 1835–1888; A Political Economy of Race." PhD diss., University of Melbourne.

Stevenson, David. 1999. "War by Timetable? The Railway Race before 1914." *Past and Present*, no. 162, 163–94.

Stevenson, Lisa. 2014. *Life beside Itself: Imagining Care in the Canadian Arctic*. Berkeley: University of California Press.

Stoler, Ann Laura. 2016. *Duress: Imperial Durabilities in Our Times*. Durham, NC: Duke University Press.

Stoler, Ann Laura. 2018. "Colonial Toxicities in a Recursive Mode." *Postcolonial Studies* 21 (4): 542–47.

Stoltzfus, Nathan. 2016. *Hitler's Compromises: Coercion and Consensus in Nazi Germany*. New Haven, CT: Yale University Press.

Stratfor. 2011a. "Above the Tearline: Emergency Evacuation Plans." March 16, 2011. https://worldview.stratfor.com/article/above-tearline-emergency -evacuation-plans.

Stratfor. 2011b. "How to Travel Safely: Tips from a Former Agent." July 14, 2011. https://worldview.stratfor.com/article/how-travel-safely-tips-former -agent.

Stratton, Jon. 1989. "Deconstructing the Territory." *Cultural Studies* 3 (1): 38–57.

Stretton, Leonard E. B. 1939. *Report of the Royal Commission to Inquire into the Causes of and Measures Taken to Prevent the Bush Fires of January, 1939, and to Protect Life and Property and the Measures to Be Taken to Prevent Bush Fires in Victoria and to Protect Life and Property in the Event of Future Bush Fires*. Melbourne, VIC: Judge Leonard E. B. Stretton.

Stroud, John. 2019. "The AFM's 'Defense Estate': A Legacy of British Military Architecture." *OnParade*, October 2019, 35–37. https://www.um.edu.mt /library/oar/bitstream/123456789/49389/1/The%20AFM%27s%20De-fence%20Estate.pdf.

Sunshine, Gregory, and Aila Hoss. 2015. "Emergency Declarations and Tribes: Mechanisms under Tribal and Federal Law." *Michigan State International Law Review* 24 (1): 33–44.

Survey. 1911a. "The Common Welfare: A Fire College for New York." 25 (January 7, 1911): 519–20.

Survey. 1911b. "The Common Welfare: Prevention of Factory Fires." 26 (April 8, 1911): 81.

Suzuki, Michiko. 2023. *Reading the Kimono in Twentieth-Century Japanese Literature and Film*. Honolulu: University of Hawai'i Press.

Szerszynski, Bronislaw. 2016. "Planetary Mobilities: Movement, Memory and Emergence in the Body of the Earth." *Mobilities* 11 (4): 614–28.

Tateishi, John, and Roger Daniels. 2012. *And Justice for All: An Oral History of the Japanese American Detention Camps*. Seattle: University of Washington Press.

Taylor, A. J. P. 1969. *War by Time-Table: How the First World War Began*. New York: Macdonald.

Taylor, Alan. 2015. "Zoo Security Drills: When Animals Escape." *Atlantic*, February 10, 2015. https://www.theatlantic.com/photo/2015/02/zoo-security -drills-when-animals-escape/385346/.

Taylor, Chris. 2011. "CF Operation *Mobile* Air Evacuation." Taylor Empire Airways, March 5, 2011. Accessed June 20, 2018. http://taylorempireairways .com/2011/03/cf-operation-mobile-air-evacuation/.

Tazzioli, Martina. 2018. "Spy, Track and Archive: The Temporality of Visibility in Eurosur and Jora." *Security Dialogue* 49 (4): 272–88.

Templer, John. 1974. "Stair Shape and Human Movement." PhD diss., Columbia University.

Templer, John. 1992. *The Staircase: Studies of Hazards, Falls and Safer Design*. Cambridge, MA: MIT Press.

Tenorio, Richard. 2016. "Animals Rescued from the 'World's Worst Zoo.'" *National Geographic*, September 26, 2016. https://www.nationalgeographic .com/news/2016/09/saving-animals-gaza-strip-khan-younis-zoo/.

Thiesmeyer, Lynn. 1995. "The Discourse of Official Violence: Anti-Japanese North American Discourse and the American Internment Camps." *Discourse and Society* 6 (3): 319–52.

Thorpe, Stanley W. 1919. *Fire Prevention and Protection as Exemplified in an English Factory*. "Red Books" of the British Fire Prevention Committee No. 236. London: British Fire Prevention Committee.

Thurber, James. 1933. "The Day the Dam Broke." In *My Life and Hard Times*, 40–56. London: Hamilton.

Tierney, Kathleen. 2014. *The Social Roots of Risk: Producing Disasters, Promoting Resilience.* Palo Alto, CA: Stanford University Press.

Tobin, Kathleen. 2002. "The Reduction of Urban Vulnerability: Revisiting 1950s American Suburbanization as Civil Defence." *Cold War History* 2 (2): 1–32.

Tochterman, Brian L. 2017. *The Dying City: Postwar New York and the Ideology of Fear.* Chapel Hill: University of North Carolina Press.

Todd, Ellen Wiley. 2005. "Photojournalism, Visual Culture, and the Triangle Shirtwaist Fire." *Labor: Studies in Working-Class History of the Americas* 2 (2): 9–22.

Torrie, Julia S. 2006. "'If Only Family Unity Can Be Maintained': The Witten Protest and German Civilian Evacuations." *German Studies Review* 29 (2): 347–66.

Torrie, Julia S. 2010. *"For Their Own Good": Civilian Evacuations in Germany and France, 1939–1945.* New York: Berghahn.

A Train Errant: Being the Experiences of a Voluntary Unit in France, and an Anthology from Their Magazine. 1919. Hertford, UK: Simson and Co.

Troller, Norbert, Joel Shatzky, and Susan E. Cernyak-Spatz. 2004. *Theresienstadt: Hitler's Gift to the Jews.* Chapel Hill: University of North Carolina Press.

Trotter, David. 2014. "Messages from the 29th Floor: Lifts." *London Review of Books*, July 3, 2014. https://www.lrb.co.uk/the-paper/v36/n13/david-trotter/messages-from-the-29th-floor.

Tyler, Meagan, and Peter Fairbrother. 2013. "Gender, Masculinity and Bushfire: Australia in an International Context." *Australian Journal of Emergency Management* 28 (2): 20–25.

Tyner, James A. 2017. *From Rice Fields to Killing Fields: Nature, Life, and Labor under the Khmer Rouge.* Syracuse, NY: Syracuse University Press.

Tyner, James A., Andrew Curtis, Sokvisal Kimsroy, and Chhunly Chhay. 2018. "The Evacuation of Phnom Penh during the Cambodian Genocide: Applying Spatial Video Geonarratives to the Study of Genocide." *Genocide Studies and Prevention* 12 (3): 163–76.

Tyner, James A., and Stian Rice. 2016. "To Live and Let Die: Food, Famine, and Administrative Violence in Democratic Kampuchea, 1975–1979." *Political Geography* 52 (May): 47–56.

Urry, John. 2004. "The 'System' of Automobility." *Theory, Culture and Society* 21 (4–5): 25–39.

Urry, John. 2013. *Societies beyond Oil: Oil Dregs and Social Futures.* London: Zed Books.

US House Subcommittee of the Committee on Government Operations. 1956. *Civil Defense for National Survival: Part 4.* Washington, DC: US Government Printing Office.

van der Kolk, Ruth. 2020. "Actiegroep We Gaan Ze Halen zonder vliegtuig op weg naar Lesbos: 'We doen niets illegals.'" *nederlandsdagblad*, October 5,

2020. https://www.nd.nl/nieuws/nederland/995725/actiegroep-we-gaan
-ze-halen-zonder-vliegtuig-op-weg-naar-lesbos.

Vankawala, Hemant H. 2005. "A Doctor's Message from Katrina's Front Lines."
NPR, September 7, 2005. https://www.npr.org/templates/story/story.php
?storyId=4836926&t=1568320237351.

Victoria State Government. 2014. *Leave and Live.* Video produced for the
FireReadyVictoria 2014/15 campaign, Emergency Victoria. Country
Fire Authority YouTube channel. https://www.youtube.com/watch?v
=pbSWJFPfjdo.

Victoria State Government. 2018. "Leaving Early: Bushfire Survival Planning
Template." Emergency Victoria, Country Fire Association, August 2018.
Accessed October 2018. https://cdn.cfa.vic.gov.au/documents/20143
/70643/4713_CFA_Pullout_LEAVING_web.pdf/714383d1-5570-9529
-aebc-b50c4abbfc0f?t=1536016340934.

Vidal-Naquet, Pierre. 1992. *Assassins of Memory: Essays on the Denial of the
Holocaust.* New York: Columbia University Press.

Vines, Gary. 1985. "The Historical Archaeology of Forest Based Sawmilling in
Victoria, 1855–1940." BA thesis, La Trobe University.

Vogel, Lise. 1983. *Marxism and the Oppression of Women: Toward a Unitary
Theory.* Leiden: Brill.

Vogel-Klein, Ruth. 2014. "History, Emotions, Literature: The Representation of
Theresienstadt in H. G. Adler's *Theresienstadt 1941–1945: Das Antlitz einer
Zwangsgemeinschaft* and W. G. Sebald's *Austerlitz.*" In *Witnessing, Memory,
Poetics: H. G. Adler and W. G. Sebald*, edited by Helen Finch and Lynn L.
Wolff, 180–98. Norwich, UK: Boydell and Brewer.

Wagner, Laura Rose. 2014. "Haiti Is a Sliding Land: Displacement, Community and Humanitarianism in Post-earthquake Port-au-Prince." PhD diss.,
University of North Carolina.

Wallander, Arthur, and Robert Moses. 1951. *New York City Civil Defense.* New
York: Office of Civil Defense.

Wall Street Journal. 2010. "France's Sarkozy Tries to Smooth Waters over U.S.
Role in Haiti." January 19, 2010. https://blogs.wsj.com/dispatch/2010/01
/19/frances-sarkozy-tries-to-smooth-waters-over-us-role-in-haiti/.

Wallwork, Ellery D., Kathy S. Gunn, Mark L. Morgan, and Kathryn A. Wilcox-
son. 2010. *Operation Unified Response: Air Mobility Command's Response to
the 2010 Haiti Earthquake Crisis.* Scott Air Force Base, IL: Office of History,
Air Mobility Command.

Walters, William. 2015. "Migration, Vehicles, and Politics: Three Theses on
Viapolitics." *European Journal of Social Theory* 18 (4): 469–88.

Walters, William, Charles Heller, and Lorenzo Pezzani, eds. 2021. *Viapolitics:
Borders, Migration, and the Power of Locomotion.* Durham, NC: Duke University Press.

Walters, William, and Barbara Lüthi. 2016. "The Politics of Cramped Space: Dilemmas of Action, Containment and Mobility." *International Journal of Politics, Culture, and Society* 29:359–66.

Ward, Barbara, and René Dubos. 1972. *Only One Earth: The Care and Maintenance of a Small Planet.* Harmondsworth, UK: Penguin.

Ward, Jesmyn. 2011. *Salvage the Bones.* New York: Bloomsbury.

War Relocation Authority. 1946. *WRA: A Story of Human Conservation.* Washington, DC: United States Department of the Interior, War Relocation Authority.

We Gaan Ze Halen. 2020. "Volunteers Take a Boeing 737 and Head Out to Lesbos to Bring Back Refugees." October 4, 2020. https://wegaanzehalen .nl/volunteers-take-a-boeing-737-and-head-out-to-lesbos-to-bring-back -refugees/.

Weir, Jessica K., Stephen Sutton, and Gareth Catt. 2020. "The Theory/Practice of Disaster Justice: Learning from Indigenous Peoples' Fire Management." In *Natural Hazards and Disaster Justice,* edited by Claudia Baldwin and Anna Lukasiewicz, 299–317. Berlin: Springer.

Weis, Efrat. 2005. "Gush Katif Residents Won't Go Quietly." *ynetnews,* August 14, 2005. https://www.ynetnews.com/articles/0,7340,L-3126917,00 .html.

Weisman, Alan. 2008. *The World without Us.* New York: Macmillan.

Weizman, Eyal. 2006. "Lethal Theory." *Log,* no. 7, 53–77.

Weizman, Eyal. 2007. *Hollow Land: Israel's Architecture of Occupation.* London: Verso.

Weizman, Eyal. 2011. *The Least of All Possible Evils: Humanitarian Violence from Arendt to Gaza.* London: Verso.

Weizman, Eyal. 2017. *Forensic Architecture: Violence at the Threshold of Detectability.* Cambridge, MA: MIT Press.

Welshman, John. 1998. "Evacuation and Social Policy during the Second World War: Myth and Reality." *Twentieth Century British History* 9 (1): 28–53.

Welshman, John. 1999. "Evacuation, Hygiene, and Social Policy: The *Our Towns* Report of 1943." *Historical Journal* 42 (3): 781–807.

Wermiel, Sara. 2003. "No Exit: The Rise and Demise of the Outside Fire Escape." *Technology and Culture* 44 (2): 258–84.

Wisemon, Tamar. 2006. "Living on a Prayer." *Jerusalem Post,* May 11, 2006. https://www.jpost.com/Israel/Living-on-a-Prayer-May-11.

Wivell, E. J. 1883. *Grand Fine Art Exhibition on William Strutt's Great Historical Picture: Black Thursday: An Episode of the Bush Fires in Victoria of February 6 1881.* Adelaide: McGlory and Co.

Wojcik, Daniel. 1997. *The End of the World as We Know It: Faith, Fatalism, and Apocalypse in America.* New York: NYU Press.

Wong, Julie Carrie, and Sam Levin. 2016. "North Dakota Governor Orders Evacuation of Standing Rock Protest Site." *Guardian*, November 28, 2016. https://www.theguardian.com/us-news/2016/nov/28/north-dakota.

Wright, Jessica. 2012. "Shoe in Custody as Protesters Give Gingerella the Slipper." *Sydney Morning Herald*, January 28, 2012. https://www.smh.com .au/national/shoe-in-custody-as-protesters-give-gingerella-the-slipper -20120127-1qlmu.html.

Yahil, Leny. 2000. "The Double Consciousness of the Nazi Mind and Practice." In *Probing the Depths of German Antisemitism*, edited by David Banker, 36–53. New York: Berghahn.

Yatsushiro, Toshio, Iwao Ishino, and Yoshiharu Matsumoto. 1944. "The Japanese-American Looks at Resettlement." *Public Opinion Quarterly* 8 (2): 188–201.

Young-Bruehl, Elisabeth. 2013. *Freud on Women*. London: Random House.

Zeiderman, Austin. 2020. "Danger Signs: The Aesthetics of Insecurity in Bogotá." In *Futureproof: Security Aesthetics and the Management of Life*, edited by D. Asher Ghertner, Hudson McFann, and Daniel M. Goldstein, 63–86. Durham, NC: Duke University Press.

Zelinger, Amir. 2018. "Unnatural Pet-Keeping: Pet-Custody Disputes in the Aftermath of Hurricane Katrina." *Humanimalia* 9 (2): 92–120. https://doi .org/10.52537/humanimalia.9544.

Zelinsky, Wilbur, and Leszek A. Kosinski. 1991. *The Emergency Evacuation of Cities: A Cross-National Historical and Geographical Study*. Savage, MD: Rowman and Littlefield.

Zerba, Shaio H. 2014. "China's Libya Evacuation Operation: A New Diplomatic Imperative—Overseas Citizen Protection." *Journal of Contemporary China* 23 (90): 1093–112.

Ziadah, Rafeef. 2019. "Circulating Power: Humanitarian Logistics, Militarism, and the United Arab Emirates." *Antipode* 51 (5): 1684–702.

Zipp, Samuel. 2010. *Manhattan Projects: The Rise and Fall of Urban Renewal in Cold War New York*. New York: Oxford University Press.

Žižek, Slavoj. 1989. *The Sublime Object of Ideology*. London: Verso.

Zoos Victoria. 2019. "How Do You Evacuate a Zoo." *Fauna* podcast. https:// open.spotify.com/episode/490wk9gqv78SKJNGTCMsFI.

www.ingramcontent.com/pod-product-compliance
Lightning Source LLC
Chambersburg PA
CBHW020825270326
41928CB00006B/439